Analysis of tall buildings by the force-displacement method

Analysis of tall buildings by the force-displacement method

M. Smolira
Ph.D., D.I.C., C.Eng., M.I.Struct., M.I.C.E.
Reader in Structural Engineering
the City University, London

A HALSTED PRESS BOOK

John Wiley and Sons
New York

Published in the U.S.A.
by **Halsted Press**
A Division of **John Wiley and Sons, Inc., New York**

Library of Congress Cataloging in Publication Data

Smolira, M
Analysis of tall buildings by the force-displacement method.

"A Halsted Press book."
Includes bibliographical references.
1. Tall buildings. 2. Structural dynamics.
I. Title.
TH845.S63 1974b 690 74-19011
ISBN 0–470–80620–6

PRINTED AND BOUND IN GREAT BRITAIN

To my wife Eileen

Contents

page

Notation xi

1 Analysis of Tall Building Frames 1

1.1 Introduction 1
1.2 Statical Indeterminacy, Assumptions and Requirements of Structural Analyses 2
1.3 Analysis of Frames for Lateral Forces 4
1.4 General Matrix for a Symmetrical Frame on Two Columns 9
1.5 General Matrix for an Unsymmetrical Frame on Two Columns 14
1.6 Analysis of Multi-bay Frames 18
1.7 The Effect of Distributed Gravity and Lateral Forces 24
1.8 The Effect of Settlement of Foundations on Tall Building Frames 25
1.9 The Effect of Members with Variable Sections 28
1.10 The Effect of Discontinuities in the Analysis of Tall Frames 34
1.11 The Effect of Boundary Conditions in the Analysis of Frames and Shear Walls 38
1.12 Approximate Analysis of Continuous Foundations to Tall Frames 40
1.13 Soil–Structure Interaction in Tall Buildings 43
1.14 Continuum Analysis of Tall Frames and Shear Walls by the Force-displacement Method 49
References 53

2 Analysis of Shear Walls 55

2.1 Introduction 55
2.2 Analysis of Plane Shear Walls 57
2.3 Shear Walls with Two Rows of Openings 58
2.4 Approximate Analysis of Shear Walls with Local Openings 62
2.5 Distribution of Lateral Forces in Shear Walls 67
2.6 Preliminary Analysis of Shear Walls without the Use of Computers 73
2.7 Infilled Frames 79
2.8 Effect of Perforations in the Analysis of Infilled Frames 89
2.9 Analysis of Frames with Parapet Infills 92
2.10 Effect of Contact Pressure in the Analysis of Infilled Frames 98
2.11 Stress Concentrations in Frames and Shear Walls 99
References 102

3 Principles of Three-dimensional Analysis of Tall Buildings 103

4 Analysis of Perforated Cores 111

4.1 Introduction 111
4.2 Pure Torsion in Thin Tubes 112
4.3 Bending of Perforated Cores 117
4.4 Warping Stresses in Closed Tubes 122
4.5 Shear Lag in Cores 123
4.6 Bending of Frame-tube Structures 124
References 125

5 Analysis of Indeterminate Floor Systems in Tall Buildings 127

5.1 Introduction 127
5.2 Interconnected Floor Beams 128
5.3 Beam on Elastic Foundation 128
5.4 Analysis of Vierendeel Girders 131
5.5 'Diagrid' Floors 134
5.6 'Waffle' Floors 135
5.7 Cantilever Stairs 136
5.8 Deflection of a Stepped Balcony Slab 142
References 144

6 Analysis of Thermal Stresses and Deformations 145

6.1 Introduction 145
6.2 Heat Transfer and its Effect on Concrete Structures 147
6.3 Thermal Stresses in Statically Determinate Members 149
6.4 Analysis of Thermal Deformations in Beams 156
6.5 Cantilever Shear Wall subjected to a Linear Temperature Distribution 161
6.6 Cantilever Shear Wall subjected to a Parabolic Temperature Distribution 164
6.7 Parapet Beams subjected to Temperature Changes 165
6.8 Effect of Temperature Changes on Statically Indeterminate Structures 166
6.9 The Effect of Axial Forces arising from Temperature Changes 171
6.10 Tall Building Frames and Shear Walls subjected to Temperature Changes 172
6.11 Summary of Unusual Effects of Temperature Changes on Structures 180
References 183

7 Analysis of Vibrations of Elastic Structures 184

7.1 Introduction 184
7.2 Single-Degree-of-Freedom Systems 185
7.3 Rayleigh's Method 187
7.4 Ritz Method 189
7.5 Damping of Vibrations of Tall Buildings 194
References 199

8 Secondary Effects in the Analysis of Tall Buildings 200

8.1 Introduction 200
8.2 Effect of Axial Forces on Flexibility Coefficients 201
8.3 Relationship between the Stability Functions and Flexibility Coefficients 202
8.4 Stiffness and Carry-over Factors, including Instability Functions 203
8.5 The Effect of Axial Deformations of Columns 204
8.6 The Effect of Shearing Forces in the Analysis of Shear Walls 210
8.7 The Effect of Flexible or Semi-rigid Connections in the Analysis of Frames 214
References 223

9 Elastic Instability of Frames and Shear Walls 225

9.1 General Considerations 225
9.2 Elastic Instability of Columns—General Review 227
9.3 Elastic Instability of Frames 229
References 238

10 Volume Changes in Concrete and their Effect on Structures 240

10.1 Introduction 240
10.2 Short Review of Mechanical Properties of Concrete 241
10.3 Drying Shrinkage 242
10.4 Creep in Concrete Structures 244
10.5 Approximate Analysis of the Effect of Creep on Concrete Members 247
10.6 Effect of Creep on Columns 252
10.7 Effect of Creep on Frames and Shear Walls 256
10.8 Approximate Analysis of the Effect of Shrinkage on Concrete Structures 259
10.9 Effect of Shrinkage on Parapet Beams 267
10.10 Analysis of Shrinkage Deformations 268
10.11 The Effect of Shrinkage on Frames 270
References 273

11 Analysis of Trusses by the Force-displacement Method 274

11.1 Introduction 274
11.2 Derivation of Compatibility Relationship 274
11.3 Statically Determinate Trusses 280
11.4 Internally Statically Indeterminate Trusses 284
11.5 Externally Indeterminate Trusses 288
11.6 Bending Stresses in Trusses 289
11.7 Effect of Rigid Joints in the Analysis of Trusses 291
11.8 Trusses with Continuous Chords 293

Concluding Remarks 294

Index 297

Notation

a	dimension
A	cross-sectional area of a member or a non-dimensional coefficient
b	dimension
B	non-dimensional coefficient
c	stability coefficient and specific creep
C	non-dimensional coefficient
d	total depth of a member, also length of the diagonal in brick panels
D	non-dimensional coefficient
e	eccentricity of load
E	modulus of elasticity of the material
f	stress, also flexibility coefficient
F	force
g	gravity load per unit area or unit length of a member
G	shear modulus of elasticity, also gravity load
h	storey height of a frame or shear wall
H	total height of shear wall
k	coefficient, stiffness of a member or modulus of subsoil reaction
l and L	length of a member
m and M	bending moments due to the primary and secondary effects respectively
N	axial force in a member, also a number of storeys in a tall building
p	pressure per unit length of a member
P	concentrated gravity force
r	coefficient
R	radius
s	shrinkage coefficient or stability coefficient
t	thickness of a plate, also time
T	torsional moment, also temperature
v	shear stress
V	shear force
W	wind load
x	depth of neutral axis
z	lever arm
α and β	angles
δ	axial deformation
Δ	horizontal deflection
ε	strain
ϕ	torsional coefficient

η	coefficient
λ	vertical deflection
θ	angular deformation of a member at release due to gravity or wind force

other symbols when used are defined in the text.

1 Analysis of Tall Building Frames

1.1 Introduction

This book concerns structural analyses as applied mainly to tall building frames and shear walls, although the general principles discussed are applicable to other structures. The book is intended as a reference for students, practising engineers, and researchers.

The need for more accurate analysis of tall buildings is stimulated by the continued trend to build slender structures of reduced mass, the safety and stability of which depend on a high degree of continuity of joints. The latter results in highly statically indeterminate forms, which more than old forms are vulnerable to the effect of gravity and lateral forces as well as other effects which need to be considered at the design stage.

There is no point at which a building can be considered as tall, and the term is usually given to one the design of which is affected more by lateral than by gravity forces. In addition, tall buildings, more than low buildings, can be affected by instability forces in their broad form, and by 'secondary' effects such as changes of temperature, shrinkage and creep of concrete, settlement of foundations, dynamic action of loads, etc., and these problems therefore constitute the main topics of this book.

The analysis of structures concerns the determination of the effect of these forces on stresses and deformations and is the main subject of this work. Design concerns the proportioning of sizes of members and is not considered here. The two aspects, however, are related and, in principle, cannot be separated.

The exact magnitude of stresses and deformations in structures is a function of imposed loads, but depends also on a number of other effects, such as the frequency and duration of loads, past history of loading, physical properties of materials, environmental conditions, etc. These are usually very complex parameters and only rarely can they be formulated exactly in mathematical terms. Mathematical formulation, however, is necessary for the development of the analysis, for checking the validity of assumptions, for establishing the limits of applicability of these assumptions and the theory, and also for repetitive use of the theory in the design. It is mainly because of the complexity of the parameters involved that the assessment of stresses and deformations has to be based on some simplified assumptions and idealized models. These simplifications in turn result in discrepancy between the predicted and actual response of structures which, in some cases, may be small and unimportant for practical design, but can also be very large. Even the reverse response to that predicted can take place. This discrepancy can be narrowed only by further developments in understanding of the properties of materials, by improvements in analytical formulations of problems, and by correlation with the observed response of structures.

All modern structural analyses are based on well-established energy principles, and most theories developed from these principles treat materials as linearly elastic, because beyond this elastic range many problems become extremely complex and often intractable, unless simplifying assumptions are made. Particularly difficult to formulate accurately are problems involving inelastic response of materials and time-dependent stresses.

In the past, many simplified assumptions had also to be accepted to make a structure suitable for analysis by the use of desk computers. High-speed digital computers now available not only eliminate tedious calculations but also allow us to dispense with some of the approximate assumptions made in the past, and permit the use of more powerful methods of analysis which depend on complex numerical calculations. For the same reason, matrix formulation of complex problems has been developed in recent years, and this is particularly suitable for use in conjunction with computers. Systematic matrix techniques are now well established and are presented in many textbooks.

The primary elastic analyses are now well developed and most problems can now be solved within this range, based on the assumed linearly elastic response of materials. The Stiffness, Flexibility and Finite Element Methods are basically in this range, although they can be extended to the inelastic range. The 'secondary' effects in the analysis include higher order effects, such as altered geometry of a structure, effect of axial forces, flexible connections, temperature changes, creep, shrinkage, etc. In the following, attempts are made to review some of these problems.

Generally, structural analyses are based either on methods which attempt to formulate and to solve a problem in an explicit or closed form, or on numerical methods which provide a solution to a specific problem in numerical terms only. In the first group are the stress analysis and energy methods generally, but the explicit solutions can be obtained for the simplest problems only. The second approach is based mainly on the stiffness method and provides a numerical solution of a particular problem. The latter is based on the slope–deflection relationship and is now well developed and formulated in matrix form suitable for computers. Similarly, the Finite Element Method is based on the Stiffness Method and provides only the numerical answer to a problem. These methods are now presented in many textbooks and are not reviewed here.

The classical Flexibility Method is difficult to apply to any but the simplest problems and is considered unsuitable for complex structures. Most of this work is therefore based on the concept of the force–displacement method in which the unknown values are both the forces and displacements considered simultaneously. Although the matrix is therefore a little larger than in the Stiffness Method, both values are required in modern design, particularly of tall buildings. For simple problems, the method provides an explicit solution, but for complex structures, the method must be used as numerical. The formulation of a matrix has clear physical meaning at all stages and this facilitates the setting out of the matrix and avoidance of errors.

1.2 Statical Indeterminacy, Assumptions and Requirements of Structural Analyses

Structures whose reactions or internal stresses cannot be found by using the conditions of equilibrium of forces are called statically indeterminate. Members are considered redundant if they are not essential for the stability of a structure.

In the stiffness or displacement method, statical indeterminacy is expressed in relation to the number of degrees of freedom of joints. The latter is equal to the number of translational or rotational displacements which joints can undergo. The forces are then calculated from the slope–deflection relationships. In the conventional flexibility or force method, the degree of statical indeterminacy of a structure is related to the number of redundant forces which could be removed to render a structure statically determinate. In the force–displacement method used in this text, the statical indeterminacy is related to the number of redundant forces and linear displacement of joints.

The number of degrees of freedom is usually different from the number of degrees of statical indeterminacy, e.g., a fixed-end portal frame has three degrees of freedom but five statically indeterminate values: four bending moments and one deflection.

In most parts of this text, it will be assumed that structures are made of homogeneous isotropic materials which respond linearly and elastically to the applied forces, as defined by Hooke's Law. All actions are assumed to be static, except in case of wind and earthquake forces. It is also assumed that the displacements are small, and that the analysis can be based on the undeformed configuration of a structure. This limitation will be removed in chapters 9 and 10. The assumption of linear elastic response of a structure implies that the principle of superposition is assumed to be valid. This principle assumes that, if there are two or more influences acting together, their effect on stresses and deformations can be considered as additive in any sequence of their action. For this principle to apply, the assumptions of linearly elastic response of materials and small deformations are necessary.

In most structures, deformations are sufficiently small to justify the primary analysis being based on undeformed configuration. In tall buildings, however, this may not always be true, particularly for buildings with high slenderness ratio, and the effect of large deformations may then become significant.

The equations necessary to solve statically indeterminate structures are generally derived from the imposed conditions of deflections or rotations. These are called the conditions of consistent deformations or the compatibility of deformations. These conditions can be written when the load–deflection characteristics of individual members are known and are usually derived from Hooke's law. The second group of conditions necessary for the solution of statically indeterminate problems contains the conditions of the equilibrium of forces.

With regard to the basic requirements of structural analyses, statical equilibrium of forces must always apply, whether a structure is statically determinate or indeterminate, whether forces are gravity or lateral, and for all 'secondary' effects.

All structures must deform in a compatible way, so that members fit together and form a continuous medium of a similar form to that before loads were applied. This must be true (provided cracks do not form), whatever the cause of deformations. Statically determinate structures satisfy these conditions automatically. For indeterminate structures, this requirement provides an independent set of conditions.

The last requirement of elastic analyses, that the load–deflection or the stress–strain characteristics of members must relate deformations to external forces, is usually simple and unique within the elastic range, but in the inelastic range, this relationship can be complex, particularly when the time effect has to be considered.

The equilibrium and compatibility are essentially the properties of the structural form chosen, while the load–deflection relationship is related to the properties of the materials.

Tall building frames are always highly indeterminate and require the use of all three groups of conditions stated above, while statically determinate structures require only the conditions of equilibrium and the two other groups are automatically satisfied.

In statically determinate structures, the distribution of moments and forces depends on spans and loading system and can be determined before sizes of members are known. A statically indeterminate structure must be fully proportioned before the analysis of the distribution of forces and deflections can be made. This is because the latter depend on the elastic deformations, and these in turn depend on the proportioning of members and on properties of materials used.

In principle, therefore, all structural sizes must be known or assumed, the redundants calculated, all stresses and deformations checked, and the members proportioned again. If sections thus evaluated differ substantially from those originally assumed, the process should be repeated. Furthermore, having given maximum permissible stresses, there is no assurance, when dealing with highly indeterminate structures, that stresses in all members will be fully utilized, although it is possible to arrive at a solution in which the waste of materials is small and the structure will have the required minimum of safety factors, either stress or load factor.

In preliminary calculations, all moments and forces are often found from the conditions of equilibrium of external and internal forces without trying to satisfy the conditions of compatibility of deformations, and treating each frame as a separate unit from the rest of the structure. On the other hand, a 'rigorous' solution can be obtained by the elastic theory, by setting out a great number of simultaneous equations which usually can only be solved by electronic computers. Even these methods, however, ignore inelastic behaviour of materials, formation of cracks, variation of E value with time, non-monolithic connections of joints, differential settlement, creep and shrinkage, and other secondary effects.

Because of these difficulties, many analyses are either based on some simplified assumptions or efforts are made to analyse such structures at the ultimate stress conditions, in the hope that, at this stage, the structure will become more tractable to analysis. However, simple plastic theories are not readily applicable to the design of tall building frames because of the inherent limitations of the basic assumptions, particularly that the equilibrium is based on the undeformed structure, and that the deformations are small and instability of members and of the whole structure cannot occur. These assumptions do not hold for tall buildings at ultimate stress.

In the following, the force-displacement method is used as applied to the analysis of frames and shear walls, at first for the 'primary' effects of gravity and wind forces, and this is later extended to include the 'secondary' effects.

1.3 Analysis of Frames for Lateral Forces

In the conventional flexibility method of analysis of statically indeterminate structures, the unknown values are forces (and therefore moments). The term 'force–displacement' or 'mixed flexibility' is used here to describe the procedure in which calculations include the unknown forces and translations of joints simultaneously.

This approach is particularly useful in the analysis of tall frames, shear walls, and other interconnected structures for primary as well as for secondary effects.

The general assumptions on which the analysis is based are those inherent in all elastic analyses, as discussed in section 1.2, and no further limitations are needed. However, the analysis will be later broadened by the inclusion of the effect of creep and shrinkage, large deformations, elastic instability, etc.

The sign convention in this method is not related to the clockwise direction or to the hogging or sagging moments. The sign convention follows from the use of the principle of Virtual Work, so that the displacement set should correspond to the force set, i.e., the same sign must be used for both: if a moment is assumed positive, then the corresponding deformation is also positive. In other words: moments are positive if their action closes the assumed discontinuity, and negative if it increases. The setting-out of equilibrium of forces must follow the assumed pattern of bending moments. The sign 'minus', if obtained from the solution, will then mean that the assumed direction of this moment must be reversed. It should be noted that no other sign convention is suitable in this method.

Any lack of symmetry in the geometry of the structure or in the system of loading, whether gravity or lateral, can be taken into account, including changes of the form of construction at the ground floor level, frequently imposed by the architectural planning. Any symmetry of members in the structure is also reflected in a similar symmetry of the matrix coefficients.

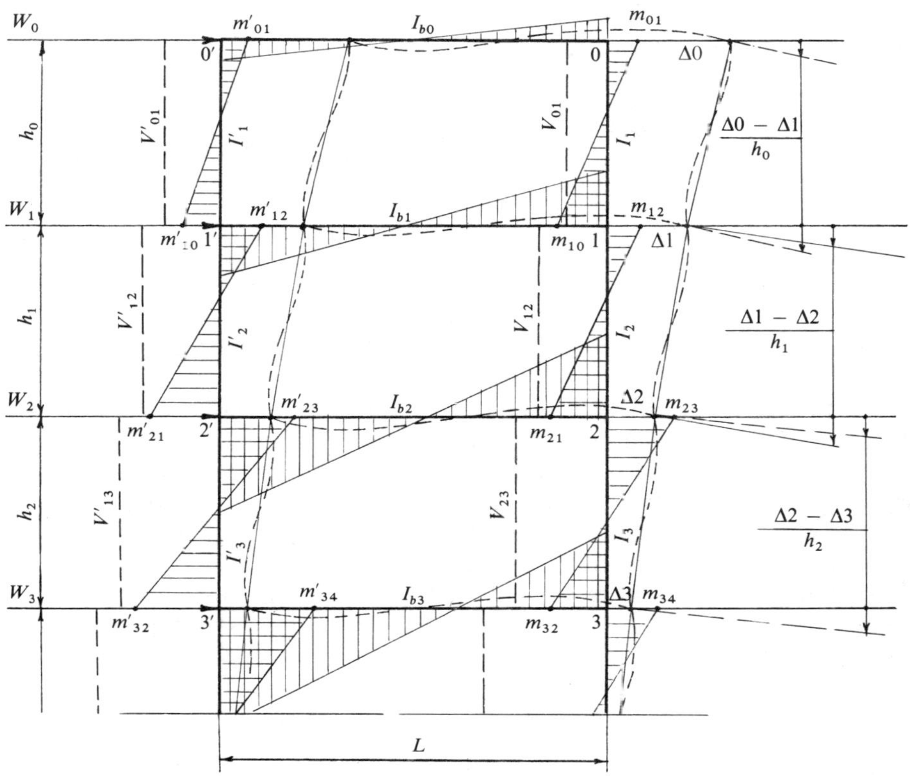

Fig. 1.1 Analysis of frames by the force-displacement method

The procedure in the analysis is as follows: First, the structure is assumed to deflect to its equilibrium position defined by the unknown deflections Δ0, Δ1, Δ2, etc., as shown in Fig. 1.1. The distribution of bending moments can, at this stage, be assumed of any form, but the rest of the procedure must closely correspond to this assumption. If the assumed directions of moments are not correct, but the rest of the analysis is correctly carried out, the final solution obtained from the matrix will still be correct, although the sign of some moments may need to be changed, or the relative values of deflections may be different from those assumed.

Next, releases are introduced at all junctions in their final position, so that all members become statically determinate and stresses are zero (when forces are applied at the nodal points only). In this released configuration all the discontinuities between the respective members meeting at each joint are marked, e.g., in Fig. 1.1, the discontinuity at the nodal point No. 1, between the beam 1′–1 and column 12 is equal to $(\Delta 1-\Delta 2)/h_2$, and between column 12 and column 10 the discontinuity is equal to $(\Delta 0-\Delta 1)/h_1-(\Delta 1-\Delta 2)/h_2$. In a similar way, discontinuities at all other nodal points can be defined.

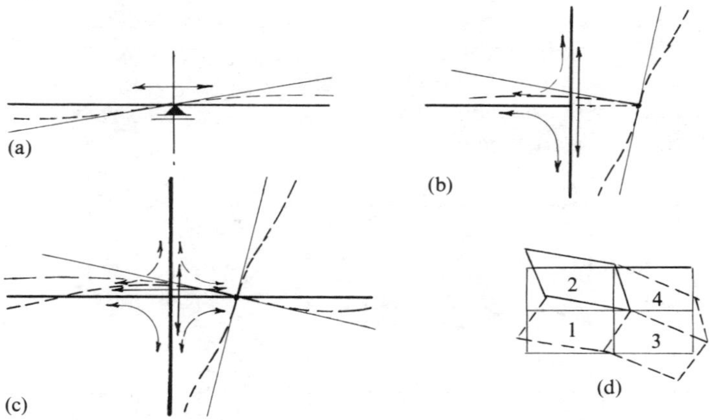

Fig. 1.2 Compatibility of deformations

Next, statically indeterminate bending moments are reapplied as assumed earlier, of such magnitude as are necessary to restore the structure to full compatibility of deformations at all the joints simultaneously. These conditions mean that strains or deflections in the adjoining elements must be compatible, e.g., in Fig. 1.2(a), (b), and (c), the tangent to the deflected lines of the adjoining beams or columns must be the same on either side of members. Similarly, in Fig. 1.2(d), the two-dimensional element 1 can deform freely, while elements 2 and 3 are restrained in one direction by common lines with the element 1. The deformation of element 4 is completely determined by the deformations of elements 1, 2, and 3. In other words, all members in linear structures and all elements in two or three-dimensional structures must deform in a compatible way, so that parts fit together without discontinuities at all stages of loading and deflections. An exception to this comes from the flexible connections of joints, cracks, stress concentrations, and in inelastic analysis. This will be discussed later.

The setting out of the conditions of compatibility of deformations requires a knowledge of the stress–strain or the load–deflection characteristics of each element.

The number of independent conditions of compatibility of deformations depends on the number of members meeting at a joint, e.g., in Fig. 1.2(a) with two members, only one such condition is required; in Fig. 1.2(b) with three members, two independent conditions can be set out as indicated by arrows, while the condition indicated by the broken arrow is not independent and is automatically satisfied by the two previous conditions. Similarly, in Fig. 1.2(c), with four members intersecting, three independent conditions can be chosen in any preferred pattern, e.g., as shown by the arrows. Two other conditions, shown by broken arrows, are automatically satisfied and do not form independent conditions.

Referring to Fig. 1.1, the compatibility conditions at the level 1–1′ are as follows:

$$2\,1\,1'. \quad -m_{21}\cdot\frac{h_1}{6EI_2} + m_{12}\cdot\frac{h_1}{3EI_2} + (m_{10} + m_{12})\cdot\frac{L}{3EI_{b_1}} - (m'_{10} + m'_{12})\cdot\frac{L}{6EI_{b_1}} = \frac{\Delta 1 - \Delta 2}{h_1}$$

$$2'\,1'\,1. \quad -m'_{21}\cdot\frac{h_1}{6EI'_2} + m'_{12}\cdot\frac{h_1}{3EI'_2} + (m'_{10} + m'_{12})\frac{L}{3EI_{b_1}} - (m_{10} + m_{12})\cdot\frac{L}{6EI_{b_1}} = \frac{\Delta 1 - \Delta 2}{h_1}$$

$$0\,1\,2. \quad -m_{01}\frac{h_0}{6EI_1} + m_{10}\cdot\frac{h_0}{3EI_1} - m_{12}\frac{h_1}{3EI_2} + m_{21}\frac{h_1}{6EI_2} = \frac{\Delta 0 - \Delta 1}{h_0} - \frac{\Delta 1 - \Delta 2}{h_1} \tag{1.1}$$

$$0\,1'\,2'. \quad -m'_{01}\frac{h_0}{6EI'_1} + m'_{10}\frac{h_0}{3EI'_1} - m'_{12}\frac{h_1}{3EI'_2} + m'_{21}\frac{h_1}{6EI_2} = \frac{\Delta 0 - \Delta 1}{h_0} - \frac{\Delta 1 - \Delta 2}{h_1}$$

The second set of conditions which a structure must satisfy are the conditions of equilibrium of forces. These mean that all members (and the whole structure) must be in equilibrium with the external forces. These conditions can be written in two ways; (a) at a junction, when the sum of all internal shearing forces must be equal to the applied force at this junction, or (b) at any section through the columns, the sum of shearing forces in columns must be equal to the sum of all the forces applied above that section. For example, at the level 1–1′ we can write the equilibrium conditions as:

$$\text{E.C. } 1\text{–}1'. \quad \left[\frac{m_{12} + m_{21}}{h_1} - \frac{m_{01} + m_{10}}{h_0}\right] + \left[\frac{m'_{12} + m'_{21}}{h_1} - \frac{m'_{01} + m'_{10}}{h_0}\right] = W_1$$

or in the form: (1.2)

$$\text{E.C. } 1\text{–}2. \quad \frac{m_{12} + m_{21}}{h_1} + \frac{m'_{12} + m'_{21}}{h_1} = V_{12} + V'_{12}$$

When the number of unknown statically indeterminate values is small (three or four), then a solution in a closed form can be obtained. However, for complex arrangements of frames, with many unknowns, the method must be used numerically.

With regard to the interconnected shear walls, the question arises as to what extent do respective flexural, shearing, and axial deformation affect the behaviour of such systems. While it is difficult to study this problem in an explicit form, some light is thrown by many analyses performed and measurements on models. It appears that slender frames are very little affected by shearing deformations, while the effect of axial deformations, loss of stiffness of compression members, and elastic instability may become important factors.

On the other hand, if members are of low height to width ratio, as in shear walls, the shear deformations should be included while the effect of axial deformations becomes small.

It is important to note that the overall height to width ratio is the criterion of the two actions, and not the ratio of the individual floor height to width of the member.

The following examples are chosen to illustrate the procedure and to discuss the implications of the results.

Example 1.1

Figure 1.3 shows a portal frame subjected to the lateral force W applied at a nodal point 0. The four compatibility conditions at the junctions 0, 0′, 1, and 1′, and the equilibrium condition can be written as follows. (The column on the left is used to identify the compatibility or equilibrium condition):

$$\text{0.} \quad m_{01}\cdot\left(\frac{L}{3EI_b}+\frac{h}{3EI_1}\right)-m_{10}\cdot\frac{h}{6EI_1}-m'_{01}\cdot\frac{L}{6EI_b}=\frac{\Delta_0}{h}$$

$$\text{0}'. \quad -m_{01}\cdot\frac{L}{6EI_b}+m'_{01}\cdot\left(\frac{L}{3EI_b}+\frac{h}{3EI_2}\right)-m'_{10}\cdot\frac{h}{6EI_2}=\frac{\Delta_0}{h}$$

$$\text{1.} \quad -m_{01}\cdot\frac{h}{6EI_1}+m_{10}\cdot\frac{h}{3EI_1}=\frac{\Delta_0}{h}$$

$$\text{1}'. \quad -m'_{01}\cdot\frac{h}{6EI_2}+m'_{10}\cdot\frac{h}{3EI_2}=\frac{\Delta_0}{h}$$

$$\text{E.C.} \quad \frac{m_{01}+m_{10}}{h}+\frac{m'_{01}+m'_{10}}{h}=W$$

Multiplying compatibility conditions by $6EI_1/h^2$ and equilibrium conditions by h, we obtain the following general matrix in terms of the non-dimensional parameters:

$$\begin{bmatrix} 2(r_1+1) & -r_1 & -1 & 0 & -1 \\ -r_1 & 2(r_1+r_2) & 0 & -r_2 & -1 \\ -1 & 0 & 2 & 0 & -1 \\ 0 & -r_2 & 0 & 2r_2 & -1 \\ 1 & 1 & 1 & 1 & 0 \end{bmatrix} \begin{Bmatrix} m_{01} \\ m'_{01} \\ m_{10} \\ m'_{10} \\ \Delta_0 \end{Bmatrix} = \begin{Bmatrix} 0 \\ 0 \\ 0 \\ 0 \\ W.h \end{Bmatrix}$$

in which $$r_1=\frac{LI_1}{hI_b} \quad \text{and} \quad r_2=\frac{I_1}{I_2}$$

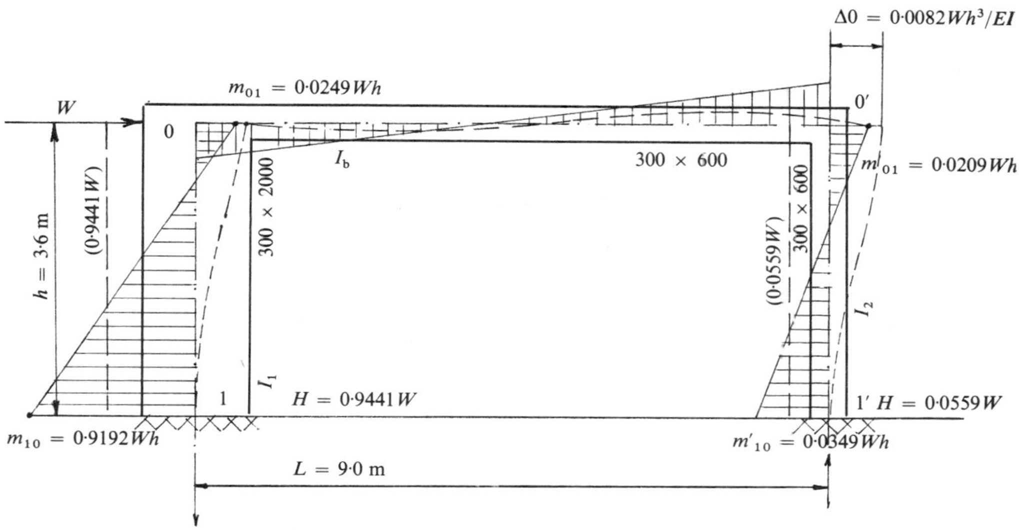

Fig. 1.3 Example of the analysis of a portal frame by the force-displacement method

or, generally, in the form:

$$[F]\begin{Bmatrix} M \\ \Delta \end{Bmatrix} = \{W\}$$

in which: $[F]$ is the force–displacement matrix
$\{M\}$ is the force–displacement vector, and
$\{W\}$ is the load vector.

In the example, the values of r_1 and r_2 are:

$$r_1 = \frac{9 \times 2^3}{3{\cdot}6 \times 0{\cdot}6^3} = 92{\cdot}59, \qquad r_2 = 37{\cdot}04$$

Based on the above matrix, we obtain bending moments and deformation $\Delta 0$ as shown in Fig. 1.3. It should be noted that bending moments are in terms of $W.h$ and deflections in terms of $W.h^3/EI_1$. The solution is therefore in terms of non-dimensional parameters.

1.4 General Matrix for a Symmetrical Frame on Two Columns

For a symmetrical frame on two columns as shown in Fig. 1.4 a general matrix can be set out in an explicit form in terms of the non-dimensional parameter:

$$r = \frac{L.I_c}{h.I_b} \tag{1.3}$$

Proceeding as shown in section 1.3 and rearranging terms, a general matrix is set out as shown in Table 1.1. This matrix is set out for a ten-storey frame, but can be easily adjusted for a frame of any given number of floors by taking an appropriate number of element matrices F^e. In this matrix, the values of V represent the shearing

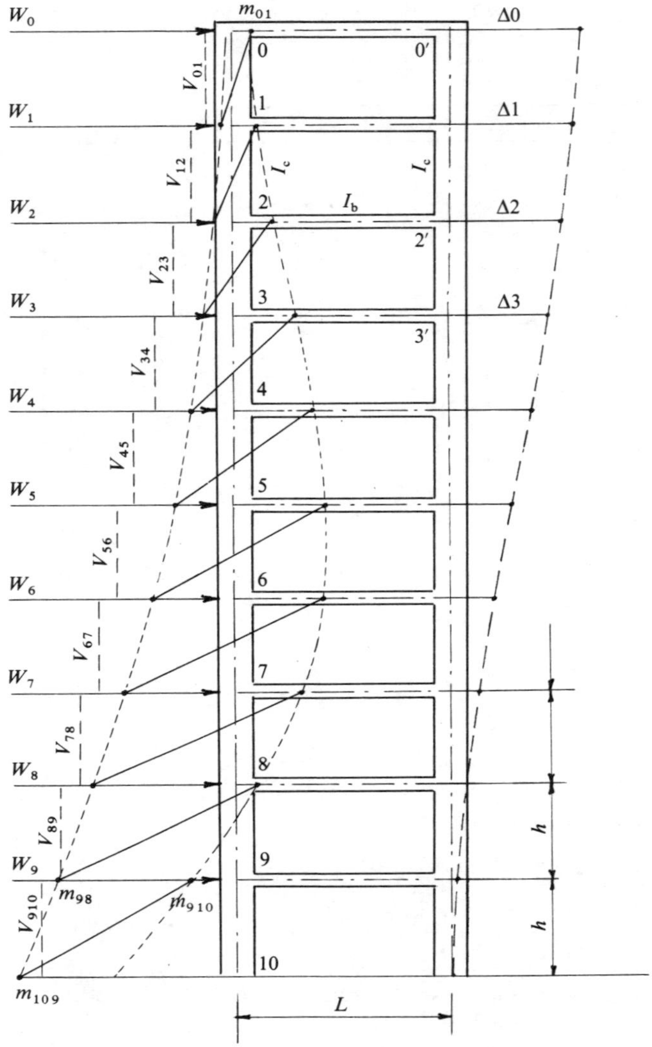

Fig. 1.4 Analysis of a symmetrical frame on two columns

forces in columns, i.e., $V_0=\frac{1}{2}W_0$, $V_1=\frac{1}{2}(W_0+W_1)$, etc. For uniformly distributed lateral forces $W=wh$, $V_0=0{\cdot}25W$, $V_1=0{\cdot}75W$, $V_2=1{\cdot}25W$, etc.

It should be noted that the bending moments obtained from the matrix are expressed in terms of Wh, and deflections in terms of $Wh^3/6EI_c$.

Example 1.2

For a ten-storey frame as shown in Fig. 1.5, $r=2{\cdot}5$ and the solution of the matrix gives the distribution of bending moments and deflections shown in Fig. 1.5. Shearing and axial forces are calculated using the usual conditions of statics. Free cantilever bending moments are indicated for comparison of the results or for static checks,

Table 1.1 General matrix for the analysis of symmetrical frames on two columns ($r = L.I_c/h.I_b$)

	0		1			2			3			4			5			6			7			8			9			10	
	0 1	Δ0	1 0	1 2	Δ1	2 1	2 3	Δ2	3 2	3 4	Δ3	4 3	4 5	Δ4	5 4	5 6	Δ5	6 5	6 7	Δ6	7 6	7 8	Δ7	8 7	8 9	Δ8	9 8	9 10	Δ9	10 9	
1 0 0′	$r+2$	−1	−1	0	1																										0
0–1	1	0	1	0	0																										$V_{cr}.h$
2 1 1′	0	0	r	$r+2$	−1	−1	0	1																							0
0 1 2	−1	−1	2	−2	2	1	0	−1																							0
1–2	0	0	0	1	0	1	0	0																							$V_{12}.h$
3 2 2′																															0
1 2 3							F^e																								0
2–3																															$V_{23}.h$
4 3 3′																															0
2 3 4										F^e																					0
3–4																															$V_{34}.h$
5 4 4′																															0
3 4 5													F^e																		0
4–5																															$V_{45}.h$
6 5 5′																															0
4 5 6																F^e															0
5–6																															$V_{56}.h$
7 6 6′																															0
5 6 7																			F^e												0
6–7																															$V_{67}.h$
8 7 7′																															0
6 7 8																						F^e									0
7–8																															$V_{78}.h$
9 8 8′																															0
7 8 9																									F^e						0
8–9																															$V_{89}.h$
10 9 9′																															0
8 9 10																															0
9–10																															$V_{910}.h$
10 9																												−1	−1	2	

Note: all bending moments are in terms of $W.h$ and deflections in terms of $W.h^3/6EI_c$.

e.g., at any horizontal section just above the joint we can write that for equilibrium:

$$2m_n + N_n . L = M_{n,\text{cantilever}} \tag{1.4}$$

For example, at joint 5:

$$2 \times 0{\cdot}9182\,Wh + 1{\cdot}1848\,Wh \times 9{\cdot}0 = 12{\cdot}5\,Wh$$

Equation (1.4) can be modified by expressing axial force N_n at any floor level in terms

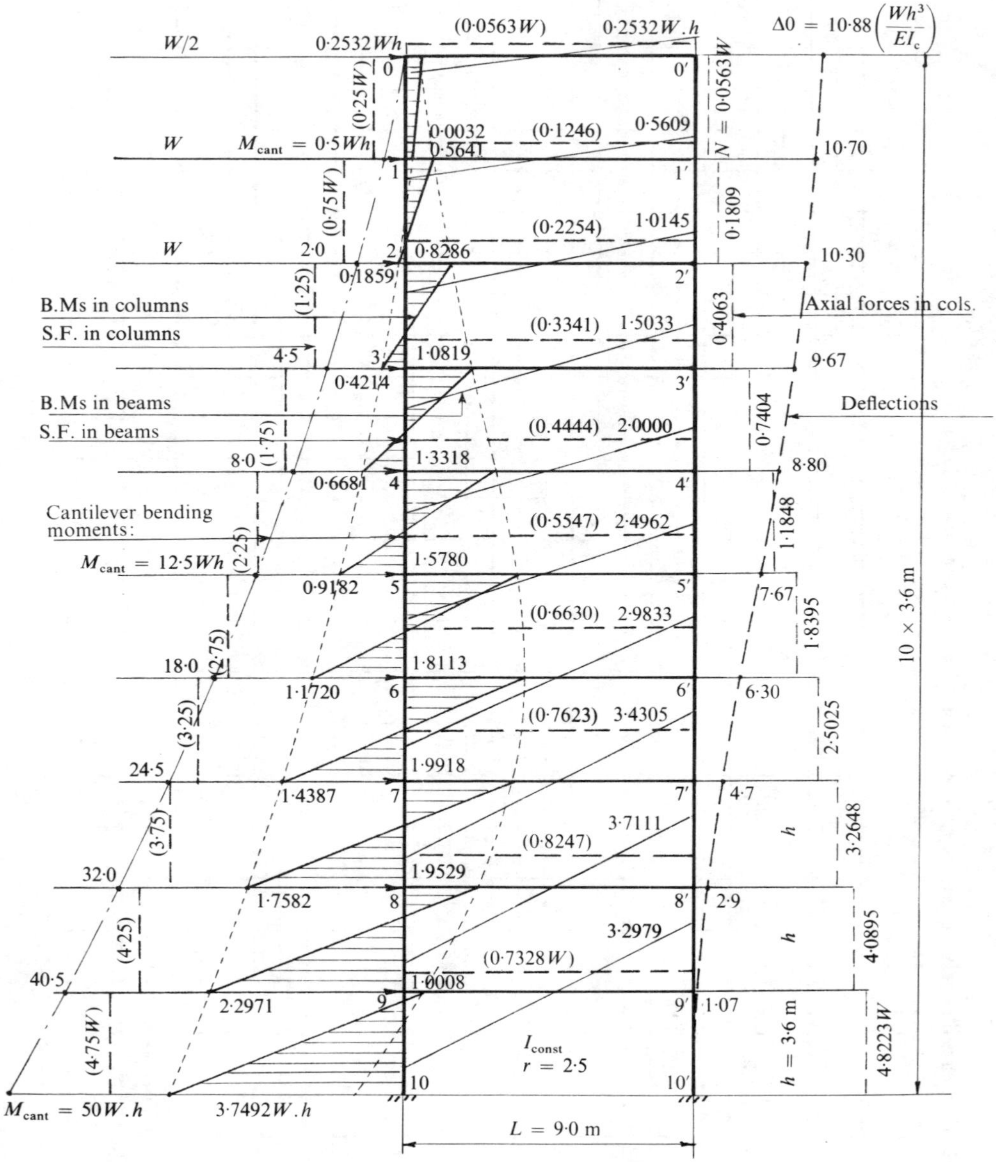

Fig. 1.5 Example of the analysis of a tall frame

of shear forces in beams or bending moments in beams. We obtain:

$$m_n = \tfrac{1}{2} M_{n,\,\text{can.}} - \sum_{0}^{n} m_{\text{beams}} \tag{1.5}$$

e.g., $m_{56} = \tfrac{1}{2} 12{\cdot}5Wh - [4{\cdot}9962 + 2{\cdot}0 + 1{\cdot}5033 + 1{\cdot}0145 + 0{\cdot}5609 + 0{\cdot}2532]Wh$
$= 1{\cdot}5780Wh$

which is as in Fig. 1.5.

The effect of shearing and axial deformations will be discussed later in the text.

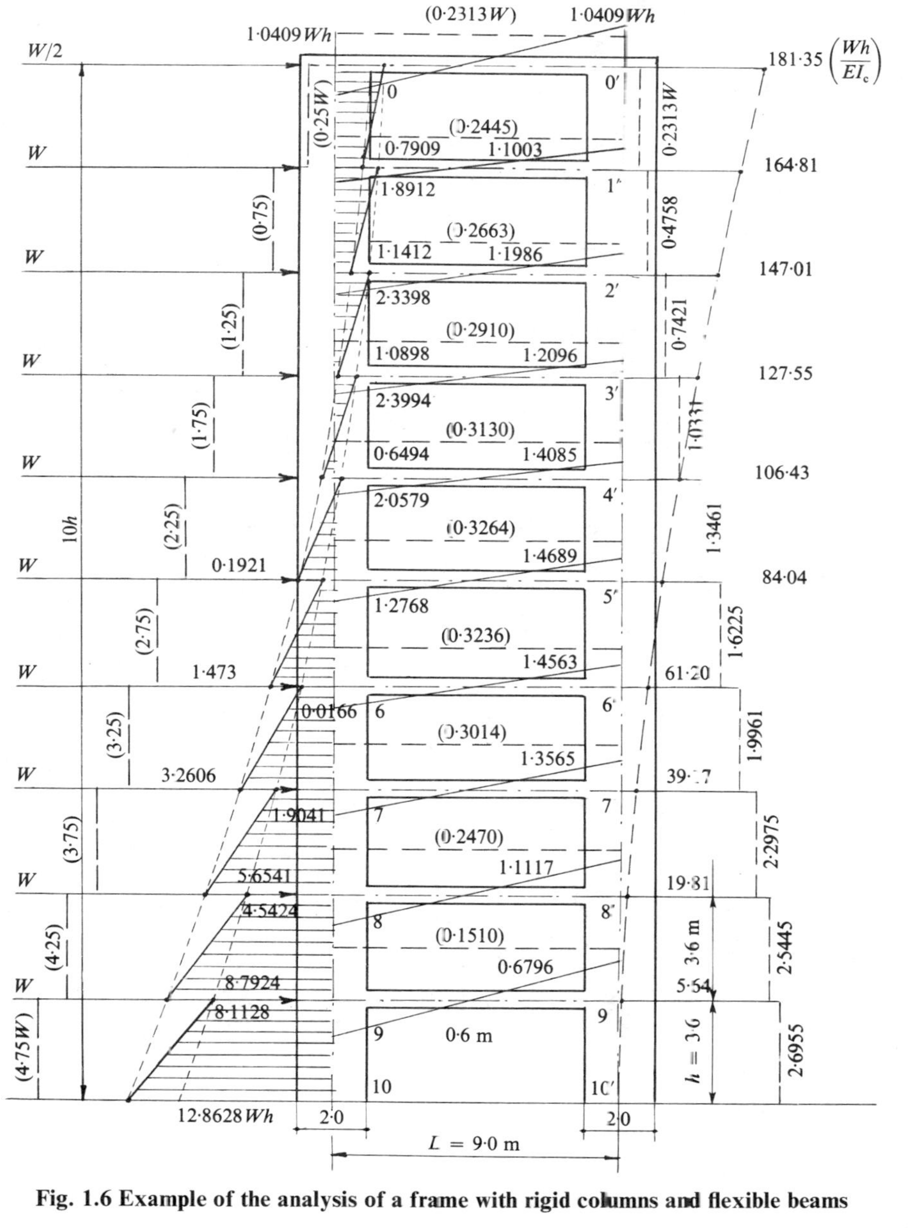

Fig. 1.6 Example of the analysis of a frame with rigid columns and flexible beams

Example 1.3

This example is set out to show that in frames in which the ratios of the stiffnesses of columns to beams differ considerably, the overall pattern of bending moments can also be radically altered.

For a frame shown in Fig. 1.6 (the value of $r = 92{\cdot}5925$) the overall pattern of bending moments is such that only two points of contraflexure exist. In such rigid frames, large 'cantilever' bending moments develop in the lower part of frames, beams absorb small moments, and deflections often do not have a point of contraflexure as in flexible frames.

1.5 General Matrix for an Unsymmetrical Frame on Two Columns

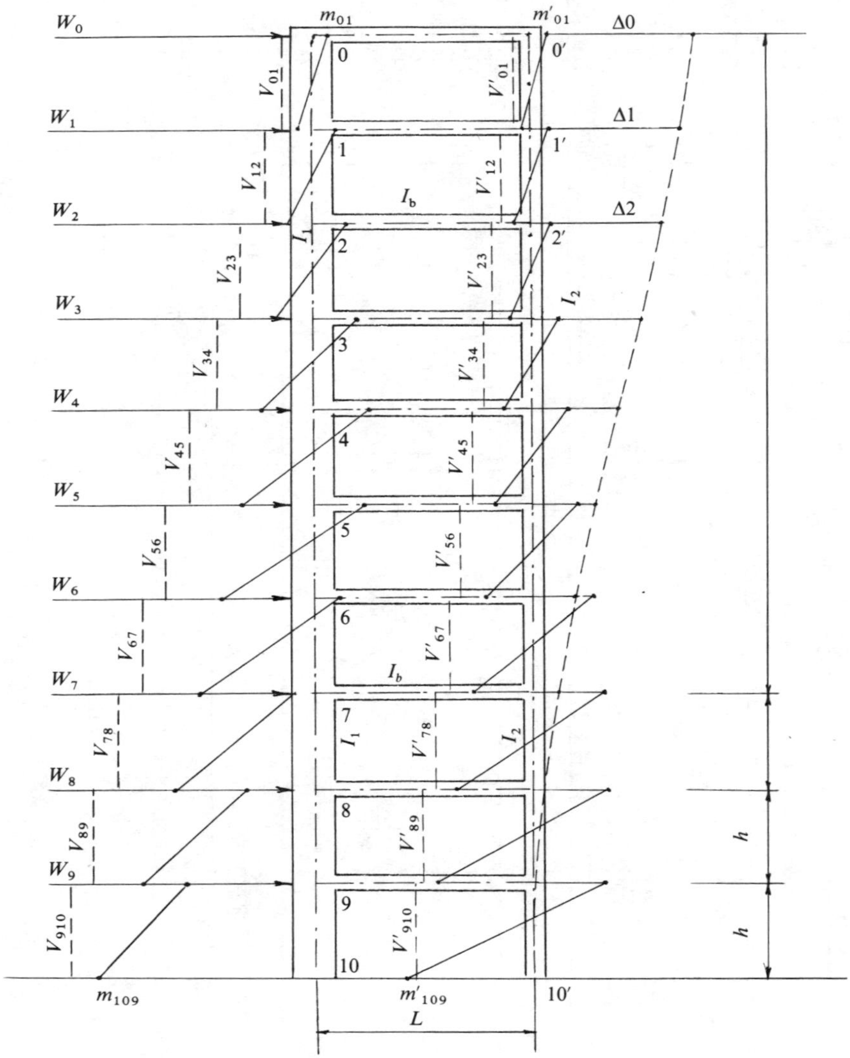

Fig. 1.7 Analysis of unsymmetrical frame on two columns

Table 1.2(a) General matrix for the analysis of unsymmetrical frames on two columns

$$\left(r_1 = \frac{L.I_1}{h.I_b},\quad r_2 = \frac{L.I_2}{h.I_b}\right)$$

	0′	0		1′		1			2′		2			3′		3			4′	…							
	0′ 1′	0 1	Δ0	1′ 0′	1′ 2′	1 0	1 2	Δ1	2′ 1′	2′ 3′	2 1	2 3	Δ2	3′ 2′	3′ 4′	3 2	3 4	Δ3	4′ 3′						n′, n′−1	n, n−1	$\frac{EI_1}{6h^2}$
1′ 0′ 0	$2\left(\frac{r_1}{r_2}+r_1\right)$	$-r_1$	−1	$-\frac{r_1}{r_2}$	0	0	0	+1																			0
1 0 0′	$-r_1$	$2(r_1+1)$	−1	0	0	−1	0	+1																			0
0–1	1	1	0	1	0	1	0	0																			$h(V_{01}+V'_{01})$
2′ 1′ 1′	0	0	0	$2r_1$	$2\left(\frac{r_1}{r_2}+r_1\right)$	$-r_1$	$-r_1$	−1	$-\frac{r_1}{r_2}$	0	0	0	1														0
2 1 1′	0	0	0	$-r_2$	$-r_2$	$2r_1$	$2(r_1+r_2)$	−1	0	0	−1	0	1														0
0′ 1′ 2′	$-\frac{r_1}{r_2}$	0	−1	$2\frac{r_1}{r_2}$	$-2\frac{r_1}{r_2}$	0	0	2	$\frac{r_1}{r_2}$	0	0	0	−1														0
0 1 2	−1	0	−1	0	0	2	−2	2	0	0	1	0	−1														0
1–2	0	0	0	0	1	0	1	0	1	0	1	0	0														$h(V_{12}+V'_{12})$
3′ 2′ 2					0	0	0	0	$2r_1$	$2\left(\frac{r_1}{r_2}+r_1\right)$	$-r_1$	$-r_1$	−1	$-\frac{r_1}{r_2}$	0	0	0	−1									0
3 2 2′					0	0	0	0	$-r_2$	$-r_2$	$2r_1$	$2(r_1+r_2)$	−1	0	0	−1	0	1									0
1′ 2′ 3′					$-\frac{r_1}{r_2}$	0	0	−1	$2\frac{r_1}{r_2}$	$-2\frac{r_1}{r_2}$	0	0	2	$\frac{r_1}{r_2}$	0	0	0	−1									0
1 2 3					−1	0	0	−1	0	0	2	−2	2	0	0	1	0	−1									0
2–3					0	0	0	0	0	1	0	1	0	1	0	1	0	0									$h(V_{23}+V'_{23})$
4′ 3′ 3																											
4 3 3′														F^e													
																			F^e								
																											0
																											0
																											0
																											0
																											$h(V_{n-1,n}+V'_{n-1,n})$
n′, n′−1																					$-\frac{r_1}{r_2}$	0	0	−1	$2\frac{r_1}{r_2}$	0	0
n, n−1																					0	0	−1	−1	0	2	0

Table 1.2(b) Flow chart for two-column shear walls

$$r = \frac{L.I_1}{h.I_b}, \quad r_c = \frac{I_1}{I_2}, \quad s_1 = \frac{h_1}{h}, \quad s_2 = \frac{h_2}{h}, \quad \text{etc.} \quad (h\text{—any reference length})$$

Design	Zeros	Element matrices														Zeros	Forces and displs.	Load vector
$1'\,0'\,0$							$2(s_1r_c+r)$		$-r$	$-1/s_1$	$-s_1r_c$	0	0	0	$1/s_1$	$5N-8$	$0'\,1'$	0
$1\,0\,0'$							$-r$		$2(s_1+r)$	$-1/s_1$	0	0	$-s_1$	0	$1/s_1$		$0\,1$	0
$E_c.\,0\,1$							$1/s_1$		$1/s_1$	0	$1/s_1$	0	$1/s_1$	0	0		$\Delta 0$	$V_{01}.h$
$2'\,1'\,1$		0		0	0	$2r$	$2(s_2r_c+r)$	$-r$	$-r$	$-1/s_2$	$-s_2r_c$	0	0	0	$1/s_2$	$5N-13$	$1'\,0'$	0
$2\,1\,1'$		0		0	0	$-r$	$-r$	$2r$	$2(s_2+r)$	$-1/s_2$	0	0	$-s_2$	0	$1/s_2$		$1'\,2'$	0
$0'\,1'\,2'$		$-s_1r_c$		0	$-1/s_1$	$2s_1r_c$	$-2s_2r_c$	0	0	$1/s_1+1/s_2$	s_2r_c	0	0	0	$-1/s_2$		$1\,0$	0
$0\,1\,2$		0		$-s_1$	$-1/s_1$	0	0	$2s_1$	$-2s_2$	$1/s_1+1/s_2$	0	0	s_2	0	$-1/s_2$		$1\,2$	0
$E_c.\,1\,2$		0		0	0	0	$1/s_2$	0	$1/s_2$	0	$1/s_2$	0	$1/s_2$	0	0		$\Delta 1$	$V_{12}.h$
$3'\,2'\,2$		0	0	0	0	$2r$	$2(s_3r_c+r)$	$-r$	$-r$	$-1/s_3$	$-s_3r_c$	0	0	0	$1/s_3$	$5N-18$	$2'\,1'$	0
$3\,2\,2'$		0	0	0	0	$-r$	$-r$	$2r$	$2(s_3+r)$	$-1/s_3$	0	0	$-s_3$	0	$1/s_3$		$2'\,3'$	0
$1'\,2'\,3'$		$-s_2r_c$	0	0	$-1/s_2$	$2s_2.r_c$	$-2s_3r_c$	0	0	$1/s_2+1/s_3$	s_3r_c	0	0	0	$-1/s_3$		$2\,1$	0
$1\,2\,3$		0	0	$-s_2$	$-1/s_2$	0	0	$2s_2$	$-2s_3$	$1/s_2+1/s_3$	0	0	s_3	0	$-1/s_3$		$2\,3$	0
$E_c.\,2\,3$		0	0	0	0	0	$1/s_3$	0	$1/s_3$	0	$1/s_3$	0	$1/s_3$	0	0		$\Delta 2$	$V_{23}.h$
									F^e									
	$5N-11$								F^e									
$N',\,N'-1$	$5N-6$	$-s_N.r_c$	0	0	$-1/s_N$	$2s_N.r_c$		0		F^e							$N',\,N'-1$	0
$N,\,N-1$		0	0	$-s_N$	$-1/s_N$	0		$2s_N$									$N,\,N-1$	0

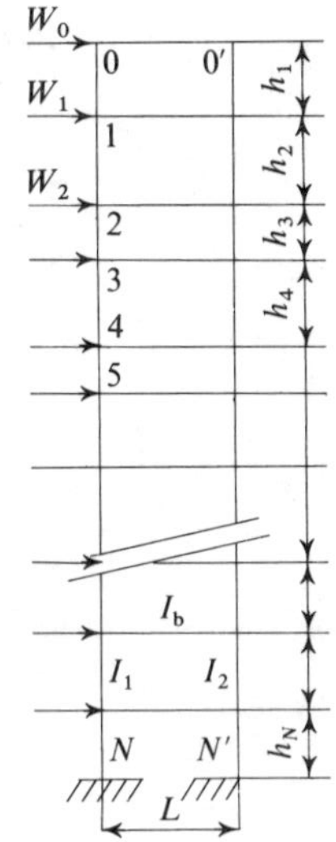

Bending moments are in terms of $W.h$ and deflections in terms of $W.h^3/6EI$.

Table 1.2(c) Flow chart for three-column symmetrical shear walls

$$r = \frac{L.I_1}{h.I_b}, \qquad r_c = \frac{I_1}{I_2}$$

Design	Zeros	Element matrices														Zeros	Forces and displs.	Load vector
1 0 0′																	0′ 1′	0
																		0
1′ 0′ 1									F^e							5N−8	0 1	0
																		0
E_c. 0 1																	Δ1	$V_{01}.h$
2 1 1′																	1′ 0′	0
0 1 2																	1′ 2′	0
2′ 1′ 1									F^e							5N−13	1 0	0
0′ 1′ 2′																	1 2	0
E_c. 1 2																	Δ1	$V_{12}.h$
3 2 2′		0	0	0	0	$-0{\cdot}5r$	$-0{\cdot}5r$	$2r$	$2(r+1)$	−1	0	0	−1	0	1		2′ 1′	0
1 2 3		0	0	−1	−1	0	0	2	−2	2	0	0	1	0	−1		2′ 3′	0
3′ 2′ 2		0	0	0	0	r	$2r_c+r$	$-r$	$-r$	−1	$-r_c$	0	0	0	1	5N−18	2 1	0
1′ 2′ 3′		$-r_c$	0	0	−1	$2r_c$	$-2r_c$	0	0	2	r_c	0	0	0	−1		2 3	0
E_c. 2 3		0	0	0	0	0	0	0	2	0	0	0	2	0	0		Δ2	$V_{23}.h$
									F^e									
																		0
																		0
	5N−1								F^e									0
																		0
																		$V_{N-1}.h$
$N, N-1$																	$N, N-1$	0
	5N−6								F^e									
$N', N'-1$																	$N', N'-1$	0

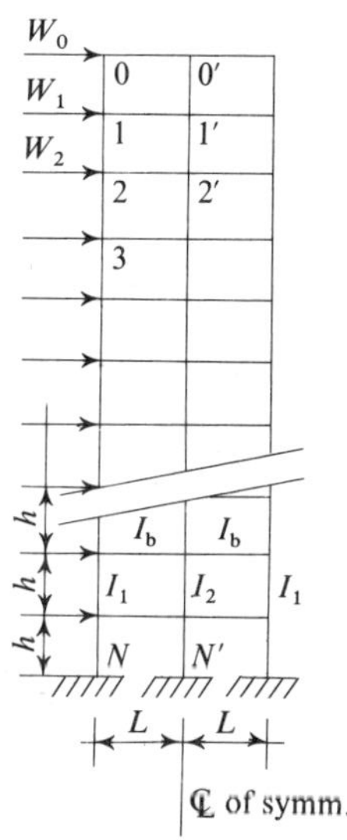

Bending moments are in terms of $W.h$ and deflections in terms of $W.h^3/6EI_1$.

For an unsymmetrical frame on two columns as shown in Fig. 1.7 a general matrix can also be set out in an explicit form in a similar way as for a symmetrical frame. Denoting the non-dimensional parameters:

$$r_1 = \frac{L.I_1}{h.I_b} \quad \text{and} \quad r_2 = \frac{L.I_2}{h.I_b} \tag{1.6}$$

the matrix is as shown in Table 1.2a.

The element matrix F^e is repetitive, provided that the column sizes repeat throughout. If, however, column sizes change at any floor level then the appropriate values of the parameters r_1 and r_2 must also be altered in the matrix.

It should be noted that at the top and foot of the frame, where discontinuities occur, the element matrices are also affected as indicated in Table 1.1 and Table 1.2a.

Based on a similar matrix to that shown in Table 1.2a, Tables 1.2b and 1.2c are prepared showing the flow charts for two- and three-column frames or shear walls. These charts show how all element matrices can be made of identical form even at the extreme boundaries. In addition, for multi-storey frames such charts are more compact than actual matrices and are more suitable for programming.

1.6 Analysis of Multi-bay Frames

In the previous sections, two-column, single-bay frames were used to illustrate the application of the force–displacement method and the setting out of the element and main matrices. The analysis of multi-bay frames for the effect of lateral or gravity forces (Fig. 1.8) does not differ in principle from the previously described procedure. The number of unknown moments, however, increases in proportion to the number of columns introduced. Each junction with four members will have also one unknown bending moment in the beam, while the bending moment in the second beam at this junction can be related, so that the sum of bending moments at each junction is equal to zero.

The main force–displacement matrix always remains diagonal and the element

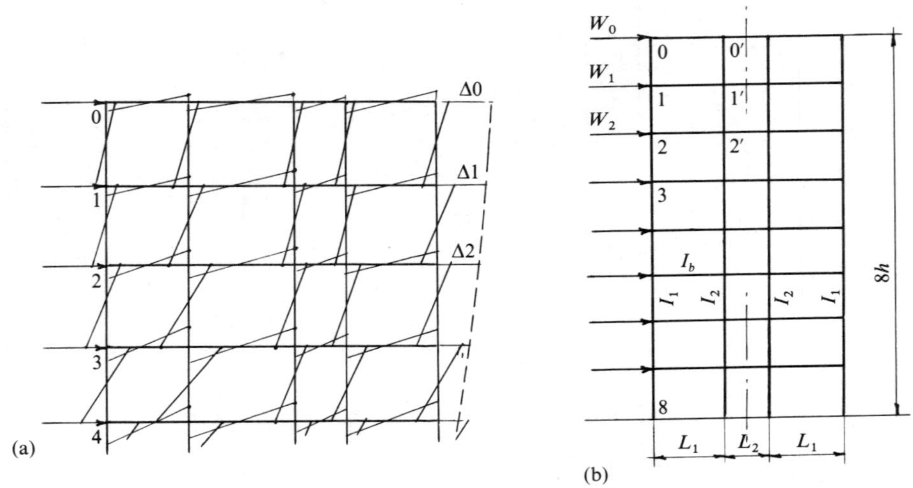

Fig. 1.8 Analysis of multi-bay frames

matrices F^e are repetitive throughout. It is usually more convenient to analyse the effect of lateral forces separately from the effect of gravity forces, but if required both effects can be included at the same time. Similarly, the effect of any arrangement of the superimposed load can be treated separately, taking advantage of the fact that the main matrix remains constant, and only the load vector changes.

To reduce the size of the main matrix, a simplified assumption is sometimes made by which the unknown moments are accepted in one or two columns only, and these are then distributed to all columns in proportion to their respective stiffnesses. This approximation may be adequate for some practical cases but it suffers from the following disadvantages:

(a) not all beam stiffnesses can be accounted for in such analysis,
(b) moments in all columns at a certain floor level are assumed to be of a similar form, while an exact analysis often indicates a great difference in the form and the distribution of moments and shear forces as will be shown later, e.g., in Fig. 2.5 (pages 65 and 66).

A simple example is used to indicate the procedure in the analysis of multi-storey, multi-bay frames by the force–displacement method. Figure 1.8(b) shows an eight-storey, four-column symmetrical frame subjected to the effect of lateral forces. The matrix can be set out in general terms by using the ratios of the member lengths and their second moments of area.

Denoting:

$$r_1 = \frac{L_1 I_1}{h.I_b}, \qquad r_2 = \frac{L_2 I_1}{h I_b} \quad \text{and} \quad r_c = \frac{I_1}{I_2}$$

the matrix can be set out as shown in Table 1.3(a). This matrix is diagonally repetitive with the exception of the first two elements, which are affected by roof discontinuities, and the last two, affected by the discontinuities at the foundation level.

For tall frames, the size of the matrix may become large, although it is always diagonally repetitive and is confined to a narrow band with zero values on either side of the diagonal as shown in Table 1.3(a). However, the setting out of the complete matrix can be avoided by preparing a flow chart to indicate a sequence of operations and their interdependence upon one another. In such a chart, all diagonal matrices (as well as the values of zeros) can be grouped in three vertical columns as shown in Table 1.3(b). All element matrices F^e remain repetitive in vertical column, while 'zero' values increase on the left-hand column and decrease on the right in a regular pattern which depends on the number of columns in the frame and their symmetry. This feature can be utilized in computing, without drawing the complete chart. However, as mentioned earlier, the first and the last two element matrices are slightly modified by the change of the boundary conditions at the roof and foundations respectively and this must be taken into account.

If the column or beam sizes change at any floor level, then the corresponding values of the ratios r must also be changed in the matrix and in the chart. A different chart, however, must be drawn for frames with a different number of columns.

Example 1.4

For a ten-storey frame with four columns, as shown in Fig. 1.9, with constant I

Table 1.3(a) Matrix for the analysis of symmetrical frames on four columns (see Fig. 1.8(b))

$$r_1 = \frac{L_1 I_1}{h I_b}, \quad r_2 = \frac{L_1 I_1}{h I_b}, \quad r_c = \frac{I_1}{I_2}$$

	0		0′		1			1′			2			2′			3			3′	
	0 1	Δ0	0′ 1′	0′ 0	1 0	1 2	Δ1	1′ 0′	1′ 2′	1′ 1	2 1	2 3	Δ2	2′ 1′	2′ 3′	2′ 2	3 2	3 4	Δ3	3′ 2′	3′ 4′
1 0 0′	$2(1+r_1)$	−1	0	$-r_1$	−1	0	1	0													
0 0′ 1′	$-r_1$	−1	$2r_c$	$2r_1$	0	0	1	$-r_c$													
0 0′ 0″	$-r_1$	0	$-r_2$	$2r_1+r_2$	0	0	0	0													
E. 0 1	1	0	1	0	1	0	0	1													
2 1 1′	0	0	0	0	$2r_1$	$2(1+r_1)$	−1	0	0	$-r_1$	−1	0	1	0							
0 1 2	−1	−1	0	0	2	−2	2	0	0	0	1	0	−1	0							
1 1′ 2′	0	0	0	0	$-r_1$	$-r_1$	−1	0	$2r_c$	$2r_1$	0	0	1	$-r_c$							
0′ 1′ 2′	0	−1	$-r_c$	0	0	0	2	$2r_c$	$-2r_c$	0	0	0	−1	r_c							
1 1′ 1″	0	0	0	0	$-r_1$	$-r_1$	0	$-r_2$	$-r_2$	$2r_1+r_2$	0	0	0	0							
E.1–2	0	0	0	0	0	1	0	0	1	0	1	0	0	1							
3 2 2′						0	0	0	0	0	$2r_1$	$2(1+r_1)$	−1	0	0	$-r_1$	−1	0	1	0	
1 2 3						−1	−1	0	0	0	2	−2	2	0	0	0	1	0	−1	0	
2 2′ 3′						0	0	0	0	0	$-r_1$	$-r_1$	−1	0	$2r_c$	$2r_1$	0	0	1	$-r_c$	
1′ 2′ 3′						0	−1	0	$-r_c$	0	0	0	2	$2r_c$	$-2r_c$	0	0	0	−1	r_c	
2 2′ 2″						0	0	0	0	0	r_1	$-r_1$	0	$-r_2$	$-r_2$	$2r_1+r_2$	0	0	0	0	
E. 2–3						0	0	0	0	0	0	1	0	0	0	0	1	0	0	1	
4 3 3′																					
2 3 4																					
3 3′ 4′																				F^e	
2′ 3′ 4′																					
3 3′ 3″																					
E. 3–4																					
5 4 4′																					
3 4 5																					
4 4′ 5′																					
3′ 4′ 5′																					
4 4′ 4″																					
E. 4–5																					
6 5 5′																					
4 5 6																					
5 5′ 6′																					
4′ 5′ 6′																					
5 5′ 5″																					
E. 5–6																					
7 6 6′																					
5 6 7																					
6 6′ 7′																					
5′ 6′ 7′																					
6 6′ 6″																					
E. 6–7																					
8 7 7′																					
6 7 8																					
7 7′ 8′																					
6′ 7′ 8′																					
7 7′ 7″																					
E. 7–8																					
8 7																					
8′ 7′																					

4		4′			5			5′			6			6′			7			7′			8	8′	$\frac{EI_1}{6h^2}$
4 5	Δ4	4′ 3′	4′ 5′	4′ 4	5 4	5 6	Δ5	5′ 4′	5′ 6′	5′ 5	6 5	6 7	Δ6	6′ 5′	6′ 7′	6′ 6	7 6	7 8	Δ7	7′ 6′	7′ 8′	7′ 7	8 7	8′ 7′	
																									0
																									0
																									0
																									$0{\cdot}25Wh$
																									0
																									0
																									0
																									0
																									0
																									$0{\cdot}75Wh$
																									0
																									0
																									0
																									0
																									0
																									$1{\cdot}25Wh$
																									0
																									0
																									0
																									0
																									0
																									$1{\cdot}75Wh$
																									0
																									0
		F^e																							0
																									0
																									0
																									$2{\cdot}25Wh$
																									0
																									0
								F^e																	0
																									0
																									0
																									$2{\cdot}75Wh$
																									0
																									0
														F^e											0
																									0
																									0
																									$3{\cdot}25Wh$
											0	0	0	0	0	$2r_1$	$2(1+r_1)$	-1	0	0	$-r_1$	-1	0	0
											-1	-1	0	0	0	2	-2	2	0	0	0	1	0	0
											0	0	0	0	0	$-r_1$	$-r_1$	-1	0	$2r_c$	$2r_1$	0	$-r_c$	0
											0	-1	0	$-r_c$	0	0	0	2	$2r_c$	$-2r_c$	0	0	r_c	0
											0	0	0	0	0	$-r_1$	$-r_1$	0	$-r_2$	$-r_2$	$2r_1+r_2$	0	0	0
											0	0	0	0	0	0	1	0	0	1	0	1	1	$3{\cdot}75Wh$
																	-1	-1	0	0	0	2	0	0
																	0	-1	0	$-r_c$	0	0	$2r_c$	0

throughout and of similar dimensions to the frame in Fig. 1.5, the required ratios are:

$$r_1 = \frac{L_1 I_1}{h \, . \, I_b} = 2{\cdot}5, \qquad r_2 = \frac{L_2 I_1}{h \, . \, I_b} = 2{\cdot}5, \qquad r_c = 1$$

These values can now be introduced into the general matrix shown in Table 1.3(a) or in the corresponding chart, Table 1.3(b). The distribution of statically indeterminate

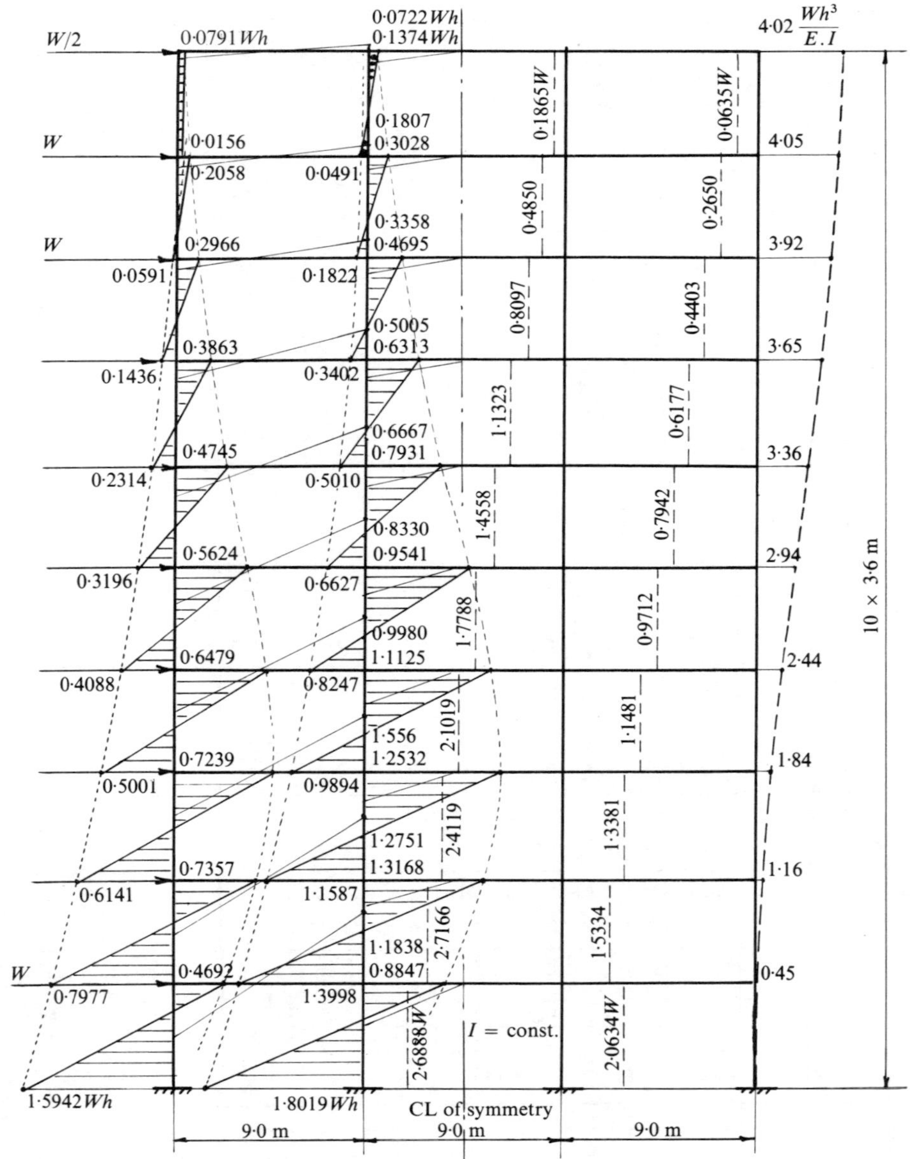

Fig. 1.9 Analysis of ten-storey, three-bay frame

Table 1.3(b) Flow chart for four-column symmetrical shear walls

$$r_1 = \frac{L_1 . I_1}{h . I_b}, \quad r_2 = \frac{L_2 . I_1}{h . I_b}, \quad r_c = \frac{I_1}{I_2}$$

Design	Zeros	Element matrices															Zeros	Forces and displs.	Load vector
1 0 0′								$2(r_1+1)$	-1	0		$-r_1$	-1	0	1	0		0 1	0
0 0′ 1′								$-r_1$	-1	$2r_c$		$2r_1$	0	0	1	$-r_c$	$6N-8$	Δ0	0
0 0′ 0″								$-r_1$	0	$-r_2$		$2r_1+r_2$	0	0	0	0		0′ 1′	0
E_c. 0 1								1	0	1		0	-1	0	0	1		0′ 0	$V_{01}.h$
2 1 1′																		1 0	0
0 1 2																		1 2	0
1 1′ 2′									F^e								$6N-14$	Δ1	0
0′ 1′ 2′																		1′ 0′	0
1 1′ 1″																		1′ 2′	0
E_c. 1 2																		1′ 1	$V_{12}.h$
3 2 2′		0	0	0	0	0	$2r_1$	$2(r_1+1)$	-1	0	0	$-r_1$	-1	0	1	0		2 1	0
1 2 3		-1	-1	0	0	0	2	-2	2	0	0	0	1	0	-1	0		2 3	0
2 2′ 3′		0	0	0	0	0	$-r_1$	r_1	1	0	$2r_c$	$2r_1$	0	0	1	$-r_c$	$6N-20$	Δ2	0
1′ 2′ 3′		0	-1	0	$-r_c$	0	0	0	2	$2r_c$	$-2r_c$	0	0	0	-1	$-r_c$		2′ 1′	0
2 2′ 2″		0	0	0	0	0	$-r_1$	$-r_1$	0	$-r_2$	$-r_2$	$2r_1+r_2$	0	0	0	0		2′ 3′	0
E_c. 2 3		0	0	0	0	0	0	1	0	0	1	0	1	0	0	1		2′ 2	$V_{23}.h$
									F^e										
																			0
																			0
	$6N-13$								F^e										0
																			0
																			0
																			$V_{N-1}.h$
N, N–1	$6N-7$	-1	-1	0	0	0			2	0									0
N' N'–1		0	-1	0	$-r_c$	0			2	$2r_c$									0

Bending moments are in terms of $W.h$ and deflections in terms of $W.h^3/6EI_1$.

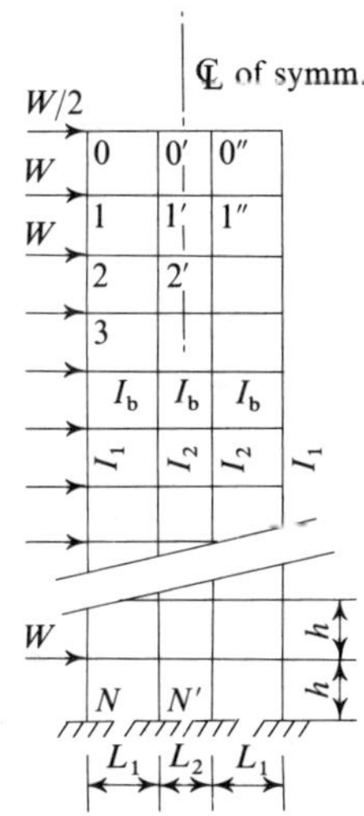

bending moments and deflections as well as the shearing forces in columns is shown in Fig. 1.9. By comparing Fig. 1.9 with the values in Fig. 1.5 for a similar frame but on two columns it can be seen that both the bending moments and deflections are greatly reduced. It is interesting to note that in this case the intermediate columns absorb a larger proportion of the horizontal shear forces and develop higher bending moments than the external columns. The latter, however, are subjected to much higher axial forces. This is because shear forces in beams at the intermediate joints partly cancel each other and only the difference accumulates as the axial force in the columns.

1.7 The Effect of Distributed Gravity and Lateral Forces

In the analysis of tall frames for gravity loads, the main matrix can remain practically the same as in the case of lateral forces, and the main difference is only in the setting-out of the load vector. The latter will now contain the values of angular deformations θ which are the function of the applied loads, e.g., for uniformly distributed load, this value is:

$$\theta = \frac{w.L_1{}^3}{24EI_b} \tag{1.7}$$

and, for a concentrated load P:

$$\theta_1 = \frac{P.a.b(L + a)}{6L.EI_b} \quad \text{and} \quad \theta'_1 = \frac{P.a.b(L + b)}{6L.EI_b} \tag{1.8}$$

The internal shearing forces must be in equilibrium at each floor level so that the sum of all shearing forces must now be equal to zero, and not W as in the case of wind loading.

Since the main matrix remains constant, and unaffected by gravity loads, and only the load vector changes with the assumed pattern of loading, any combination of live loads can be easily taken into account, if required. Live loads are usually smaller than dead loads, and only seldom need an alternate pattern of loading be considered. However, if this is required in multi-bay frames, it can be taken into account without difficulty, because computers are particularly well suited for this type of repetitive calculation.

If the frame as well as the gravity loading is symmetrical, then only half of the frame need be analysed. However, if lateral forces are simultaneously included in such frames, then the system must be considered as unsymmetrical and the matrix must include the whole frame. This is in spite of the fact that if these loads were taken into account separately, then the symmetry would apply and the matrix would need to be set out only for half of the frame in each case. This apparent inconsistency results from the fact that the lateral forces W_0, W_1, W_2 are not truly symmetrically applied to a frame, although this has no effect on the matrix, provided that these loads are taken as concentrated at the joints. For true distributed loads acting on windward side only (Fig. 1.10), the frame is not symmetrical and an exact analysis requires that the frame should be treated as unsymmetrically loaded. This, however, is of little practical importance, except for a portal frame.

To illustrate this the frame in Fig. 1.5 has been analysed on the assumption that the lateral forces are uniformly distributed along one side of the frame. Figure 1.10

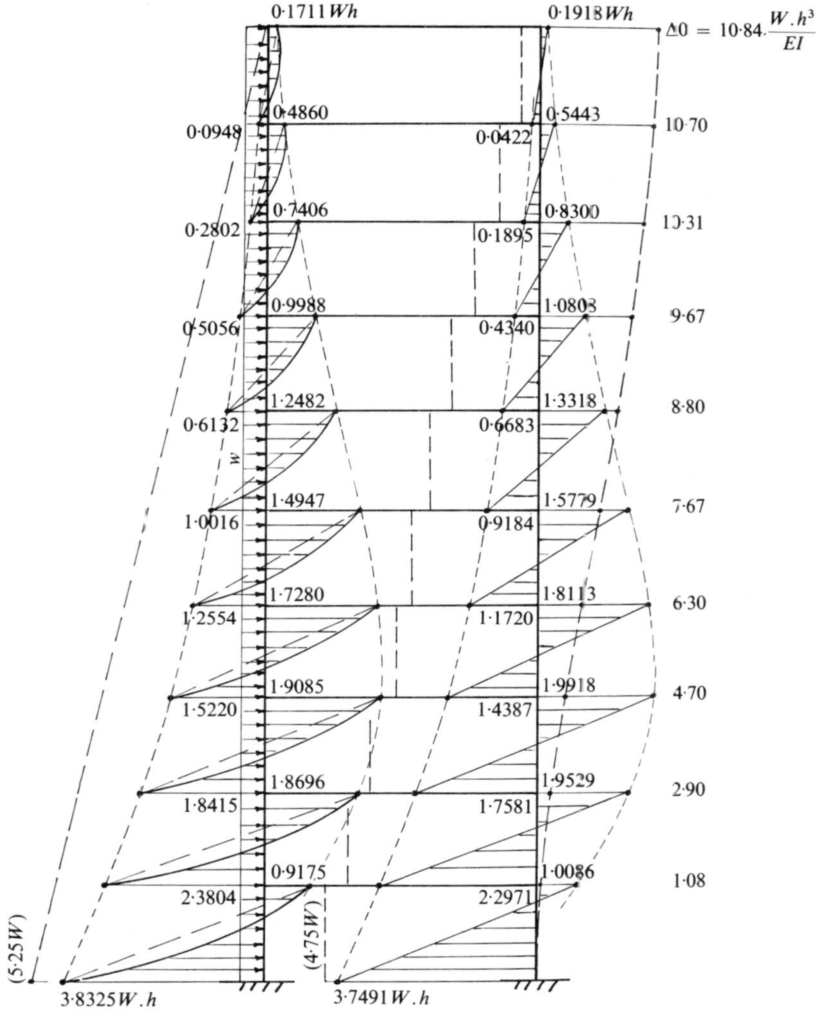

Fig. 1.10 Effect of distributed lateral forces on the frame shown in Fig. 1.5

shows the resulting diagram of bending moments, shear forces and deflections. It can be seen, when comparing this figure with Fig. 1.5 that, as could be expected, the differences in the magnitude of bending moments and deflections are small.

1.8 The Effect of Settlement of Foundations on Tall Building Frames

In the design of foundations for tall buildings, very little settlement is usually tolerated, particularly differential settlement. However, it is seldom possible to eliminate settlement entirely unless foundations are on rock, and the effect of differential settlement may need to be considered in some cases. The amount of possible differential settlement can either be predicted from soil properties or be checked after completion when, of course, little can be done to prevent its effect.

There are, generally, three forms of settlement of foundations which can affect the superstructure: horizontal, vertical, and angular, and these will be briefly reviewed.

1.8.1 *The Effect of Horizontal Settlement of Foundations*

It is well recognized and can be shown analytically that the effect of horizontal settlement of foundations (or yielding of supports) can be very serious, particularly in portal frames and in the lower storeys of tall buildings, and can affect stability of the whole structure. Fortunately, foundations of tall buildings are nearly always tied by beams, slabs, or rafts, and the possibility of this form of settlement is either completely eliminated or reduced to a negligible extent.

The effect of horizontal settlement treated separately is felt most seriously in the lower floors, and diminishes rapidly in the upper floors as shown in Fig. 1.11(b).

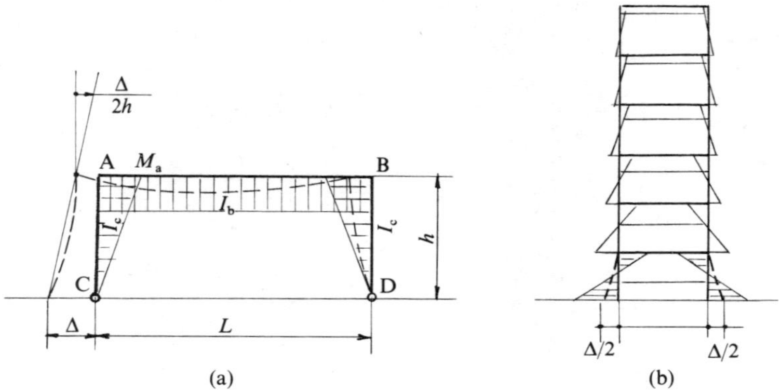

Fig. 1.11 Effect of lateral settlement of foundations on frames

For a single-storey portal frame, subjected to horizontal settlement of supports, as shown in Fig. 1.11(a), we can write the condition of compatibility of deformations at point A as follows:

$$\text{at A,}\quad M_a\left(\frac{L}{2EI_b}+\frac{h}{3EI_c}\right)=\frac{\Delta}{2h}$$

from which, denoting $n = I_b/I_c$, we obtain the well-known expression:

$$M_a = \frac{3EI_b.\Delta}{h(3L + 2nh)} \tag{1.9}$$

In the case of a multi-storey frame (Fig. 1.11(b)), the force–displacement matrix can be set out in a similar way to that used for wind loading. The shear forces are self-equilibrating and this is stated in setting out the conditions of equilibrium of forces in the load vector.

1.8.2 *The Effect of Vertical Settlement of Foundations*

Any structure can settle bodily, vertically, or by rotation with respect to some point such as point A in Fig. 1.12(a). In the latter case, vertical settlement would not result in any primary stresses at all except for those arising from the instability effect. In the limit, a structure would overturn as a rigid body, as shown diagrammatically in Fig. 1.12(b).

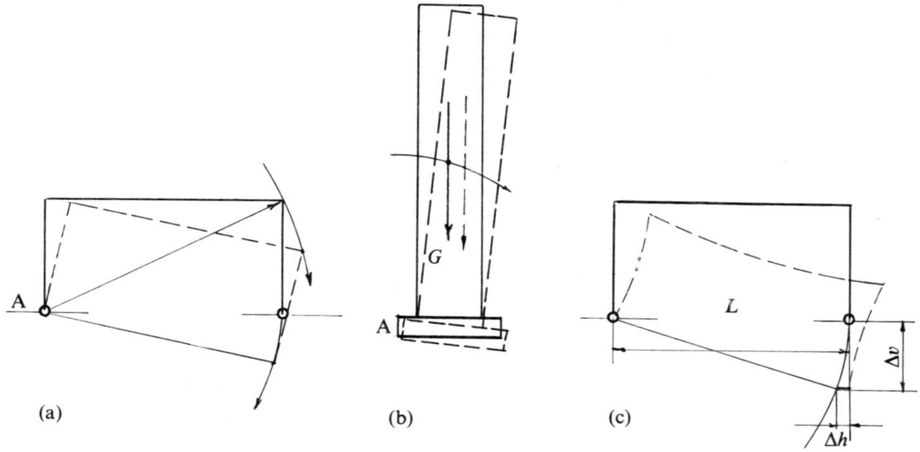

Fig. 1.12 'Body' settlement of frames

However, if one foundation settles vertically only with respect to the other, as shown in Fig. 1.12(c), then primary stresses occur only from its horizontal component Δh. This value can be approximately calculated by geometry from:

$$\frac{\Delta h}{\Delta v} = \frac{\Delta v/2}{L} \quad \text{or} \quad \Delta h = \frac{\Delta v^2}{2L} \tag{1.10}$$

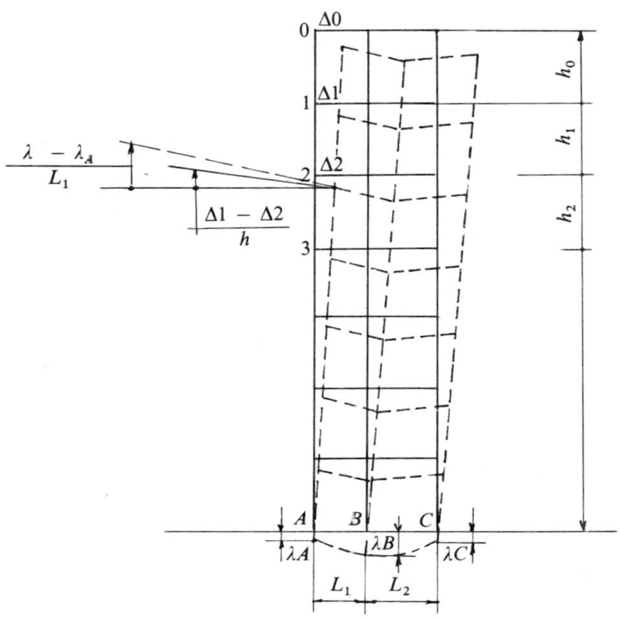

Fig. 1.13 Differential settlement of foundations

This value is usually very small in comparison with Δv, and indicates that this form of vertical settlement can result in considerably smaller stresses than from horizontal settlement, provided that its magnitude is such that instability does not take place.

In tall buildings, vertical differential settlement can be felt throughout the structure and can seriously affect distribution of stresses as well as stability. The force–displacement matrix F can be set out in a similar way as for gravity or wind forces, and only the load vector needs to be considered, e.g., at the point 1 in Fig. 1.13, between column 2–1 and beam 1–1′, the discontinuity at release becomes:

$$\frac{\lambda B - \lambda A}{L_1} - \frac{\Delta 1 - \Delta 2}{h_1} \tag{1.11}$$

in which λA, λB, etc., are given settlements of supports, and $\Delta 1$, $\Delta 2$, etc., are the unknown translations of joints.

1.8.3 *The Effect of Angular Rotations of Supports* (*Fig. 1.14*)

Angular settlements of supports are seldom known and therefore receive little attention. Their effect, however, can be very high, particularly in the lower parts of the frame, and diminishing rapidly in the upper floors. The analysis, however, could be made without difficulty if the value of the rotation of foundations, ϕ, was known.

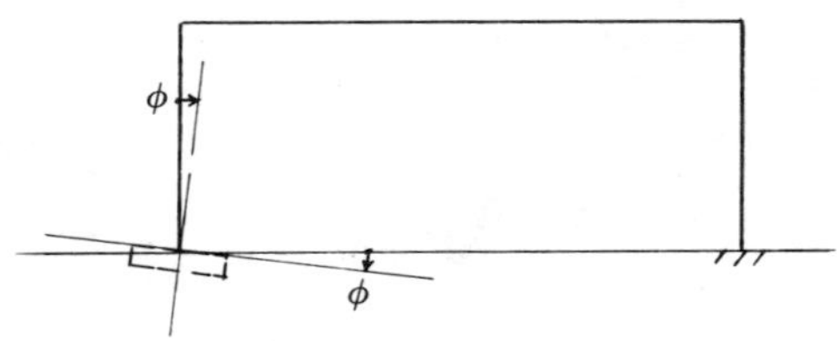

Fig. 1.14 Effect of rotation of foundation

1.9 The Effect of Members with Variable Sections

Members with variable second moments of area frequently occur in tall building frames and in floors, and their load–deflection characteristics can best be found by using Unit Load method. These characteristics can be expressed in terms of stiffness or flexibility coefficients for individual members. In the following, only the flexibility coefficients will be considered. The procedure is best illustrated by numerical examples,

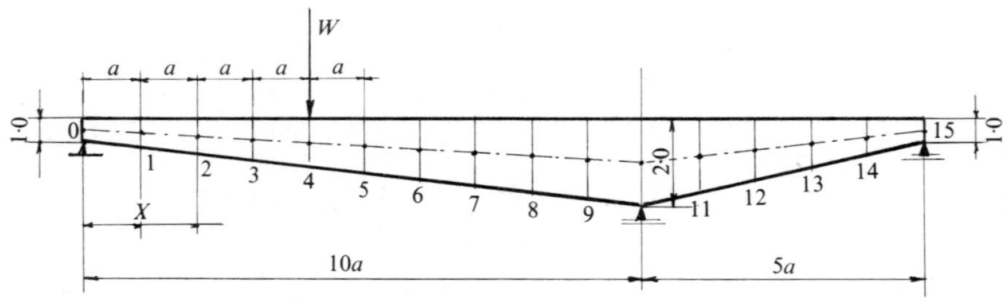

Fig. 1.15 Beam with variable second moments of area

as follows: Figure 1.15 shows a two-span beam subjected to a point load with all the relevant dimensions given.

First, beams are divided into elements, preferably of the same length, and the values required in the Unit Load expression are tabulated, as shown in Fig. 1.16.

In this example, the flexibility coefficients are expressed in terms of the second moment of area of the beam at the nodal point 0, and are as follows:

$$f_{00} = \int_0^{10} \frac{M.m.\mathrm{d}x}{EI_n} = \frac{a}{E}\sum\frac{m_a}{I_n}; \qquad \frac{a}{EI_0}\sum_1^{10} n.m_a^{\,2} = 2{\cdot}4956\frac{a}{EI_0}$$

$$f_{0,10} = \frac{a}{EI_0}\cdot\sum_1^{10} n.m_a.m_b = 0{\cdot}5648\cdot\frac{a}{EI_0} \tag{1.12}$$

$$f_{10,10} = \frac{a}{EI_0}\cdot\sum n.m_b^{\,2} = 0{\cdot}7459\cdot\frac{a}{EI_0}; \quad n = (d_1/d_n)^3$$

Similarly, for the beam between nodal points 10 and 15, we have:

$$f_{10,10} = \frac{a}{EI_0} \times 0{\cdot}4055$$

$$f_{15,10} = \frac{a}{EI_0} \times 0{\cdot}3977$$

$$f_{15,15} = \frac{a}{EI_0} \times 1{\cdot}5701$$

Now, the load function θ is found in the same way. The relevant values are shown in Fig. 1.15(c), from which:

$$\theta_{10,0} = \int_0^{10} \frac{M_x.m_b.\mathrm{d}x}{EI} = \frac{W.a}{EI_1}\cdot\sum_1^{10} n.M_x.m_b = 1{\cdot}6519 \times \frac{W.a}{EI_0}$$

The statically indeterminate bending moment at the nodal point 10 can now be calculated by setting out the condition of compatibility of deformations at this point, as follows:

$$M_{10}.(f_{10,10}^{10,0} + f_{10,10}^{10,15}) = \theta_{10,0}$$

or

$$M_{10}.(0{\cdot}7459 + 0{\cdot}4055) = 1{\cdot}6518W.a$$

from which: $M_{10} = 1{\cdot}4346W.a$
and $M_4 = 1{\cdot}8262W.a.$

The values in parentheses in Fig. 1.17 would be obtained, assuming that the beam is of constant average section, corresponding to the depth at point 5. The average value of deflection at point 4 can be calculated from the compatibility condition:

$$M_4\cdot\left(\frac{4a}{3EI_{av}} + \frac{6a}{3EI_{av}}\right) - M_{10}\cdot\frac{6a}{6EI_{av}} = \frac{\Delta 4}{4a} + \frac{\Delta 4}{6a} \tag{1.13}$$

from which:

$$\Delta 4 = 12{\cdot}9264W.a^3/EI_{av}$$

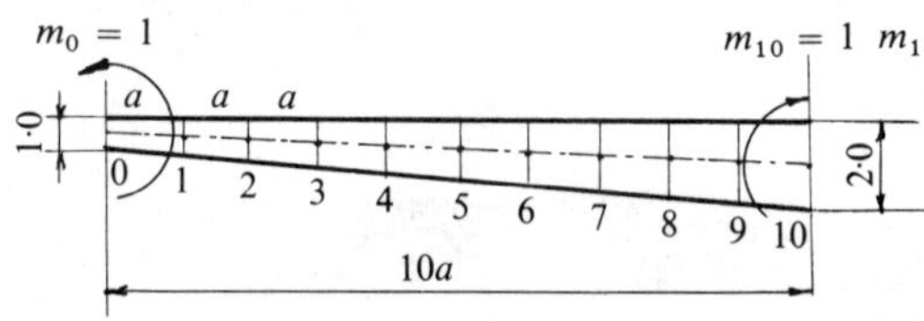

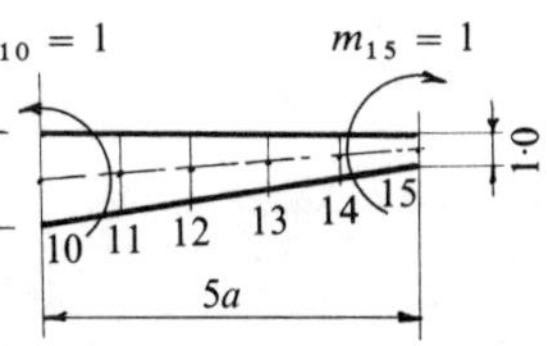

Node	d	d^3	$n=\left(\frac{d_0}{d_n}\right)^3$	$m_0=1$	$n.m_0{}^2$	$m_{10}=1$	$n.m_{10}^2$	$n.m_0.m_{10}$
0	1·0	1·000	1·0000	1·0	1·0000	0	0	0
1	1·1	1·331	0·7513	0·9	0·6085	0·1	0·0075	0·0676
2	1·2	1·628	0·6142	0·8	0·3931	0·2	0·0246	0·0983
3	1·3	2·197	0·4552	0·7	0·2230	0·3	0·0410	0·0956
4	1·4	2·744	0·3644	0·6	0·1312	0·4	0·0583	0·0874
5	1·5	3·375	0·2963	0·5	0·0741	0·5	0·0741	0·0741
6	1·6	4·096	0·2441	0·4	0·0391	0·6	0·0879	0·0586
7	1·7	4·913	0·2035	0·3	0·0183	0·7	0·0997	0·0427
8	1·8	5·832	0·1715	0·2	0·0068	0·8	0·1097	0·0274
9	1·9	6·859	0·1458	0·1	0·0015	0·9	0·1181	0·0131
10	2·0	8·000	0·1250	0	0	1·0	0·1250	0
Σ	—	—	—	—	2·4956	—	0·7459	0·5648

(a)

Node	d	d^3	$n=\left(\frac{d_0}{d_n}\right)^3$	$m_{10}=1$	$n.m_{10}^2$	$m_{15}=1$	$n.m_{15}^2$	$n.m_{10}.m_{15}$
10	2·0	8·000	0·1250	1·0	0·1250	0	0	0·1250
11	1·8	5·832	0·1715	0·8	0·1097	0·2	0·0068	0·0279
12	1·6	4·096	0·2441	0·6	0·0879	0·4	0·0390	0·0586
13	1·4	2·744	0·3644	0·4	0·0583	0·6	0·1312	0·0879
14	1·2	1·628	0·6142	0·2	0·0246	0·8	0·3931	0·0903
15	1·0	1·0	1·0	0	0	1·0	1·0	0
Σ	—	—	—	—	0·4055	—	1·5701	0·3977

(b)

Node	$n=\left(\frac{d_0}{d_n}\right)^3$	M_{0x}	$m_{10}=1$	$n.M_x.m_{10}$
0	1·0	0	0	0
1	0·7513	0·6Wa	0·1	0·0451Wa
2	0·6142	1·2Wa	0·2	0·1474
3	0·4552	1·8Wa	0·3	0·2458
4	0·3644	2·4Wa	0·4	0·3498
5	0·2963	2·0Wa	0·5	0·2963
6	0·2441	1·6Wa	0·6	0·2343
7	0·2035	1·2Wa	0·7	0·1709
8	0·1715	0·8Wa	0·8	0·1097
9	0·1458	0·4Wa	0·9	0·0525
10	0·1250	0	1·0	0
Σ	—	—	—	1·6519 $\frac{Wa}{EI_0}$

Node	d	d^3	$n=\left(\frac{d_6}{d_n}\right)^3$	$m_0=1$	$n.m_0^2$	$n.m_0.m_{12}$
0	2·0	8·0	0·1250	1·0	0·1250	0
1	1·83	6·1618	0·1623	0·916	0·1364	0·0124
2	1·6	4·6291	0·2160	0·83	0·1500	0·0300
3	1·5	3·3750	0·2963	0·75	0·16	0·0555
4	1·3	2·3702	0·4219	0·6	0·1875	0·0937
5	1·16	1·5876	0·6299	0·583	0·2141	0·1530
6	1·0	1·0	1·0	0·5	0·25	0 2500
7	1·16	1·5876	0·6299	0·416	0·1094	0·1531
8	1·3	2·3702	0·4219	0·3	0·0468	0·0937
9	1·5	3·3750	0·2963	0·25	0·0185	0·0555
10	1·6	4·6291	0·2160	0·16	0·0060	0·0300
11	1·83	6·1618	0·1623	0·083	0·0011	0·0124
12	2·0	8·0	0·1250	0	0	0
Σ	—	—	—	—	1·4114	0·9393

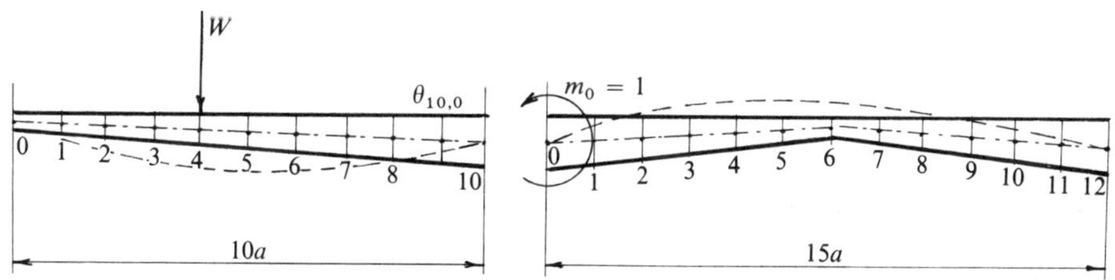

(c) (d)

Fig. 1.16 Example of calculation of flexibility coefficients of beams with variable second moments of area

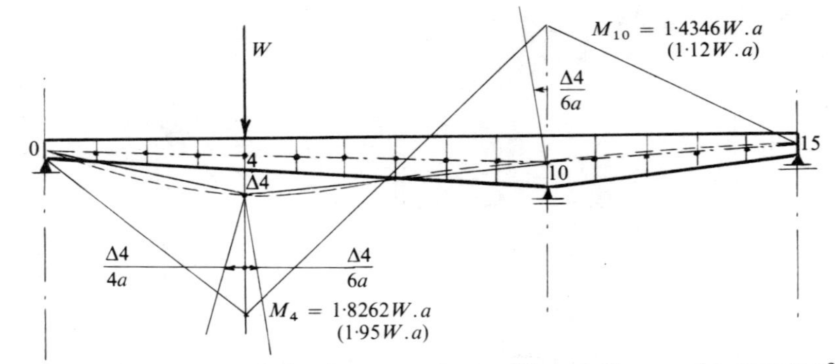

Fig. 1.17 Example of the analysis of two-span beam with variable second moments of area

An alternative analysis by the force–displacement method could be made by introducing releases at all the nodal points simultaneously as shown in Fig. 1.18. In this way, moments and deflections at all nodal points can be calculated simultaneously by setting out conditions of compatibility of deformations and equilibrium of forces at all the nodal points, e.g., a typical compatibility condition at point 2, based on Fig. 1.18, can be written as follows:

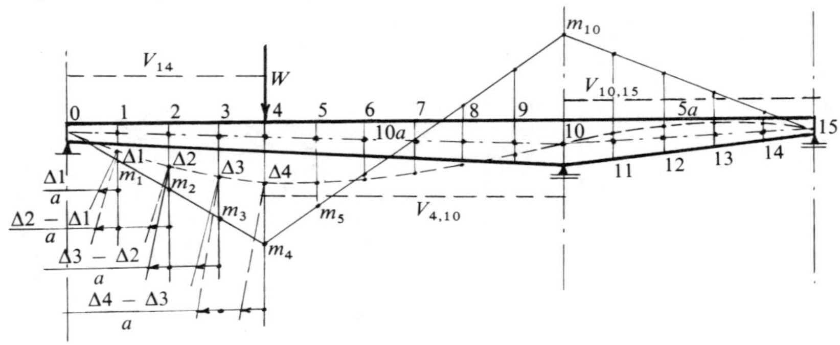

Fig. 1.18 Finite segments analysis of a beam with variable second moments of area

2. $$m_1 \cdot \frac{a}{6EI_1} + m_2 \cdot \frac{2a}{3EI_2} + m_3 \cdot \frac{a}{6EI_3} = \frac{\Delta 2 - \Delta 1}{a} - \frac{\Delta 3 - \Delta 2}{a} \qquad (1.14)$$

or $$m_1 \cdot \frac{n_1}{6} + m_2 \cdot \frac{2n_2}{3} + m_3 \cdot \frac{n_3}{6} + (\Delta 1 - 2\Delta 2 + \Delta 3) \cdot \frac{EI_0}{a^2} = 0$$

or $$0{\cdot}1252m_1 + 0{\cdot}4096m_2 + 0{\cdot}0759m_3 + (\Delta 1 - 2\Delta 2 + \Delta 3)\frac{EI_0}{a^2} = 0$$

The condition of equilibrium of forces at point 2 becomes:

S.C. 2. $$\frac{m_2 - m_1}{a} - \frac{m_3 - m_2}{a} = 0$$

or $$-m_1 + 2m_2 - m_3 = 0 \qquad (1.15)$$

In a similar way, these conditions can be set out at all nodal points. Table 1.4 shows the setting-out of the general matrix for calculation of moments and deflec-

Table 1.4 Force–displacement matrix for the analysis of beam in Fig. 1.17

Node	m_1	Δ1	m_2	Δ2	m_3	Δ3	m_4	Δ4	m_5	Δ5	m_6	Δ6	m_7	Δ7	m_8	Δ8	m_9	Δ9	m_{10}	m_{11}	Δ11	m_{12}	Δ12	m_{13}	Δ13	m_{14}	Δ14	
1	0·5008	−2	0·1024	+1																								0
S.C. 1	+2	0	−1	0																								0
2	0·1252	+1	0·4096	−2	0·0758	+1																						0
S.C. 2	−1	0	+2	0	−1	0																						0
3			0·1024	+1	0·3032	−2	0·0607	+1																				0
S.C. 3			−1	0	+2	0	−1	0																				0
4					0·0759	+1	0·2692	−2	0·0495	+1																		0
S.C. 4					−1	0	+2	0	−1	0																		$1{\cdot}0Wu$
5							0·0673	+1	0·1976	−2	0·0407	+1																0
S.C. 5							−1	0	+2	0	−1	0																0
6									0·0494	+1	0·1628	−2	0·0339	+1														0
S.C. 6									−1	0	+2	0	−1	0														0
7											0·0407	+1	0·1356	−2	0·0286	+1												0
S.C. 7											−1	0	+2	0	+1	0												0
8													0·0339	−1	0·1140	+2	0·0243	−1										0
S.C. 8													−1	0	−2	0	+1	0										0
9															0·0285	−1	0·0972	+2	0·0208									0
S.C. 9															−1	0	+2	0	−1									0
10																	0·0243	−1	0·0832	0·0285	+1							0
11																			0·0208	0·1140	−2	0·0407	+1					0
S.C. 11																			−1	+2	0	−1	0					0
12																				0·0285	+1	0·1628	−2	0·0673	+1			0
S.C. 12																				−1	0	+2	0	−1	0			0
13																						0·0407	+1	0·2692	−2	0·1024	+1	0
S.C. 13																						−1	0	+2	0	+1	0	0
14																								0·0673	+1	0·4096	−2	0
S.C. 14																								−1	0	+2	0	0

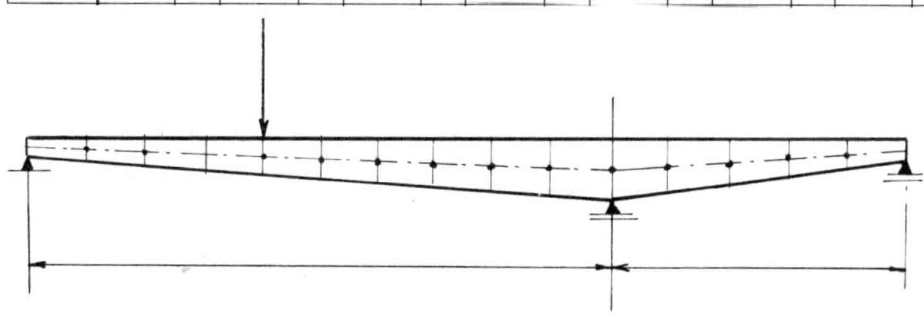

tions at all nodal points simultaneously, and Fig. 1.19 shows the distribution of these values throughout the beam.

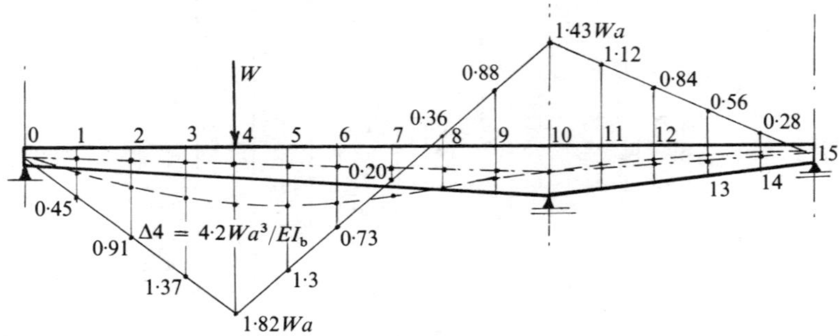

Fig. 1.19 Distribution of bending moments and deflections of a beam shown in Fig. 1.17

1.10 The Effect of Discontinuities in the Analysis of Tall Frames

The 'discontinuities' in the geometrical arrangement of frames or shear walls are sometimes introduced as a result of planning requirements and can take various forms, some of which are shown in Fig. 1.20. In these, either beams or columns change their normal pattern at certain levels, when some shear walls may not extend into foundations and instead are carried by a special arrangement of frames. Such frames must carry not only the accumulation of column loads but also the full value of the wind force acting at eccentricity. The supporting frames may have inclined columns or beams with constant or variable cross-sections, as shown in Fig. 1.20.

Structures containing such discontinuities can be analysed either as one unit with the lowest frame or, approximately, by separately analysing the upper and the lower part.

There is no basic difference in the setting out of the matrix as far as the 'primary' moments and deflections are concerned. The high concentration of stresses at some points, however, may need special consideration. Another factor which needs to be considered in some frames is the effect of vertical translation of joints. This is illustrated by the following examples.

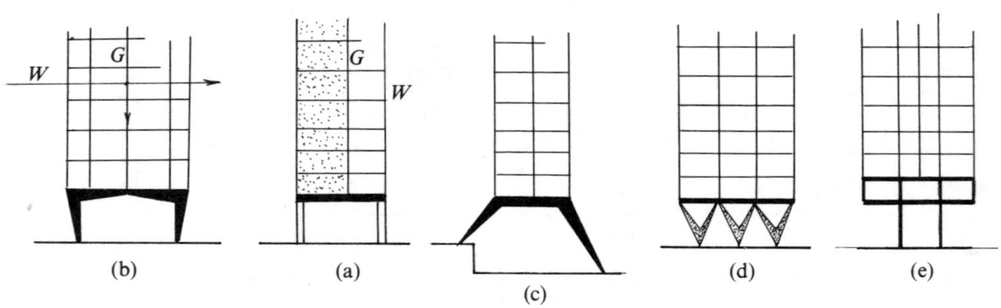

Fig. 1.20 Discontinuities in tall building frames

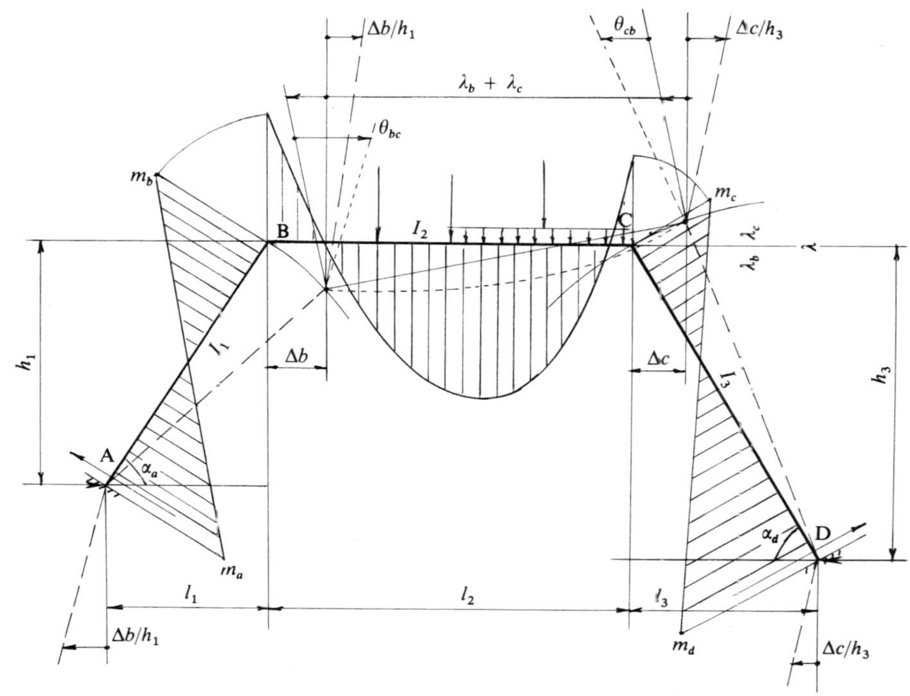

Fig. 1.21 Analysis of frames with inclined columns

1.10.1 *Frames with inclined columns*

The analysis of frames with inclined columns and with constant or variable sections can be made based on Fig. 1.21 after the flexibility coefficients are evaluated. In Fig. 1.21 the frame is shown in its final deflected configuration when, because of the inclination of columns both the horizontal and vertical deflections take place.

Assuming that the deformations are small, $\Delta b = \Delta c = \Delta$, vertical deformations λ can be expressed in terms of the horizontal deflections as follows:

$$\lambda_b = \Delta \cdot \frac{l_1}{h_1} \quad \text{and} \quad \lambda_c = \Delta \cdot \frac{l_3}{h_3}$$

so that $$\lambda = \lambda_b + \lambda_c = \Delta \left(\frac{l_1}{h_1} + \frac{l_2}{h_2} \right) \tag{1.16}$$

For a frame with constant second moments of area in all members, the equations of compatibility and equilibrium become:

A. $$-m_a \frac{h_1}{3EI_1 . \sin \alpha_a} + m_b \cdot \frac{h_1}{6EI_1 . \sin \alpha_a} = \frac{\Delta_b}{h_1}$$

B. $$-m_a \cdot \frac{h_1}{6EI_1 . \sin \alpha_a} + m_b \left(\frac{h_1}{3EI_1 . \sin \alpha_a} + \frac{l_2}{3EI_2} \right) + m_c \cdot \frac{l_2}{6EI_2} = \theta_{bc} - \frac{\lambda_b + \lambda_c}{l_2} - \frac{\Delta_b}{h_1}$$

C. $$m_b \cdot \frac{l_2}{6EI_2} + m_c\left(\frac{l_2}{3EI_2} + \frac{h_3}{3EI_3 . \sin\alpha_d}\right) - m_d \frac{h_3}{6EI_3 \sin\alpha_d}$$

$$= \theta_{cb} + \frac{\lambda_b + \lambda_c}{l_2} + \frac{\Delta_c}{h_3}$$

D. $$-m_c \cdot \frac{h_3}{6EI_3 \sin\alpha_d} + m_d \frac{h_3}{3EI_3 . \sin\alpha_d} = \frac{\Delta c}{h_3}$$

E.C. $$\frac{m_a + m_b}{h_1/\sin\alpha_a} \sin\alpha_a - \frac{m_c + m_d}{h_3/\sin\alpha_d} \cdot \sin\alpha_d = 0$$

Multiplying compatibility equations by $6EI_1/h_1$ and denoting

$$r_1 = \frac{l_2 . I_1}{h_1 . I_2}, \qquad r_2 = \frac{l_1}{l_2} + \frac{l_3}{l_2} \cdot \frac{h_1}{h_3}, \qquad r_3 = \frac{h_3 I_1}{h_1 I_3}$$

the matrix can be written in the form:

	m_a	m_b	m_c	m_d	$\Delta \cdot \frac{6EI_1}{h_1^2}$	
A	-2	$+1$	0	0	$-\sin\alpha_a$	0
B	-1	$2(1 + r_1 . \sin\alpha_a)$	$r_1 . \sin\alpha_a$	0	$r_2 + 1$	$\theta_{bc} \cdot \frac{6EI_1}{h_1} \cdot \sin\alpha_a$
C	0	$r_1 . \sin\alpha_d$	$2(r_1 \sin\alpha_d + r_2)$	$-r_2$	$-\left(r_2 + \frac{h_1}{h_3}\right) \sin\alpha_d$	$\theta_{cb} \cdot \frac{6EI_1}{h_1} \cdot \sin\alpha_d$
D	0	0	$-r_3$	$2r_3$	$-\frac{h_1}{h_3} \cdot \sin\alpha_d$	0
E.C.	$\sin^2\alpha_a$	$\sin^2\alpha_a$	$-\frac{h_1}{h_3} \sin^2\alpha_d$	$-\frac{h_1}{h_3} \sin^2\alpha_d$	0	0

For members with variable cross-sections the above matrix must be adjusted by introducing the appropriate values of the flexibility coefficients.

1.10.2 *Frames with inclined beams*

In tall buildings the use of frames with inclined beams is rare because of their large deformations in both vertical and horizontal directions. However, the analysis represents an interesting application of the force–displacement method. Figure 1.22 shows a general case of an unsymmetrical frame subjected to any system of loading. The vertical deflection λ of the joint C can be either related to the lateral deflections Δb and Δd or, alternatively, can be included in the matrix as an unknown value. In the latter case, the additional equation required is derived from the equilibrium of vertical forces at point C. However, to eliminate the value of λ, releases are introduced at all joints in the frame, when, with no external loads, the point C is displaced vertically by λ and the points B and D are displaced horizontally by $\Delta 1$ and $\Delta 2$ respectively so

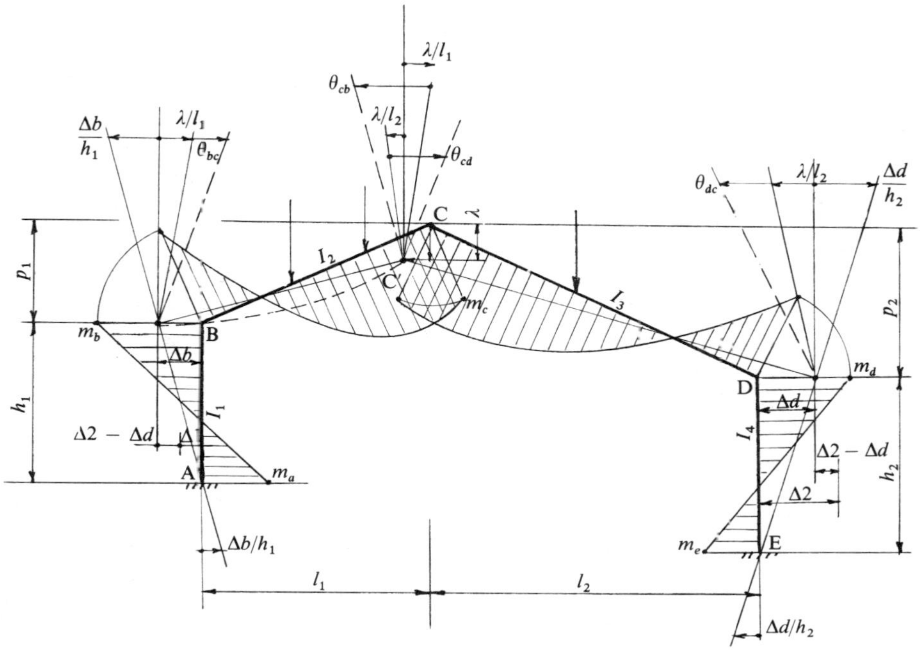

Fig. 1.22 Analysis of frames with inclined beams

that:

$$\Delta 1 = \lambda \frac{p_1}{l_1} \quad \text{and} \quad \Delta 2 = \lambda \frac{p_2}{l_2}$$

When loads are applied, the joints B and D will be displaced by the unknown values Δb and Δd and the following relationship can be written, based on Fig. 1.22:

$$\Delta b = \Delta 1 + (\Delta 2 - \Delta d)$$

from which

$$\lambda = (\Delta d - \Delta b) \cdot \frac{l_1 . l_2}{p_1 . l_2 + p_2 . l_1}$$

If $l_1 = l_2 = l$ and $p_1 = p_2$, then:

$$\lambda = (\Delta d - \Delta b) \frac{l}{2p}$$

Now the equations of compatibility and equilibrium can be written in the usual way except that the effect of inclination of members must be considered. These equations are as follows:

A. $$m_a \cdot \frac{h_1}{3EI_1} - m_b \cdot \frac{h_1}{6EI_1} = \frac{\Delta b}{h_1}$$

B. $$-m_a \cdot \frac{h_1}{6EI_1} + m_b \cdot \left(\frac{h_1}{3EI_1} + \frac{l_1}{3EI_2 \cos \alpha_a} \right) - m_c \frac{l_1}{6EI_2 \cos \alpha_a} = \frac{\Delta b}{h_1} + \frac{\lambda}{l_1} + \theta_{bc}$$

C. $$-m_b \frac{l_1}{6EI_2 \cos \alpha_a} + m_c\left(\frac{l_1}{3EI_2 \cos \alpha_a} + \frac{l_2}{3EI_3 \cos \alpha_d}\right) - m_d \cdot \frac{l_2}{6EI_3 \cos \alpha_d} = \theta_{cb} + \theta_{cd} - \frac{\lambda}{l_1} - \frac{\lambda}{l_2}$$

D. $$-m_c \cdot \frac{l_2}{6EI_3 \cos \alpha_d} + m_d\left(\frac{l_2}{3EI_3 \cos \alpha_d} + \frac{h_2}{3EI_4}\right) - m_e \frac{h_2}{6EI_4} = \frac{\Delta d}{h_2} + \frac{\lambda}{l_2} + \theta_{dc}$$

E. $$-m_d \cdot \frac{h_2}{6EI_4} + m_e \cdot \frac{h_2}{3EI_4} = \frac{\Delta d}{h_2}$$

E.C.AE. $$\frac{m_a + m_b}{h_1} - \frac{m_d + m_e}{h_2} = 0$$

E.C.C. $$V_{cb}^0 - \frac{m_b + m_c}{l_2/\cos \alpha_a} + V_{cd}^0 - \frac{m_c + m_d}{l_2/\cos \alpha_d} = 0$$

in which $$\lambda = (\Delta d - \Delta b)\frac{l}{2p}$$

If the members are of variable sections, the flexibility coefficients must be adjusted accordingly as discussed in section 1.9.

1.11 The Effect of Boundary Conditions in the Analysis of Frames and Shear Walls

In flexible frames (Figs. 1.23(a) and 1.5) beams absorb comparatively large bending moments which are the sum of moments in the column above and below the junction. As a result of this, large shearing forces develop in beams and therefore axial forces in columns as shown in Fig. 1.5.

In rigid frames, however, beams are usually very flexible relative to columns, and can balance only a small proportion of column bending moments (Figs. 1.23(b) and

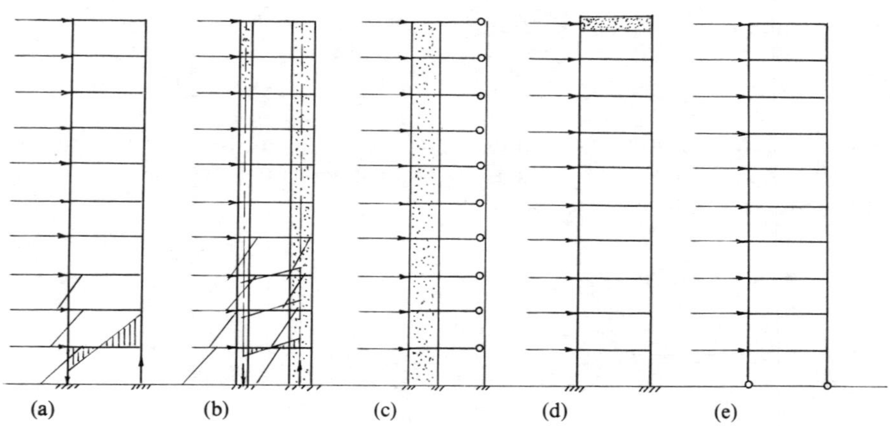

Fig. 1.23 Effect of boundary conditions in frames and shear walls

1.6). Small shearing forces then develop in beams and small direct forces in shear walls. In the limit, when either the beam sizes become too small to absorb any bending moment, or their connections to walls are hinged, the moments in columns approach those of a free cantilever.

Shear walls, in which one column is very flexible and practically cannot absorb any bending moment are sometimes used (Fig. 1.23(c)) for aesthetic reasons. Such columns, however, can usually transmit a direct force and are beneficial in reducing bending moments in the shear wall provided that the beams are rigidly connected at the far end to the shear wall.

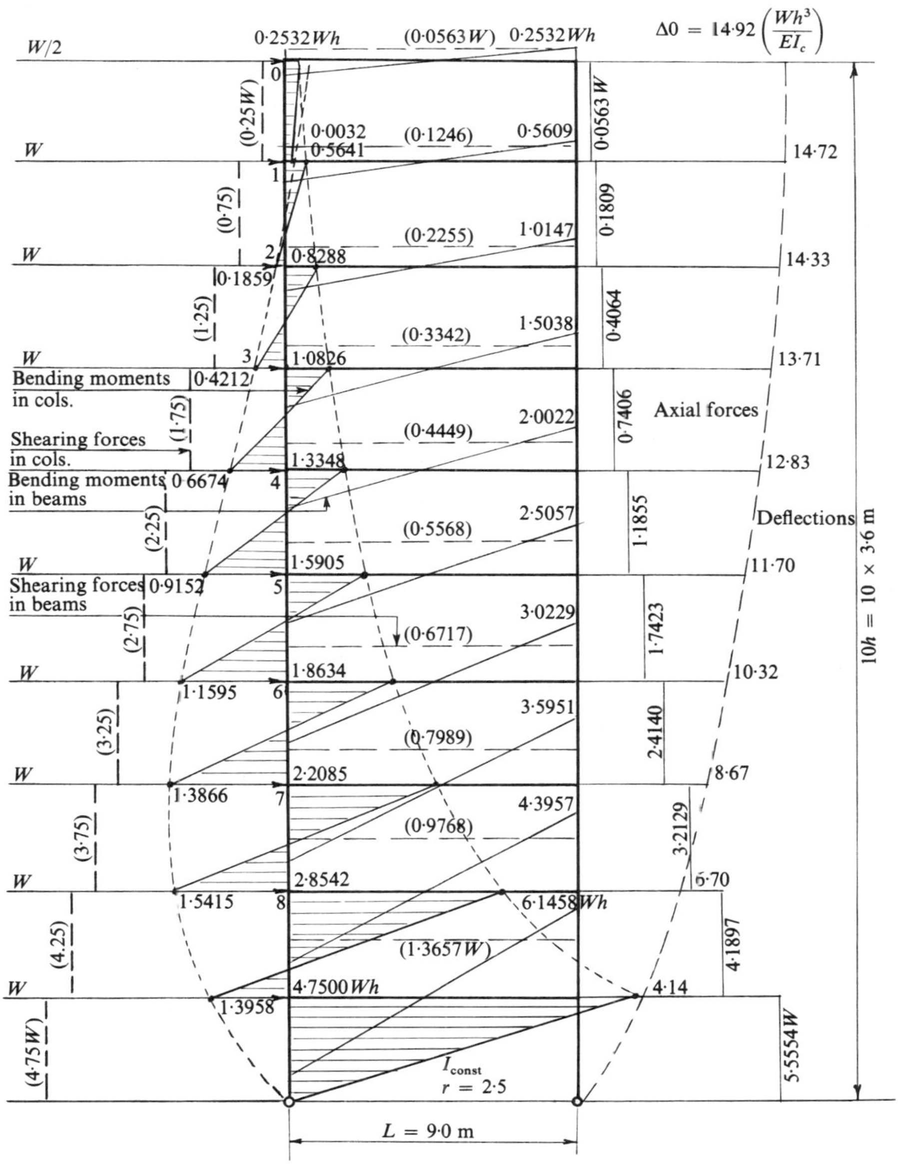

Fig. 1.24 Example of the analysis of a hinged frame

The arrangement of large beams and very flexible columns is not practical, but a very rigid roof beam is sometimes possible, and the effect of such a beam (Fig. 1.23(d)) is very beneficial in reducing the overall deflections of a frame. However, the second-last beam at the top of the frame often carries an exceptionally high bending moment, shearing and direct forces. An exact analysis in such cases is essential in avoiding cracks and redistribution of moments and deformations.

A deliberate introduction of hinges in tall buildings or shear walls at the base or at intermediate points is very rare (Fig. 1.23(e)). However, cracks resulting from differential settlement or dynamic action of wind and earthquake forces can result in redistribution of bending moments and deflections. To illustrate this effect, the frame in Fig. 1.5, previously analysed as fixed at the base, is now analysed assuming that full hinges develop at the base. The matrix remains almost identical to that shown in Table 1.1 except that the bending moment at the point 10 is taken as zero, and therefore the size of the matrix is 29×29. Figure 1.24 shows the resulting distribution of bending moments, forces, and deflection. It can be seen, upon comparison with Fig. 1.5, that the presence of hinges can seriously affect the magnitude of bending moments particularly in the lower floors, as well as forces and deflections. The latter became almost constant throughout many upper floors and diminish rapidly only in the lowest floors. Hinges or cracks can therefore result in considerable increase of stresses in some members and can adversely affect deflections as well as vibrations of frames.

1.12 Approximate Analysis of Continuous Foundations to Tall Frames

Tall building frames and shear walls are usually supported by rigid foundations and it is then assumed that they settle uniformly or as a rigid body, and that the pressure distributed can be calculated on the basis of the column formula.

However, if the foundation beam is elastic, then differential settlement may occur, the pressure may not be linear throughout, and bending moments may be seriously

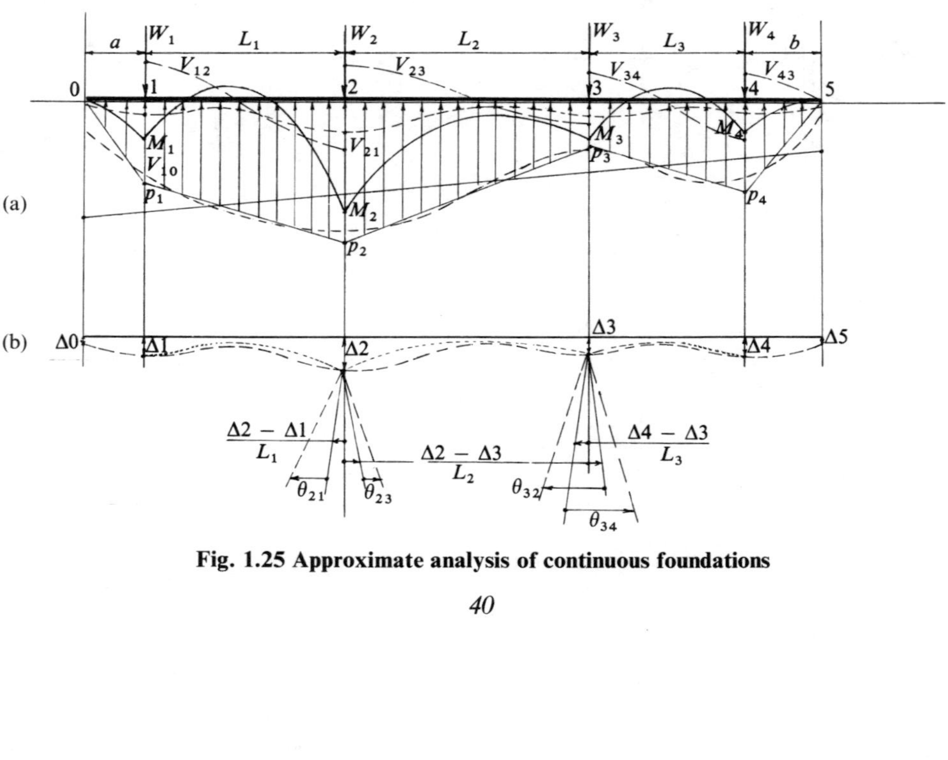

Fig. 1.25 Approximate analysis of continuous foundations

affected. This problem has been investigated by many authors and a considerable amount of literature is available on this subject. An approximate analysis can also be made, by the force–displacement method, based on the following assumptions:

(a) that the pressure varies linearly between the points of application of concentrated loads,
(b) that the soil mass responds elastically and can be represented by its modulus of subsoil reaction, k, and
(c) that the settlements Δ at these points are proportional to the corresponding pressures, i.e., $\Delta = p/k$.

Although in soils, the pressure distribution cannot change rapidly at the points of concentrated loads, and a smooth distribution must take place, assumption (a) may give a closer answer than a calculation based on the column formula.

The unknown values at each point load are: The pressure p_i, deflection Δ_i, and bending moment M_i (see Fig. 1.25). These values must be calculated simultaneously so that the conditions of compatibility of deformations and equilibrium of forces are satisfied at all points at the same time. The procedure which is adopted in the analysis is as follows:

First, releases are introduced so that the beam is reduced to a series of simply supported elements between columns, e.g., in Fig. 1.25(b), these are taken at points 2 and 3. Next, the values of angular discontinuities are calculated at these points, including the effect of cantilevers and of differential settlement. The following equations of compatibility of deformations at points 2 and 3 can be written:

$$\text{at point 2}\quad M_2\left(\frac{L_1}{3EI}+\frac{L_2}{3EI}\right)+M_3\frac{L_2}{6EI}=\theta_{21}+\theta_{23}+\frac{\Delta 2-\Delta 1}{L_1}+\frac{\Delta 2-\Delta 3}{L_2} \tag{1.17}$$

$$\text{at point 3}\quad M_2\frac{L_2}{6EI}+M_3\left(\frac{L_1}{3EI}+\frac{L_3}{3EI}\right)=\theta_{32}+\theta_{34}-\frac{\Delta 2-\Delta 3}{L_2}+\frac{\Delta 3-\Delta 4}{L_3}$$

in which the values of θ can be expressed in terms of the unknown pressures as follows:

$$\begin{aligned}\theta_{21}&=\frac{7}{8}\cdot\frac{p_1L_1{}^3}{45EI}+\frac{p_2 . L_1{}^3}{45EI}-M_1\cdot\frac{L_1}{6EI}\\ \theta_{23}&=\frac{p_2L_2{}^3}{45EI}+\frac{7}{8}\cdot\frac{p_3 . L_2{}^3}{45EI}\end{aligned} \tag{1.18}$$

The values of settlements Δ can be eliminated by expressing them in terms of pressures and the soil modulus k by assuming that:

$$\Delta_i=\frac{p_i}{k} \tag{1.19}$$

The equilibrium conditions at points 1, 2, 3, and 4 in Fig. 1.25(a) can be expressed in terms of shear forces as:

$$\begin{aligned}&\text{at point 1}\quad \cdots V_{10}+V_{12}=W_1\\ &\text{at point 2}\quad \cdots V_{21}+V_{23}=W_2\\ &\text{at point 3}\quad \cdots V_{32}+V_{34}=W_3\\ &\text{at point 4}\quad \cdots V_{43}+V_{45}=W_4\end{aligned} \tag{1.20}$$

in which the values of shear forces can be expressed in terms of pressures and bending moments, i.e.,

$$\begin{aligned} V_{10} &= \tfrac{1}{2}p_1 . a \\ V_{12} &= \tfrac{1}{3}p_1 L_1 + \tfrac{1}{6}p_2 . L_1 + M_1/L_1 \\ V_{23} &= \tfrac{1}{3}p_2 . L_2 + \tfrac{1}{6}p_3 L_2 + M_2/L_1 + (M_2 - M_3)/L_2 \qquad (1.21) \\ &\text{etc.} \end{aligned}$$

It should be noted that the cantilever bending moments can be expressed directly in terms of pressures and do not enter as separate unknown values, i.e.,

$$M_1 = (\tfrac{1}{2}p_1 . a) \cdot \frac{a}{3}$$

Example 1.5

In a beam shown in Fig. 1.26, the end pressures based on the conventional column formula are 233·0 and 138·0 kN/m. However, assuming that the pressure varies linearly between column loads, the following matrix can be set out:

	m_3	m_4	p_2	p_3	p_4	p_5	
3	5·0	1·0	−14·175	−21·0	−4·2	0·25	0
4	1·0	3·0	0	−4·2	−4·4	−0·4417	0
S.C. 2	−0·1	0	3·5185	1·5	0	0	1000
3	0·2$\dot{7}$	−0·16	1·4815	5·0	1·0	0	2000
4	−0·1$\dot{6}$	0·$\dot{5}$	0	2·0	2·0	0·$\dot{4}$	200
5	0	−0·$\dot{3}$	0	0	1·0	2·5	500

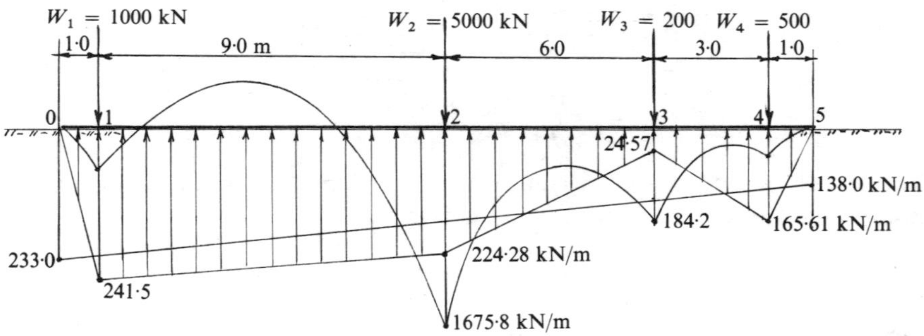

Fig. 1.26 Example of the analysis of continuous foundation

The resulting distribution of bending moments and pressures is shown in Fig. 1.26. The settlements can now be calculated from eq. (1.19).

In the above analysis, only the effect of column loads on foundation beams has been considered. However, in tall buildings, it is usual to assume that columns are fixed at the base level, and therefore absorb bending moments. These in turn must be resisted by the foundation beam and will produce additional bending moments and shear forces, and can also affect the distribution of pressures and deflections. In Fig. 1.27, it is assumed that column bending moments M_1, M_2, etc., are given, and that the required values are bending moments in the foundation beam m_2, m_3, etc.

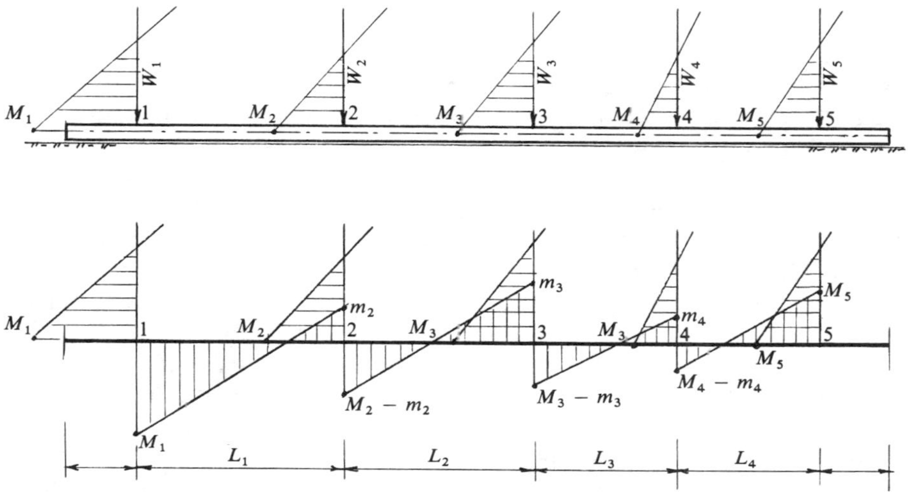

Fig. 1.27 Effect of bending moments in columns on foundations

A typical compatibility condition at point 2 can now be written in the form:

$$-M_1\frac{L_1}{6EI}+m_2\frac{L_1}{3EI}-(M_2-m_2)\frac{L_2}{3EI}+m_3\frac{L_2}{6EI}=0 \quad (1.22)$$

A similar condition can be written at other points and statically indeterminate bending moments m_2, m_3, etc., can thus be determined.

A possible local deformation of the soil mass and the change of pressure distribution resulting from flexural deformations of the beam are high order effects and can be expected to be negligible in most practical cases.

1.13 Soil–Structure Interaction in Tall Buildings

The superstructure in tall buildings is usually analysed independently from foundations and the soil medium on the assumption that no interaction takes place. However, modern tall buildings can be very slender, and although their overall mass tends to be reduced, their foundations are often small and concentration of loads can be very high. High pressures induced in the soil mass can cause considerable deformation of the soil, even if piled foundations are used, and this may affect deflections of the superstructure and sometimes even the distribution of stresses. The reverse effect also takes place when the deflection of the superstructure and the forces transferred at the foundation level may affect the deformations of the soil. This relationship is referred to as the soil–structure interaction, which shows itself in increased deformations and stresses when compared with those obtained on the assumption of unyielding foundations.

The main factors affecting such interaction are: the magnitude and eccentricity of loads, the stiffness of the superstructure, the soil, and the nature of the lateral forces. The deformations of the soil mass directly under the foundation depend mainly on the properties of the soil and the magnitude of forces but also on the type of foundation and the stiffness of the superstructure. The settlements of foundations can be translational or rotational and can be partly elastic and partly permanent. In this analysis only the effect of elastic rotational deformations is considered.

The exact analysis of soil–structure interaction is complicated by the dynamic nature of forces and by the inertia of the superstructure, foundations, and soil. Although the soil is a powerful source of damping of vibrations in tall buildings, deformations of the soil may increase deflections and vibrations of the superstructure to such an extent that they may become perceptible to occupants or cause damage to non-structural finishes, partition walls, and services; for example, Meyerhof[10] suggests that the limiting angular distortion of foundations should not exceed 1/300 for framed buildings and 1/1000 for shear walls. Skempton and MacDonald[11] suggest that structural damage occurs from angular distortions above 1/150 and for rafts 1/300.

The bearing capacity of the soil and its deformation are related to shearing strength which, if exceeded, may result in rupture of soil, loss of shear transfer between soil particles and slip along a curved surface of rupture, and the structure may then tilt permanently. However, usual settlement of soil does not involve shear rupture, but gradual reduction in the voids ratio of the soil. A part of the resulting deformation is recoverable and can, for practical purposes, be treated as elastic. This can be related to the E value of the soil or to its modulus of subgrade reaction k. Although soils in general do not obey Hooke's Law, and show strains which are only partly recoverable, Terzaghi[12] has shown that it is possible to assign an E value to soils, related to a particular loading condition.

Similarly, the modulus of subsoil reaction, k, can be used in the analysis. Average values are suggested by Terzaghi and Peck[12] and can also be determined from soil exploration. This mathematical equivalent of soil properties varies with the depth, soil, and loading conditions, and is therefore difficult to assess accurately. However, analysis shows that even large errors in the assessment of this value have little effect on the response of the superstructure.

In the following analysis, only the effect of 'static' wind forces is considered as the source of soil–structure interaction, and soil is represented by its modulus, k. It is assumed that the foundation is sufficiently rigid to settle as a rigid body, so that only its rotational component of deformation is taken as significant to the response of the superstructure. The analysis is based on the force–displacement method.

Referring to Fig. 1.28, it is assumed that the shear wall is subjected to gravity forces G and to the lateral wind forces W, and is supported by the soil of stiffness k. The following notation is used:

Δ	the elastic deformation of the superstructure
λ	the deflection of the frame due to soil settlement only
δ	$\Delta+\lambda$, total lateral deflection of the frame
ϕ	rotation of the rigid foundation due to soil deformation
e	total eccentricity of the resultant force on the underside of the base
e_s	eccentricity due to the 'static' resultant of wind and gravity forces

a and *b* plan dimensions of the foundation to the shear wall
l the position of the resultant wind force from the foundation.

From Fig. 1.28 we can write that:

$$e = \Delta + \lambda + e_s \tag{1.23}$$

and for rigid foundation:

$$\lambda = l.\phi$$

in which

$$\phi = \frac{y_1 - y_2}{a}$$

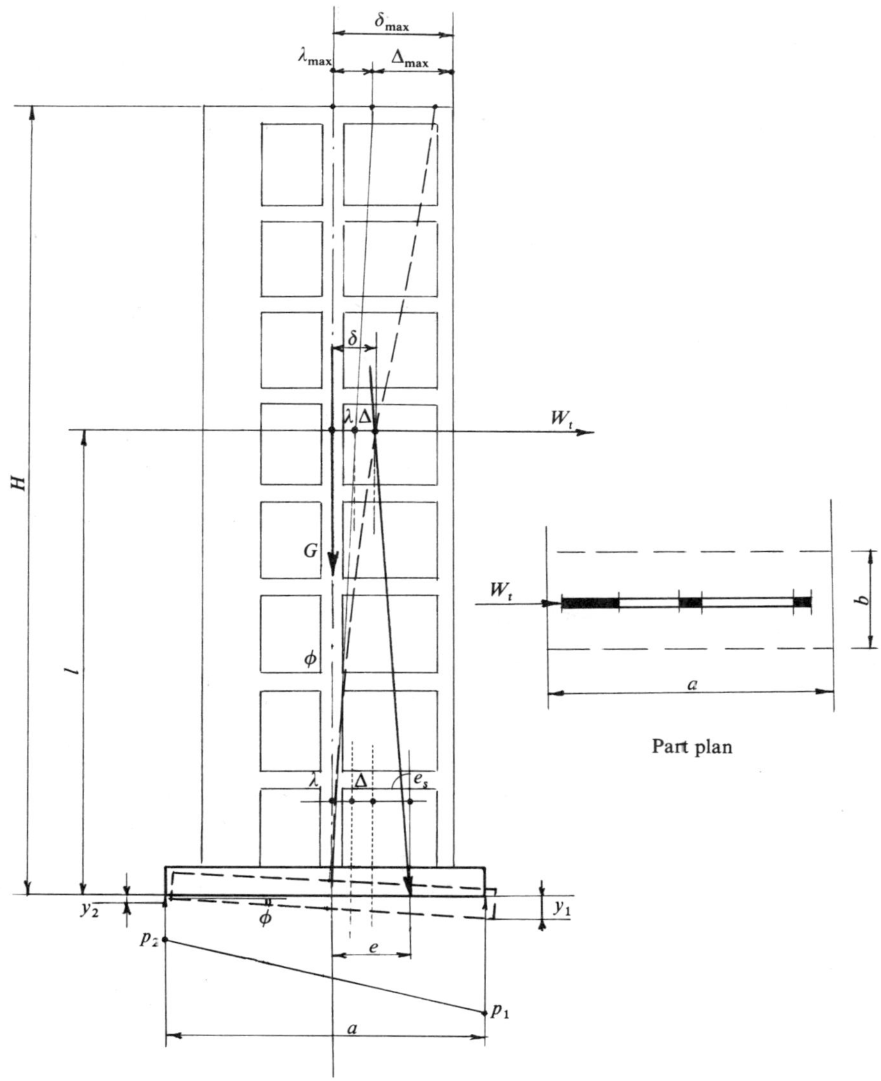

Fig. 1.28 Soil–structure interaction

Assuming that $y = p/k$ we obtain:

$$\phi = \frac{p_1 - p_2}{k.a}$$

and using the column formula for pressures induced in the soil by a rigid slab we obtain:

$$\phi = \frac{126}{kba^3}(\Delta + \lambda + e_s) \qquad (1.24)$$

from which:

$$\lambda = l.\frac{W/G + \Delta/l}{kI/G.l - 1} \qquad (1.25)$$

I is the second moment of area of the part of foundation supporting the shear wall.

If the elastic deformations Δ were known, e.g., from primary analysis, then the effect of soil deformation could be added based on eq. (1.25). This is illustrated by the following example.

Example 1.6

In the analysis of the frame shown in Fig. 1.29 the following data are assumed: $G = 2000$ kN, $W_t = 216$ kN, $k = 10\,000$ kN/m^3, and $E_c = 28 \times 10^6$ kN/m^2. Other data are indicated in Fig. 1.29.

It can be shown that the primary analysis gives the following deflections of the frame at points 0 and 5 respectively

$$\Delta_0 = 137\frac{Wh}{EI} = 70 \text{ mm}$$

$$\Delta 5 = 87\frac{Wh}{EI} = 44{\cdot}8 \text{ mm}$$

From eq. (1.25), the additional deflections resulting from soil deformations are:

$$\lambda_0 = 37\frac{216/2000 + 0{\cdot}044/19}{(10\,000 \times 864)/(2000 \times 19) - 1} = 18{\cdot}0 \text{ mm}$$

and $\lambda_5 = 9{\cdot}6$ mm

The deflection at the top of the frame in this example is therefore increased by approximately 26 per cent as the result of the deformation of the soil.

In rigid tall buildings, deflections caused by soil deformation may exceed those of the elastic deflection, while stresses may not be affected. However, in flexible buildings, both the deflections and stresses may be significantly affected and these can be obtained by considering simultaneously the effect of soil settlement and the deformation of the superstructure. Equation (1.25) can still be used in setting out of the matrix. The procedure is illustrated by the numerical example in which the same frame is used as shown in Fig. 1.29. The simple geometry of the frame used is to facilitate the setting out of the matrix and to avoid unnecessary arithmetical complications.

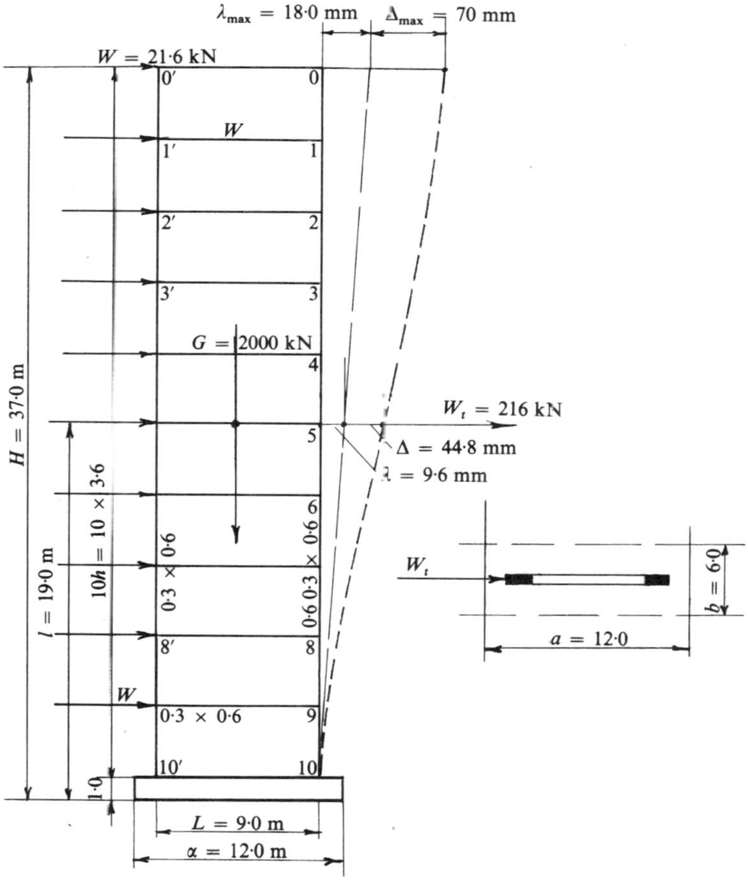

Fig. 1.29

Example 1.7

Referring to Fig. 1.30 the element force–displacement matrix is set out at level 1–1′. The two compatibility conditions at 211′ and 012 are:

$$211'. \quad -m_{21}\frac{h}{6EI_c} + m_{12}\cdot\frac{h}{3EI_c} + (m_{10} + m_{12})\frac{L}{6EI_b} = \frac{\delta_1 - \delta_2}{h} \tag{1.26}$$

$$012. \quad -m_{01}\frac{h}{6EI_c} + m_{10}\frac{h}{3EI_c} - m_{12}\frac{h}{3EI_c} + m_{21}\frac{h}{6EI_c} = \frac{\delta_0 - \delta_1}{h} - \frac{\delta_1 - \delta_2}{h}$$

and the equilibrium condition, including the effect of elastic instability, i.e., the so-called $P.\delta$ effect, becomes:

$$1\text{–}2. \qquad \frac{m_{12} + m_{21}}{h} = \tfrac{1}{2}(2W) + (2P)\frac{\delta_1 - \delta_2}{h} \tag{1.27}$$

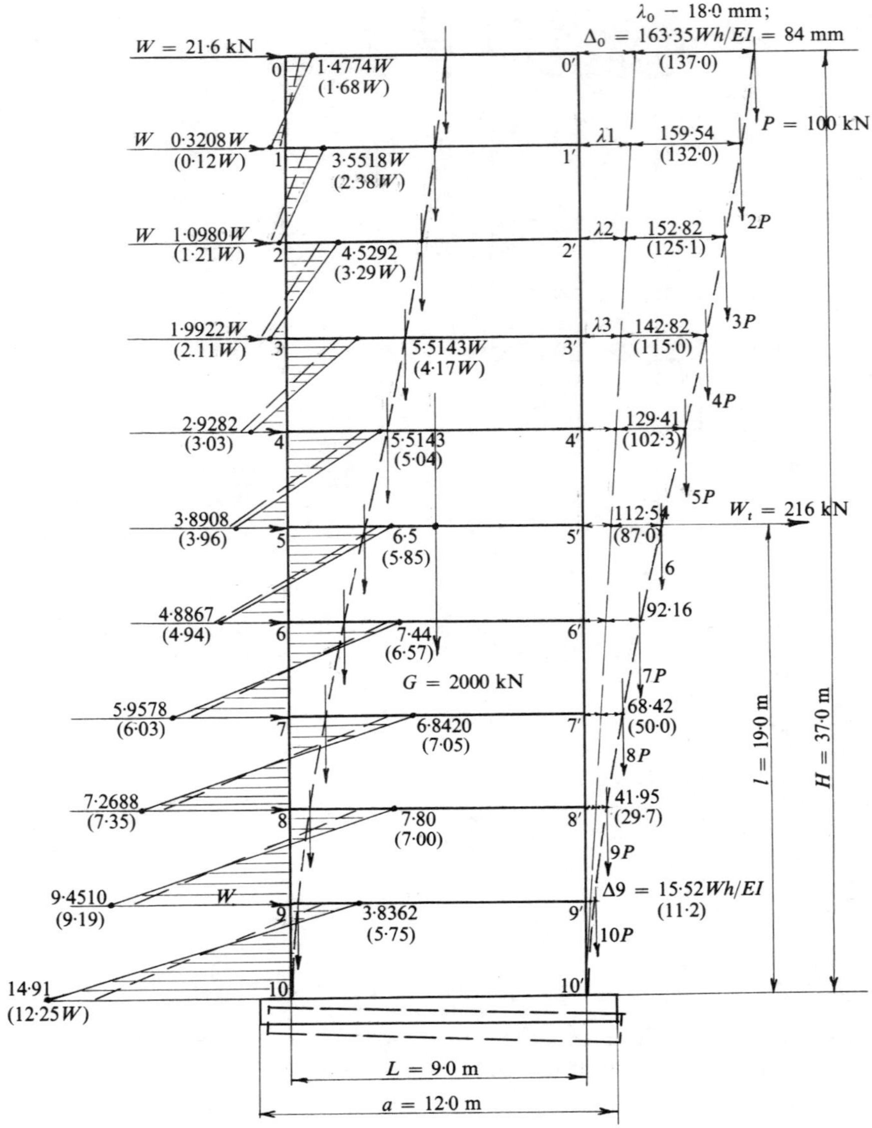

Fig. 1.30 Example of the analysis of a frame including the effect of soil–structure interaction

Introducing numerical values and with the help of eq. (1.25), the overall force–displacement matrix can now be obtained as shown in Table 1.5. The solution of this matrix gives the distribution of bending moments, forces and deflections, as shown in Fig. 1.30. The values in brackets refer to the primary analysis of the effect of wind forces only, based on the assumption of unyielding foundations, and on equilibrium being satisfied in undeformed configuration of the frame.

It can be seen from Fig. 1.30, comparing the results of the primary analysis, that both the stresses and deflections are affected by the soil–structure interaction. It is interesting to note that there is even a small increase of elastic deformations (14 mm) caused by the elastic instability initiated by the soil movement.

Although such analysis cannot be claimed to be exact, it indicates that some interaction does take place. An exact analysis is difficult on account of numerous parameters involved. Similarly, experimental verification is difficult except in ideal laboratory conditions.

Finally, while in the above analysis only the bending deformations were included, the effect of axial or shear deformations can be taken into account in setting out the matrix, with very little additional work.

1.14 Continuum Analysis of Tall Frames and Shear Walls by the Force–displacement Method

The analysis of tall frames discussed previously is based on the setting out of the matrices for moments and deflections to be evaluated at all nodal points simultaneously. This procedure requires the use of computers to solve often large matrices,

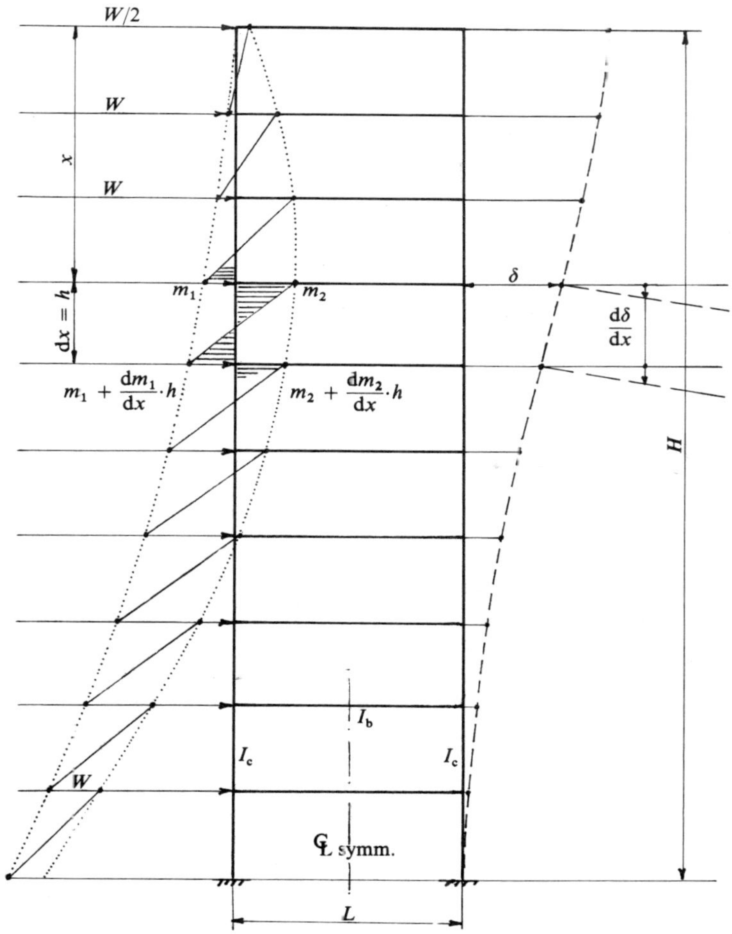

Fig. 1.31 Continuum analysis of tall frames by the force-displacement method

Table 1.5 Example of the analysis of the frame in Fig. 1.29 including the effect of soil–structure interaction

	0		1			2			3			4		
	0 1	Δ0	1 0	1 2	Δ1	2 1	2 3	Δ2	3 2	3 4	Δ3	4 3	4 5	Δ4
1 0 0′	2·7	−1·0127	−0·6	0	1·0127									
0–1	1	−0·0048	1	0	0·0048									
2 1 1′	0	0	1·5	2·7	−1·0041	−0·6	0	1·0041						
0 1 2	−0·6	−1·0041	1·2	−1·2	2·0082	0·6	0	−1·0041						
1–2	0	0	0	1	−0·0096	1	0	0·0096						
3 2 2′				0	0	1·5	2·7	−1·0041	−0·6	0	1·0041			
1 2 3				−0·6	−1·0041	1·2	−1·2	2·0082	0·6	0	−1·0041			
2–3				0	0	0	1	−0·0144	1	0	0·0144			
4 3 3′							0	0	1·5	2·7	−1·0041	−0·6	0	1·00
2 3 4							−0·6	−1·0041	1·2	−1·2	2·0082	0·6	0	−1·00
3–4							0	0	0	1	−0·0192	1	0	0·01
5 4 4′										0	0	1·5	2·7	−1·00
3 4 5										−0·6	−1·0041	1·2	−1·2	2·0
4–5										0	0	0	1	−0·02
6 5 5′													0	0
4 5 6													−0·6	−1·00
5–6													0	0
7 6 6′														
5 6 7														
6–7														
8 7 7′														
6 7 8														
7–8														
9 8 8′														
7 8 9														
8–9														
10 9 9′														
8 9 10														
9–10														
10 9														

Note: Bending moments are in terms of W and deflections in terms of $W.h/EI$.

but is necessary if an exact solution is required, and particularly when the sizes of members change at various floor levels. However, if a tall frame or a shear wall is of regular form, or if an approximate solution is required without the use of computers, then the 'continuum' analysis may be used. This analysis requires much reduced number of unknowns (three, five, or seven), and these can be presented in an explicit form ready for direct use at least for the preliminary design.

In this method, the unknown are taken as the continuous curves which form the envelopes of moments on each side of the column as well as the deformations. These are indicated by the dotted lines in Fig. 1.31, as well as in Figs. 1.5, 1.6, 1.9, 1.24, etc. Apart from regularity of frame, no further limitations are needed.

In frames, in which the sizes of members and the magnitude of lateral forces do not change rapidly, these curves form a continuous function of x and depend on the properties of members as well as on the system and the magnitude of loading.

	5		6			7			8			9			10	
	5 6	Δ5	6 5	6 7	Δ6	7 6	7 8	Δ7	8 7	8 9	Δ8	9 8	9 10	Δ9	10 9	
																0
																1·8W
																0
																0
																3·6W
																0
																0
																5·4W
																0
																0
																7·2W
6	0	1·0041														0
6	0	−1·0041														0
	0	0·024														9·0W
5	2·7	−1·0041	−0·6	0	1·0041											0
2	−1·2	2·0082	0·6	0	−1·0041											0
	1	−0·0288	1	0	0·0288											10·8W
	0	0	1·5	2·7	−1·0041	−0·6	0	1·0041								0
	−0·6	−1·0041	1·2	−1·2	2·0082	0·6	0	−1·0041								0
	0	0	0	1	−0·0336	1	0	0·0336								12·6W
				0	0	1·5	2·7	−1·0041	−0·6	0	1·0041					0
				−0·6	−1·0041	1·2	−1·2	2·0082	0·6	0	−1·0041					0
				0	0	0	1	−0·0384	1	0	0·0384					14·4W
							0	0	1·5	2·7	−1·0041	−0·6	0	1·0041		0
							−0·6	−1·0041	1·2	−1·2	2·0082	0·6	0	−1·0041		0
							0	0	0	1	−0·0432	1	0	0·0432		16·2W
										0	0	1·5	2·7	−1·0041	−0·6	0
										−0·6	−1·0041	1·2	−1·2	2·0082	0·6	0
										0	0	0	1	−0·048	1	18·0W
													−0·6	−1·0041	1·2	0

For a symmetrical, two-column frame, as shown in Fig. 1.31, at any distance x from the top of the frame, there are only three unknown values: two bending moments denoted by m_1 and m_2 and one deflection δ, and unlike the previous analysis these are independent of the number of storeys. These values form continuous curves or 'envelopes' of bending moments and deformations, and can be expressed as a continuous function of x by setting out the conditions of compatibility of deformations and equilibrium of forces in a differential form. However, all moments and deformations are valid only at the nodal points by assuming that the increment of x is equal to $\mathrm{d}x = h$, i.e., to the full storey height. The increments of moments m_1 and m_2 can be expressed respectively as

$$m_1 + \frac{\mathrm{d}m_1}{\mathrm{d}x}\cdot h \quad \text{and} \quad m_2 + \frac{\mathrm{d}m_2}{\mathrm{d}x}\cdot h \tag{1.28}$$

The two compatibility and one equilibrium conditions at x and at $x+\mathrm{d}x$ can now be written by the force–displacement method in the following form:

$$m_2\cdot\frac{h}{3EI_c} - \left(m_1 + h\cdot\frac{\mathrm{d}m_1}{\mathrm{d}x}\right)\cdot\frac{h}{6EI_c} + (m_1 + m_2)\frac{L}{6EI_b} = \frac{\mathrm{d}\delta}{\mathrm{d}x}$$

$$-m_2\cdot\frac{h}{6EI_c} + \left(m_1 + h\cdot\frac{\mathrm{d}m_1}{\mathrm{d}x}\right)\cdot\frac{h}{3EI_c} + \left(m_1 + h\cdot\frac{\mathrm{d}m_1}{\mathrm{d}x} + m_2 + h\frac{\mathrm{d}m_2}{\mathrm{d}x}\right)\frac{L}{6EI_b} = \frac{\mathrm{d}\delta}{\mathrm{d}x}$$

E.C.
$$\frac{m_1 + h.(\mathrm{d}m_1/\mathrm{d}x) + m_2}{h} = \frac{W}{2}\left(\frac{x}{h} + \frac{1}{2}\right) \tag{1.29}$$

Denoting:
$$n = \frac{I_b}{I_c}, \qquad a = \frac{L}{h} \quad \text{and} \quad r = \frac{L.I_c}{h.I_b}$$

these equations can be rearranged and the variables can be separated. The bending moment m_1 can then be expressed in the form of a second-order homogeneous differential equation with constant coefficients. This is

$$\frac{\mathrm{d}^2 m_1}{\mathrm{d}x^2} + a_1\cdot\frac{\mathrm{d}m_1}{\mathrm{d}x} + a_2.m_1 = a_3.x + a_4 \tag{1.30}$$

in which

$$a_1 = -6/r.h$$
$$a_2 = -6/r.h^2$$
$$a_3 = -3W/2r.h^2$$
$$a_4 = \frac{W}{2r.h}(r - \tfrac{3}{2})$$

For the bending moment m_2 we obtain:

$$m_2 = \frac{W}{2}\left(x + \frac{h}{2}\right) - m_1 - h\cdot\frac{\mathrm{d}m_1}{\mathrm{d}x} \tag{1.31}$$

and the deflection δ can then be evaluated from one of the conditions in the eq. (1.27).

It can be shown that the solution of eq. (1.28) is in the form:

$$m_1 = A.\mathrm{e}^{D_1 x} + B.\mathrm{e}^{D_2 x} + C_1.x + C_2 \tag{1.32}$$

in which
$$C_1 = a_3/a_2, \qquad C_2 = (a_2.a_4 - a_1.a_3)/a_2{}^2$$

$$D_{1,2} = \frac{a_1}{2}\{-1 \pm \sqrt{(1 - 4a_2/a_1{}^2)}\}$$

and the values of A, B, and C are the constants of integration which must be evaluated from the boundary conditions.

The expression for m_2 can now be derived from eqs. (1.31) and (1.32) and is in the form:

$$m_2 = -(hD_1 + 1)A.e^{D_1x} - (hD_2 + 1)B.e^{D_2x} + \left(\frac{W}{2} - C_1\right)x - (h.C_1 + C_2) \tag{1.33}$$

Similarly, from the first of the eqs. (1.29) we obtain:

$$\frac{6EI_c}{h}\cdot\delta = -\left[(r + 3).h + \frac{3}{D_1}\right]A.e^{D_1x} - \left[(r + 3)h + \frac{3}{D_2}\right]B.e^{D_2x} -$$

$$-\left[\left(\frac{r}{2} + 1\right)W + 3C_1\right]\cdot\frac{x^2}{2} +$$

$$+\left[\left(\frac{r}{2} + 1\right)\left(H - \frac{h}{2}\right)W - (r + 3)h.C_1 - 3C_2\right].x + C \tag{1.32}$$

The constants A, B, and C can be calculated from the following boundary conditions:

at $x = 0, \quad m_1 = 0$

at $x = H, \quad \delta = 0$ and $d\delta/ax = 0$.

A similar procedure can be applied to an unsymmetrical frame or to multi-bay frames, but the number of the unknown 'envelope' curves is increasing.

References

1. J. H. Argyris, *Energy Theorems and Structural Analysis*, Butterworths, London (1960).
2. H. L. Laughaar, *Energy Methods in Applied Mechanics*, John Wiley & Sons Inc., N.Y. (1962).
3. A. S. Hall and R. W. Woodhead, *Frame Analysis*, John Wiley & Sons Inc., N.Y. (1961).
4. *Steel Frames of Office Buildings*, Bulletin No. 80, Univ. of Illinois (1915 and 1932).
5. R. W. Clough, 'The Finite Element Method in Plane Stress Analysis', *Proc. 2nd A.S.C.E. Conf. on Electronic Computation*, Pittsburgh, P.A. (1960).
6. E. E. Sechler, *Elasticity in Engineering*, John Wiley & Sons (1960)
7. Zienkiewicz, O. C., *The Finite Element Method in Structural and Continuum Mechanics*, McGraw-Hill (1970).
8. M. J. Turner, R. W. Clough, H. C. Martin, and L. J. Topp, 'Stiffness and Deflection Analysis of Complex Structures', *J. Aero Sci.*, **23**, No. 9 (1956).
9. R. J. Melosh, 'Basis of Derivation of Matrices for the Direct Stiffness Method', *J. Am. Inst. Aeronautics*, **1**, No. 7 (1963).

10. G. Meyerhof, 'Some recent foundation research and its application to design', *Struct. Eng.* (Jan. 1953).
11. A. W. Skempton and D. H. MacDonald, 'Allowable settlement of buildings', *Proc. I.C.E.* (Dec. 1956).
12. K. Terzaghi and R. B. Peck, *Soil Mechanics in Engineering Practice*.

2 Analysis of Shear Walls

2.1 Introduction

Shear walls are defined as two- or three-dimensional vertical structures, solid or perforated, whose individual members, particularly those of beams, cannot be strictly treated as linear. The function of shear walls is to resist the effect of lateral and gravity forces and to provide lateral stability to a tall building. In this respect, shear walls represent one of the basic structures which can fulfil this function.

Shear walls have considerable stiffness in their own plane but, with the exception of cores, have usually very little stiffness in the perpendicular direction, and their satisfactory performance depends on the stiffening effect of floor diaphragms, which prevent buckling of walls. It is an advantage for shear walls to carry as much gravity load as possible, to minimize the danger of tension or overturning by wind forces.

The analysis of shear walls should, generally, take into account the two-dimensional nature of individual members, although there is no clearly defined demarcation line between flexible frames and shear walls, and many intermediate forms occur, which are difficult to classify into one group or another. In addition, the individual members, particularly those of columns, may be two-dimensional, even of the proportion 1 : 1, and yet the whole wall may still be very flexible and its overall response may be largely in flexure. This, however, does not apply to connecting beams, and it is conventional to treat them as linear only if the ratio of their length to depth is more than two. Even in this case, stress concentration effect can be considerable, not only locally but also on the overall response of a shear wall.

There are many forms of shear walls, and Fig. 2.1 shows some of the most frequently used types. Very broadly, these are either two or three-dimensional. The first resist forces mainly in shear and flexure, while the second group can also resist torsional moments. These are as follows: solid walls, walls with regularly spaced perforations, interconnected shear walls with flexible frames, walls interconnected with 'no-bending' columns or beams, shear walls with small openings, special forms of walls, walls with discontinuities, walls with local or irregularly spaced openings, walls with staggered openings, infilled walls, 'hinge' connected precast walls, staggered shear walls, perforated lift shafts, frame-tubes, and brick buildings.

In the analysis of shear walls, both bending and shear deformations can be significant, and may have to be included. The main problems in the analysis are to determine the distribution of internal stresses and deformations. However, other problems may also need to be considered, particularly vibrations of walls, their stability, post-elastic response, and the ultimate strength. The response of walls to secondary effects may also need to be considered, particularly the effect of dynamic action of lateral forces, temperature changes, shrinkage, creep, stress concentrations, the effect of differential settlement of foundations, etc.

In recent years, a considerable volume of research has been done into the response of shear walls to loads, and some of this work is included in the bibliography.[1–10, 16] However, an exact solution of many problems is still not available because of the

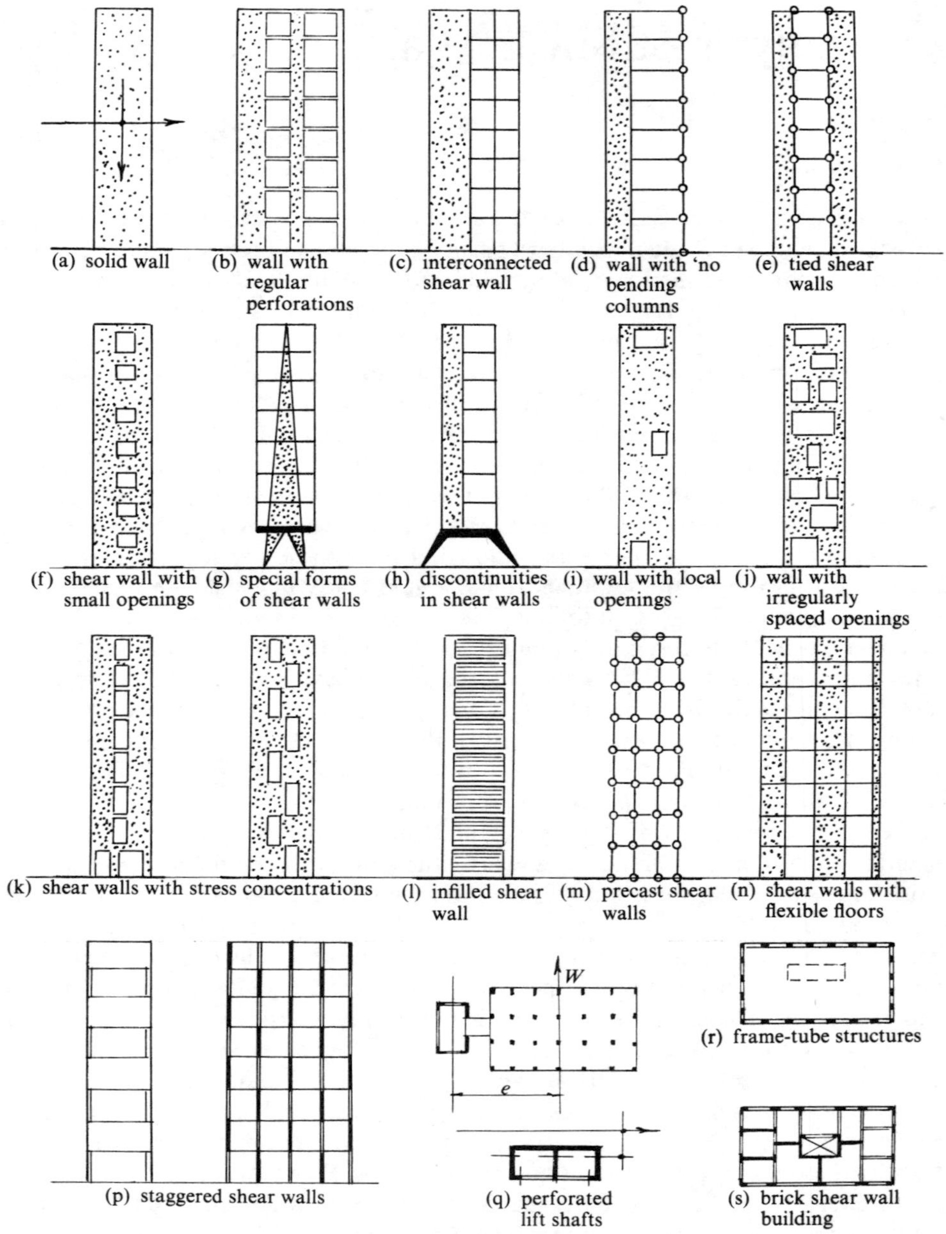

Fig. 2.1 Typical shear walls

complexity of parameters involved. For this reason, many analyses are based on simplified assumptions, mostly on the conventional linearly-elastic response of materials. In the following, only the force–displacement method is used as applied to various forms of shear walls.

2.2 Analysis of Plane Shear Walls

The force–displacement method can be used to analyse shear walls, provided that the ratio of the length of wall to its depth is more than two. However, when openings are small, and comparatively large two-dimensional areas of concrete are left, as in Fig. 2.1(f), these cannot be considered as being flexural members, and the finite elements method may give a closer solution.

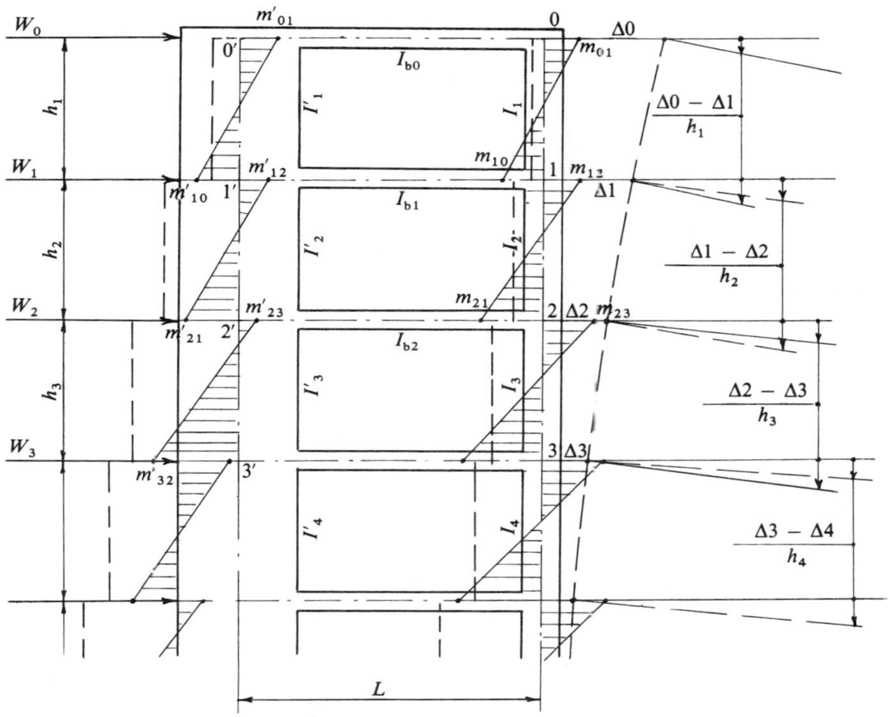

Fig. 2.2 Analysis of shear walls by the force-displacement method

The assumptions and the general procedure in setting out of the main matrix are the same as discussed in chapter 1, except that shearing deformations may have to be included in the flexibility coefficients of walls. One difficulty arises in determination of the span, L, which would give most accurate distribution of moments, forces, and deformations. This can be taken as approximately the width of opening plus twice the width of St Venant's boundary effect. More accurate analysis can only be made by assuming two-dimensional stress distribution and using the finite elements technique.

Another difficulty arises from stress concentrations at the corners, which can have an appreciable effect on the overall distribution of stresses and deformations. This problem is not yet fully resolved but reference can be made to the work shown in the bibliography.

It has been pointed out earlier that the difference in the boundary conditions at the roof and also at the foundations results in the element matrix being different at those levels to those set out at the intermediate floors. In computing it can be very useful to overcome this difficulty so that the element matrix is of the same form and can be generated by the computer in its normal form throughout the height of frame. This is possible if imaginary extensions of the shear wall are introduced at both ends as shown diagrammatically in Fig. 2.3(a). The element matrices will now be the same throughout the frame, but the overall matrix will not be square, as shown in Table 2.1.

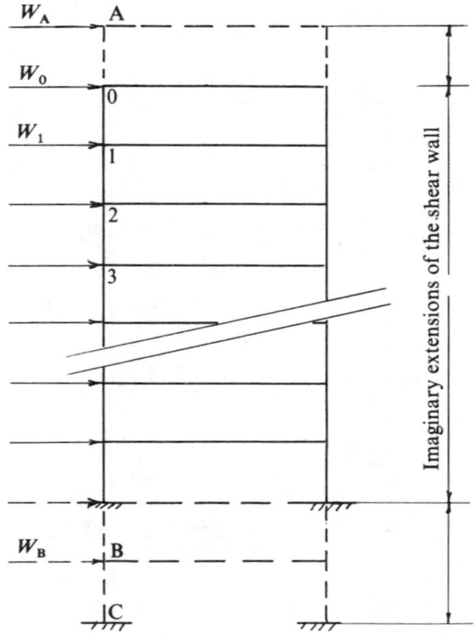

Fig. 2.3(a) The use of imaginary extensions of a frame in setting-out of matrices

Generally, such a matrix cannot be solved, but when the moments and deflections in the non-existent members are 'zero-ized' in the computing programme, the matrix then becomes square, and its eigenvalues can be obtained. Figure 2.3(b) shows the distribution of bending moments, shear, and axial forces as well as deflections based on the matrix 2.1.

2.3 Shear Walls with Two Rows of Openings

For a shear wall on three columns (with two rows of openings) as shown in Fig. 2.4, an element force–displacement matrix can be set out in an explicit form as shown in the Table 2.2. This element matrix is set out at the level 2–2″ and, for repetitive form of construction, remains diagonally repetitive throughout the main matrix. If member sizes change at various floor levels, then the corresponding coefficients will also change. The size and the form of the element matrix can only be affected by the change of the form of construction.

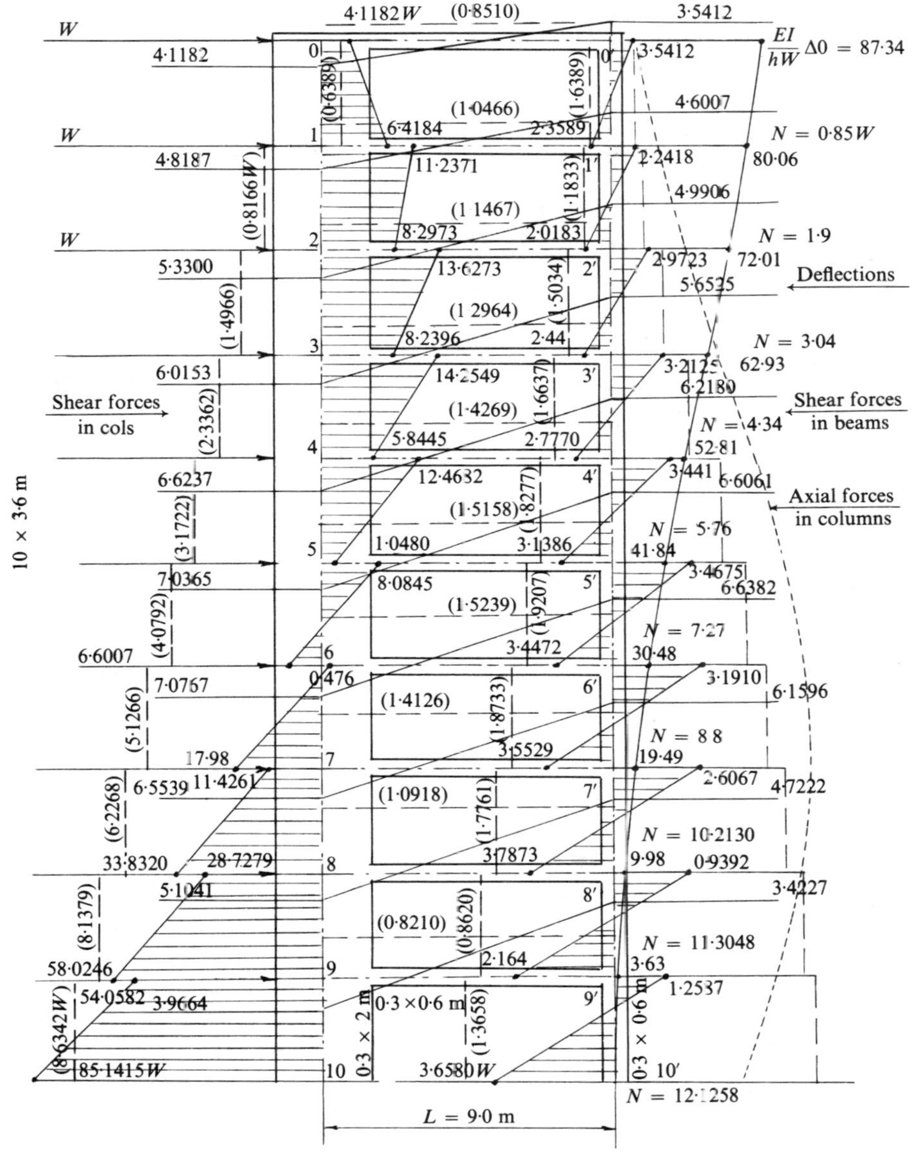

Fig. 2.3(b) Distribution of bending moments, shear and axial forces, and deformations

Example 2.1

Figure 2.4 shows a three-column shear wall subjected to the effect of lateral wind forces. All the relevant dimensions required for the analysis are indicated on the diagram. The element matrix F^e is set out at the floor level 2–2″, and follows the general pattern set out in Fig. 2.4. This is shown in Table 2.3.

The main matrix is of the order 80 × 80, and although diagonally repetitive, should

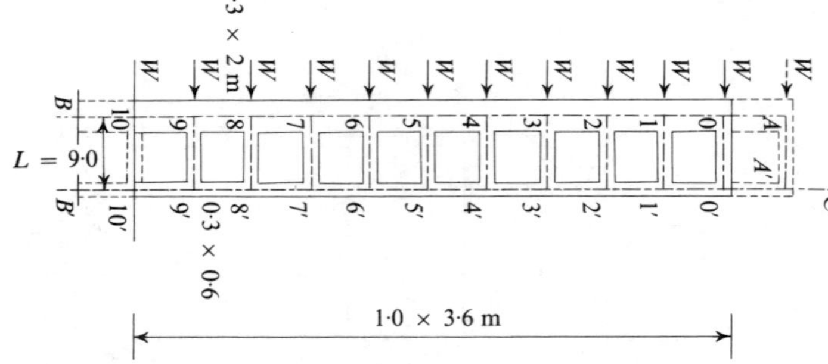

Table 2.1 Analysis of shear wall

(All moments are in terms of W, and deflections are in terms of $W.h/EI_b$.)

	A		*A'*		0			0'		1			1'		2			2'		3			3'		4			
	A 0	*A A*	*A'* 0'	*A'* 0'	0 *A*	0 1	Δ0	0' *A'*	0' 1'	1 0	1 2	Δ1	1' 0'	1' 2'	2 1	2 3	Δ2	2' 1'	2' 3'	3 2	3 4	Δ3	3' 2'	3' 4'	4 3	4 5	Δ4	4'
1 0 0'	0	0	0	0	3·0	3·0324	−1	−1·5	−1·5	−0·0162	0	+1	0															
1' 0' 0	0	0	0	0	−1·5	−1·5	−1	3·0	4·2	0	0	+1	−0·6															
A 0 1	−0·0162	−1	0	0	0·0324	−0·0324	+2	0	0	0·0162	0	−1	0															
A' 0' 1'	0	−1	0	−0·6	0	0	+2	1·2	−1·2	0	0	−1	0·6															
0–0'	−1	0	0	−1	−1	+1	0	−1	+1	+1	0	0	+1															
2 1 1'																												
2' 1' 1																												
1–2											—F^e—																	
0' 1' 2'																												
1–1'																												
3 2 2'																												
3' 2' 2																												
1 2 3																	—F^e—											
1' 2' 3'																												
2–2'																												
4 3 3'																												
4' 3' 3																												
2 3 4																						—F^e—						
2' 3' 4																												
3–3'																												
5 4 4'																												
5' 4' 4																												
3 4 5																											—F^e—	
3' 4' 5'																												
4–4'																												
6 5 5'																												
6' 5' 5																												
4 5 6																												
4' 5' 6'																												
5–5'																												
5' 6 6'																												
5' 6' 6																												
5 6 7																												
5' 6' 7'																												
6–6'																												
8 7 7'																												
8' 7' 7																												
6 7 8																												
6' 7' 8'																												
7–7'																												
9 8 8'																												
9' 8' 8																												
7 8 9																												
7' 8' 9'																												
8–8'																												
10 9 9'																												
10' 9' 9																												
8 9 10																												
8' 9' 10'																												
9–9'																												
8 10 10'																												
8' 10' 10																												
9 10 8																												
9' 10' 8'																												
10–10'																												

	5′		6			6′		7			7′		8			8′		9			9′		10			10′		C			C'	
Δ5	5′4′	5′6	6 5	6 7	Δ6	6′5′	6′7′	7 6	7 8	Δ7	7′6′	7′8′	8 7	8 9	Δ8	8′7′	8′9′	9 8	9 10	Δ9	9′8′	9′10′	10 9	10C	Δ10	10′9′	10′C'	C10	CD	AC	C'10′	
																																0
																																0
																																0
																																0
																																3·6W
																																0
																																0
																																0
																																0
																																3·6W
																																0
																																0
																																0
																																0
																																3·6W
																																0
																																0
																																0
																																0
																																3·6W
																																0
																																0
																																0
																																0
																																3·6W
																																0
																																0
																																0
																																0
																																3·6W
																																0
																																0
					F^e																											0
																																0
																																3·6W
																																0
																																0
										F^e																						0
																																0
																																3·6W
																																0
																																0
																F^e																0
																																0
																																3·6W
																																0
																																0
																					F^e											0
																																0
																																3·6W
																																0
																																0
																										F^e						0
																																0
																																3·6W

Table 2.2 Element force–displacement matrix for three-column shear wall (see Fig. 2.4)

	1			1′			1″		2		
	1 0	1 2	Δ1	1′ 0′	1′ 2′	1′–1″	1″ 0″	1″ 2″	2 1	2 3	Δ2
3 2 2′	0	0	0	0	0	0	0	0	$\frac{L_1 I_0}{3I_{b2}}$	$\frac{h_2 I_0}{3I_2} + \frac{L_1 I_0}{3I_{b2}}$	$-\frac{h_0}{h_2} \cdot \frac{EI_0}{h_0}$
1 2 3	0	$-\frac{h_1 I_0}{6I_1}$	$-\frac{h_0}{h_1} \cdot \frac{EI_0}{h_0}$	0	0	0	0	0	$\frac{h_1 I_0}{3I_1}$	$-\frac{h_2 I_0}{3I_2}$	$\left(\frac{h_0}{h_1} + \frac{h_0}{h_2}\right)\frac{EI_0}{h_0}$
2 2′ 2″	0	0	0	0	0	0	0	0	$-\frac{L_1 I_0}{6I_{b2}}$	$-\frac{L_1 I_0}{6I_{b2}}$	0
3′ 2′ 2″	0	0	0	0	0	0	0	0	0	0	$-\frac{h_0}{h_2} \cdot \frac{EI_0}{h_0}$
1′ 2′ 3′	0	0	$-\frac{h_0}{h_1} \cdot \frac{EI_0}{h_0}$	0	$-\frac{h_1 I_0}{6I'_1}$	0	0	0	0	0	$\left(\frac{h_0}{h_1} + \frac{h_0}{h_2}\right)\frac{EI_0}{h_0}$
2′ 2″ 3″	0	0	0	0	0	0	0	0	0	0	$\left(\frac{h_0}{h_1} + \frac{h_0}{h_2}\right)\frac{EI_0}{h_0}$
1″ 2″ 3″	0	0	$-\frac{h_0}{h_1} \cdot \frac{EI_0}{h_0}$	0	0	0	0	$-\frac{h_1 I_0}{6I''_1}$	0	0	$\left(\frac{h_0}{h_1} + \frac{h_0}{h_2}\right)\frac{EI_0}{h_0}$
S.C. 2–2′–2″	0	$-\frac{h_0}{h_1}$	0	0	$-\frac{h_0}{h_1}$	0	0	$-\frac{h_0}{h_1}$	$-\frac{h_0}{h_1}$	$+\frac{h_0}{h_1}$	0

generally be programmed to avoid setting out of the whole matrix. The final diagrams of bending moments, forces, and deflections are shown in Figs. 2.5 and 2.6.

This analysis shows, what was already mentioned earlier, that rigid elements absorb moments not only of higher values than flexible, but of a quite different form. The main column indicates a few points of contraflexure and in the upper floors moments are often positive, while columns in the lower floors indicate large cantilevering action.

Furthermore, the effect of a very flexible column, such as column 0‴ to 10‴ in Fig. 2.5 is still very beneficial in reducing bending moments in the whole shear wall as compared with the pure cantilever action. This is due to the ability of columns to resist axial forces. All three columns in this example indicate different patterns of moments absorbed and therefore proportioning of bending moments in relation to the respective stiffnesses of columns may lead to incorrect pattern and incorrect magnitude of moments and forces. The highest shear forces in some columns may occur in the upper floors and not, as might be expected, at the lowest level. This is due to the cantilever effect of the largest column.

2.4 Approximate Analysis of Shear Walls with Local Openings

Shear walls with local openings (Fig. 2.7) frequently occur and their exact analysis can present a difficult problem. If the size of the opening is small (15 per cent to 20 per cent) in relation to the wall area between adjacent floors, then its effect is considered as small, and may not warrant an effort of exact analysis.

2′			2″		3			3′			3″		
2′ 1′	2′ 3′	2′–2″	2″ 1″	2″ 3″	3 2	3 4	Δ3	3′ 2′	3′ 4′	3′–3″	3″ 2″	3″ 4″	
$-\frac{L_1I_0}{6I_2}$	$-\frac{L_1I_0}{6I_{b2}}$	$\frac{L_1I_0}{6I_{b2}}$	0	0	$-\frac{h_2I_0}{6I_2}$	0	$\frac{h_0}{h_2}\cdot\frac{EI_0}{h_0}$	0	0	0	0	0	0
0	0	0	0	0	$\frac{h_2I_0}{6I_2}$	0	$-\frac{h_0}{h_1}\cdot\frac{EI_0}{h_0}$	0	0	0	0	0	0
$\frac{L_1I_0}{3I_{b2}}$	$\frac{L_1I_0}{3I_{b2}}$	$-\left(\frac{L_1I_0}{3I_{b2}}+\frac{L_2I_0}{3I'_{b2}}\right)$	$\frac{L_2I_0}{6I'_{b2}}$	$\frac{L_2I_0}{6I'_{b2}}$	0	0	0	0	0	0	0	0	0
0	$\frac{h_2I_0}{3I_2}$	$\frac{L_2I_0}{3I'_{b2}}$	$-\frac{L_2I_0}{6I'_{b2}}$	$-\frac{L_2I_0}{6I'_{b2}}$	0	0	$\frac{h_0}{h_2}\cdot\frac{EI_0}{h_0}$	$-\frac{h_2I_0}{6I_2}$	0	0	0	0	0
$\frac{h_1I_1}{3I'_1}$	$-\frac{h_2I_0}{3I'_2}$	0	0	0	0	0	$\frac{h_0}{h_1}\cdot\frac{EI_0}{h_0}$	$\frac{h_2I_0}{6I'_2}$	0	0	0	0	0
0	0	$-\frac{L_2I_0}{6I'_{b2}}$	$\frac{L_2I_0}{3I'_{b2}}$	$\left(\frac{L_2I_0}{3I_{b2}}+\frac{h_2I_0}{3I''_2}\right)$	0	0	$\frac{h_0}{h_2}\cdot\frac{EI_0}{h_0}$	0	0	0	$-\frac{h_2I_0}{6I''_2}$	0	0
0	0	0	$\frac{h_1 . I_0}{3I''_1}$	$-\frac{h_2I_0}{3I''_2}$	0	0	$-\frac{h_0}{h_2}\cdot\frac{EI_0}{h_0}$	0	0	0	$\frac{h_2I_0}{6I''_2}$	0	0
$-\frac{h_0}{h_1}$	$+\frac{h_0}{h_2}$	0	$-\frac{h_0}{h_1}$	$+\frac{h_0}{h_2}$	$+\frac{h_0}{h_2}$	0	0	$+\frac{h_0}{h_2}$	0	0	$+\frac{h_0}{h_2}$	0	W_2h_0

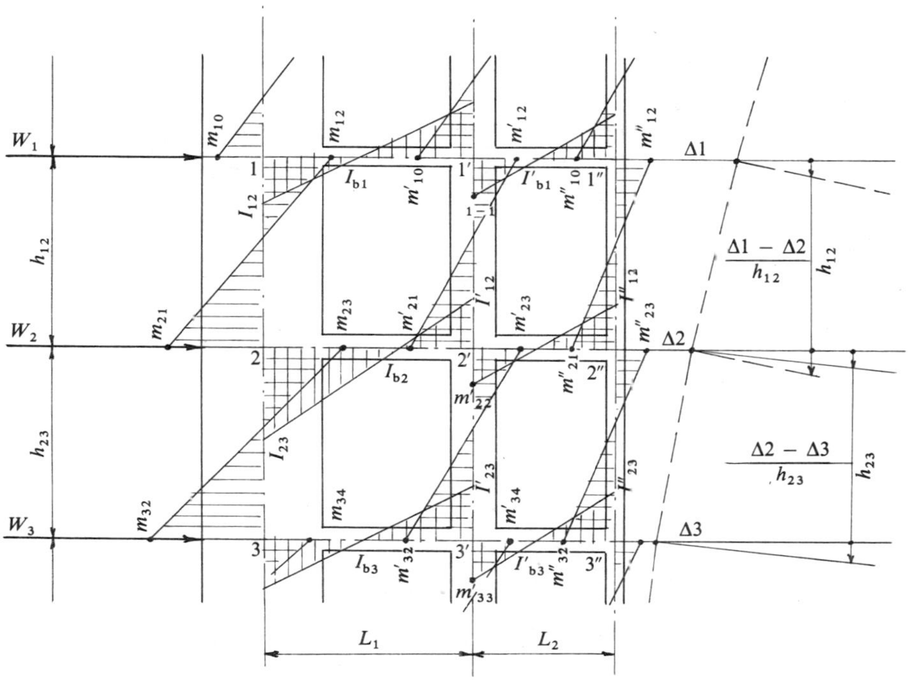

Fig. 2.4 Shear wall with two rows of openings

Table 2.3 Element force–displacement matrix for the shear wall in Fig. 2.5

	1			2			3			11			12			13			21		22		23		
	1 0	1 2	Δ1	2 1	2 3	Δ2	3 2	3 4	Δ3	11–10	11–12	11–21	12–11	12–13	12–22	13–12	13–14	13–23	21–20	21–22	22–21	22–23	23–22	23–24	
1–2–3	0	−0·0020	−1	0·0040	−0·0040	2	0·0020	0	−1	0	0	0	0	0	0	0	0	0	0	0	0	0	0	0	0
3–2–12	0	0	0	3·0	3·0040	−1	−0·0020	0	−1	0	0	0	−1·5	−1·5	1·5	0	0	0	0	0	0	0	0	0	0
2–12–22	0	0	0	−1·5	−1·5	0	0	0	0	0	0	0	3·0	3·0	−5·0	0	0	0	0	0	0	0	0	0	0
11–12–13	0	0	−1	0	0	2	0	0	−1	0	−0·0162	0	0·0324	−0·0324	0	0·0162	0	0	0	0	0	0	0	0	0
13–12–22	0	0	0	0	0	−1	0	0	−1	0	0	0	0	0·0324	0	−0·0162	0	0	0	0	−1	−1	0	0	0
12–22–23	0	0	0	0	0	−1	0	0	1	0	0	0	0	0	−1	0	0	0	0	0	2	3·2	−0·6	0	0
21–22–23	0	0	−1	0	0	2	0	0	−1	0	0	0	0	0	0	0	0	0	0	−0·6	1·2	−1·2	0·6	0	0
2–22	0	0	0	0	1	0	1	0	0	0	0	0	0	1	0	1	0	0	0	0	0	1	1	0	3·6*W*

Note: Bending moments are in terms of W and deflections in terms of $W.h/EI_b$.

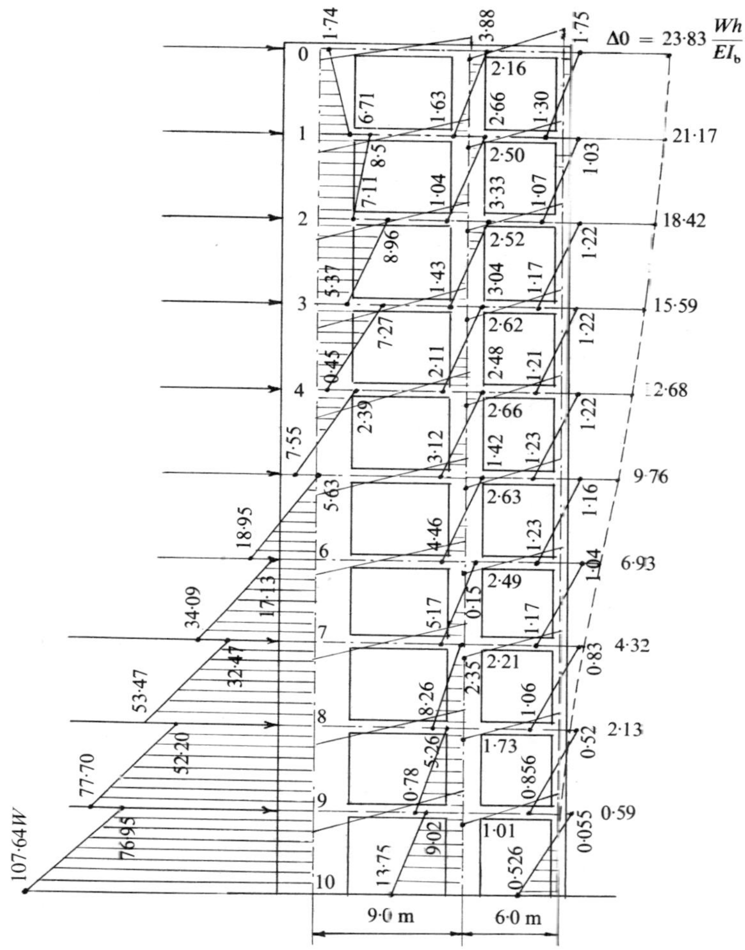

Fig. 2.5 Diagram of bending moments and deflections

If, however, the opening is large and the structure is slender, then the effect of local openings can be considerable in the response of the whole building, i.e., large deflections may occur, high local stresses, cracks may appear, and the building may vibrate excessively.

An 'exact' analysis within the elastic response can be made by the finite elements method, but this is not pursued here.

An approximate analysis can be made, assuming that the solid part works as a simple cantilever, up to the level of the opening (nodal points 2 and 2′ in Fig. 2.8), that the perforated part is then working as a frame in flexure. In Fig. 2.8, the statically indeterminate values are m_{12}, m'_{12}, m_{21}, and the deflection $\Delta 1$, while the moment m'_{21} can be expressed in terms of the cantilever bending moment M_2. This is:

$$m'_{21} = M_2 - m_{21}$$

Similarly, the deflection $\Delta 2$ can be taken as that of a cantilever wall. The four unknown values can then be calculated from the usual conditions of compatibility of

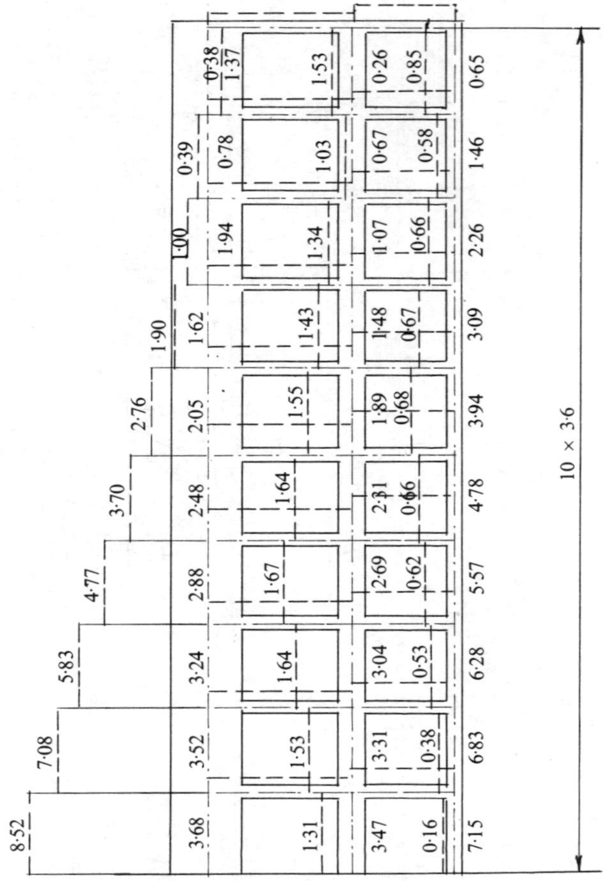

Fig. 2.6 Diagram of shear and axial forces

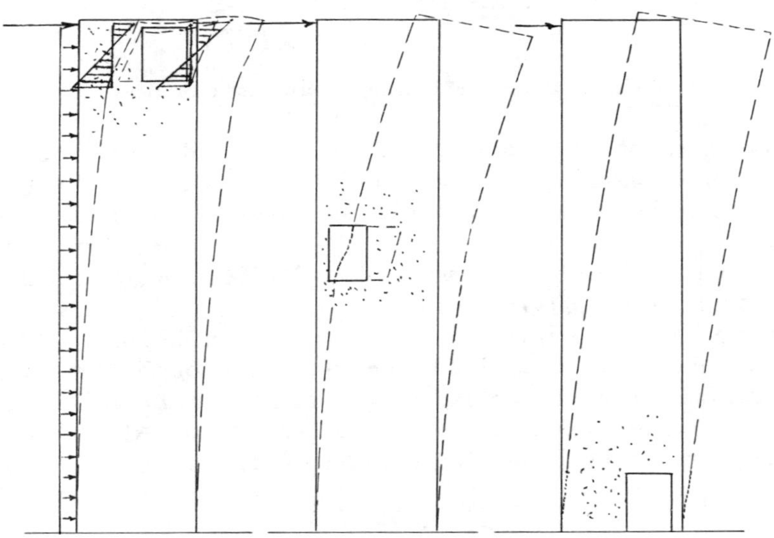

(a) (b) (c)

Fig. 2.7 Shear walls with local openings

deformations and equilibrium of forces. These are as follows:

1. $$m_{12}\left(\frac{l}{3EI_b}+\frac{h}{3EI_1}\right)-m_{21}\cdot\frac{h}{6EI_1}-m'_{12}\cdot\frac{L}{6EI_b}=\frac{\Delta 1-\Delta 2}{h}-\theta$$

1′. $$m'_{12}\left(\frac{l}{3EI_b}+\frac{h}{3EI_2}\right)-(M_2-m_{21})\frac{h}{6EI_2}-m_{12}\cdot\frac{L}{6EI_b}=\frac{\Delta 1-\Delta 2}{h}$$

2. $$m_{21}\cdot\frac{h}{3EI_1}-m_{12}\cdot\frac{h}{6EI_1}=\frac{\Delta 1-\Delta 2}{h}+\theta$$

S.C. $$\frac{m_{12}+m_{21}}{h}+\frac{m'_{12}+(M_2-m_{21})}{h}=\tfrac{1}{2}w.h$$

in which $\theta = wh^3/24EI_1$

A similar approximate analysis can be made if the opening is at an intermediate level.

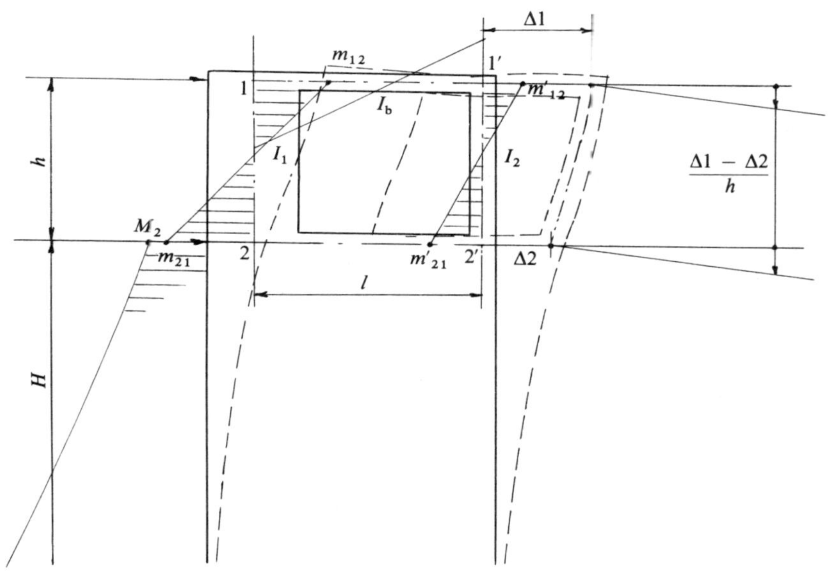

Fig. 2.8 Approximate analysis of a shear wall with a local opening

2.5 Distribution of Lateral Forces in Shear Walls

The magnitude and the form of wind pressure in buildings is stipulated by Codes of Practice and can more accurately be obtained from the wind tunnel tests. However, the distribution of the total wind force, or its storey-height parts, to the individual shear walls, presents a difficult problem when shear walls are analysed as individual two-dimensional units. The problem would not exist at all if tall buildings were analysed as three-dimensional units, but this is rarely done on account of the large amount of work involved.

The conventional distribution of wind forces is usually based on the stiffness of individual columns by using expressions such as:

$$W_i = W_t \cdot \frac{k_i}{\sum k} \qquad (2.1)$$

in which: W_t is the total wind force or the force per storey height
W_i is the force carried by any frame considered, and
k_i are the stiffnesses of columns treated as separate cantilevers.

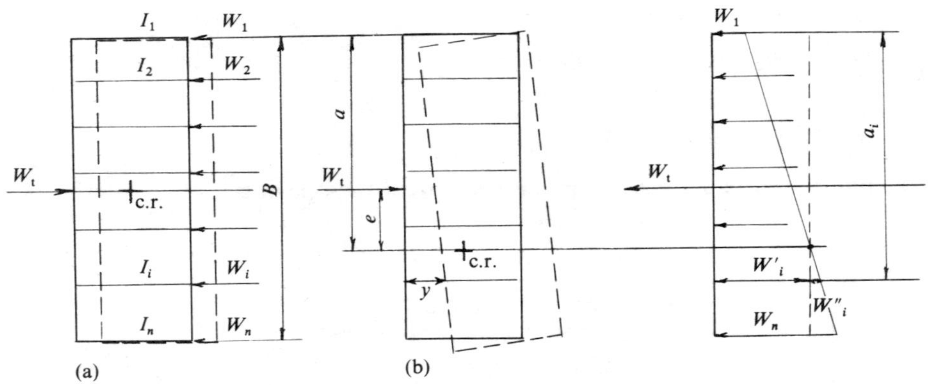

Fig. 2.9 Distribution of lateral forces in shear walls based on column formula

If the centre of pressure does not coincide with the centre of rotation of the building (Fig. 2.9(b)), then the force W_t is usually proportioned according to the concept of the eccentrically loaded column, i.e.,

$$W_i = \frac{W_t}{A} \pm \frac{W_t \cdot e \cdot y}{\sum I} \qquad (2.2)$$

Assuming linear transverse variation of forces W_i, and denoting the two extreme values of forces by W_1 and W_n as shown in Fig. 2.9(b), the latter can be calculated from the conditions of equilibrium of forces:

$$\sum W_i = W_t \quad \text{and} \quad \sum W_i \cdot (a_i - \bar{a}) = W_t \cdot e$$

The forces carried by intermediate walls are then calculated from:

$$W_i = W_n + (W_1 - W_n) \cdot a_i / B$$

Equations (2.1) and (2.2) are accurate only when the cantilevered walls are solid and free-standing or unrestrained, and do not take into account the effect of beams or perforations, frame boundary conditions, and the eccentricity of the lateral force. The error introduced by these effects can be serious in some cases, and more accurate assessment of the distribution of wind loads is therefore required.

The following suggested procedure is based on two-dimensional analysis of frames or perforated shear walls which, with the exception of torsional stiffnesses of members, include all the characteristics of members, boundary conditions, and the actual form and magnitude of the given wind pressure. In this analysis two cases are considered, namely: when the centre of rotation coincides with the centre of pressure, and a more general case, when this coincidence does not take place.

2.5.1 *Case (a). The centre of wind pressure coincides with the centre of rotation*

First, each frame or shear wall is analysed separately for the effect of unit storey forces, or forces proportional to the given pressure contour, defined by pressure coefficients. All moments, internal forces, and deflections are then calculated using any standard procedure. The force–displacement method as described previously can be conveniently used for this purpose. The deflections thus obtained contain all the characteristics of frames and reflect their stiffnesses or load/deflection properties, and can, therefore, be used for more accurate proportioning of the actual lateral forces. It is assumed that the floor diaphragms are infinitely rigid in their own planes and can equalize deflections of frames.

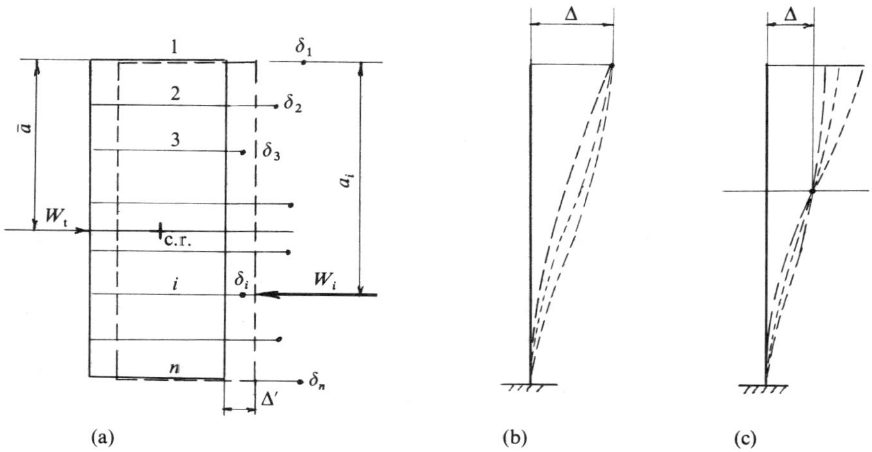

Fig. 2.10 Distribution of wind forces in shear walls

Figure 2.10(a) shows a typical floor plan with shear walls marked 1, 2, . . . , *i*, *n*. When unit loads are applied to each frame separately, their deflections at any floor level are calculated and are marked by $\delta_1, \delta_2, \ldots, \delta_i, \delta_n$. If the centre of pressure coincides with the centre of rotation, then for equilibrium we can write:

$$\sum W'_1 = W_t \tag{2.3}$$

in which W'_i is the actual force carried by the frame *i*, and W_t is the total storey wind force.

Within the elastic response of frames, we can write that:

$$W'_i = 1 \cdot \frac{\Delta'}{\delta_i} \tag{2.4}$$

in which δ_i is the deflection at any floor level due to the unit loads and Δ' is the actual deflection of all frames as equalized by the floor diaphragms.

Substituting eq. (2.3) into eq. (2.4) we obtain:

$$\Delta'_i = W_t \cdot \frac{1}{\sum 1/\delta_i} \tag{2.5}$$

and from eq. (2.4):

$$W'_i = W_t \cdot \frac{1}{\delta_i . \sum 1/\delta_i} = c . W_t \tag{2.6}$$

The centre of rotation is defined as the point in the floor area which, if the resultant wind force passes through it at any angle, translates in one or two coordinates but without any rotation. By resolving the inclined wind force into two forces perpendicular to each face of the building, the position of the centre of rotation in each direction can be calculated from the condition that:

$$W_t . \bar{a} = \sum_{1}^{n} W'_i . a_i \tag{2.7}$$

in which $\bar{a}$ is the distance of the centre of rotation from the assumed coordinate system and a_i defines the position of each shear wall in this coordinate system.

Introducing eq. (2.6) into eq. (2.7), we obtain:

$$\bar{a} = \sum \frac{a_i}{\delta_i . \sum 1/\delta_i} \tag{2.8}$$

Equations (2.5) to (2.8) assume that all the frames are 'tied' at a certain floor level, but have, theoretically, freedom to deflect differentially at all other levels, as shown in Fig. 2.10(b) and (c). This small discrepancy from the actual response is the result of the approximate assumption made, that the building can be analysed as separate two-dimensional units, instead of treating it as a three-dimensional structure. However, the analysis can be repeated at other floor levels by assuming that the frames are 'tied' by floors at other levels, as shown diagrammatically in Fig. 2.10(c), and either an average value may then be accepted or different distribution coefficients at each floor level.

The final deflection of the whole system can now be calculated from eq. (2.5). All bending moments and internal forces, as calculated for unit loads, must now be magnified or reduced by the factors $c . W_t$, as shown in eq. (2.6).

Example 2.2

In Fig. 2.11, the frames 1, 2, 3, and 4 are assumed to deflect by the values 20, 60, 40, and 30 mm respectively, under the effect of unit forces when treated separately. From eq. (2.6), we have:

$$W'_i = W_t \cdot \frac{1}{\delta_i \times 15/120}$$

so that:

$W'_1 = 6/15\ W_t$, $W'_2 = 2/15\ W_t$, $W'_3 = 3/15\ W_t$, and $W'_4 = 4/15\ W_t$

From eq. (2.5), we obtain:

$$\Delta' = W_t . 120/15 = 8{\cdot}0 W_t$$

If $W_t = 4$ units of force, then the actual deflection of all frames is: $\Delta' = 4 \times 120/15 =$ 32 mm (while the geometrical average is 37·5 mm).

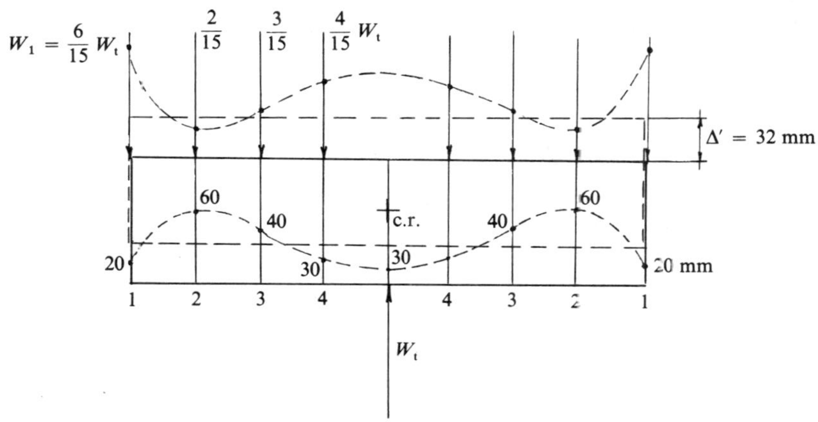

Fig. 2.11 Example of the analysis of the distribution of lateral force for symmetrically loaded building

2.5.2 *Case* (*b*). *Centre of wind pressure does not coincide with the centre of rotation*

If the centre of rotation calculated from eq. (2.8) does not coincide with the centre of pressure, then for equilibrium of moments we obtain from Fig. 2.12:

$$W_t . e = \sum_{1}^{n} W''_i . (a_i - \bar{a}) \tag{2.9}$$

in which W''_i are additional forces required for the equilibrium of moments only. From Fig. 2.12 we can write that:

$$W''_i = W''_1 . \Delta''_i / \delta_i \tag{2.10}$$

and, because of the rigidity of the floor system which can only rotate as a solid body:

$$\Delta''_i = \Delta''_n \cdot \frac{a_i - \bar{a}}{B - \bar{a}} \tag{2.11}$$

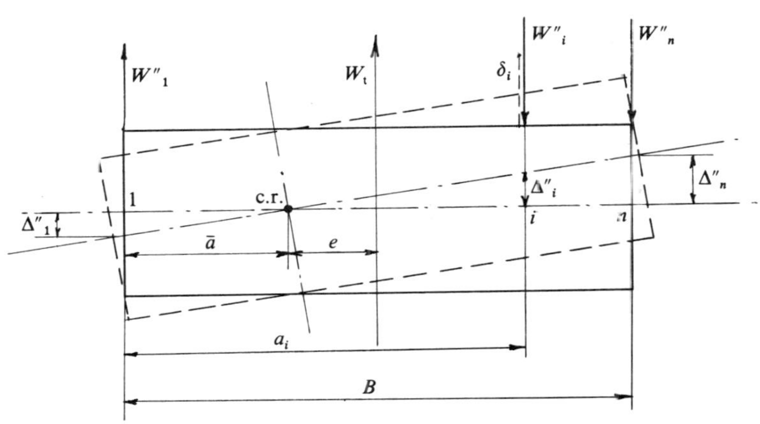

Fig. 2.12 Effect of the eccentricity of force on distribution of lateral forces

Introducing (2.11) and (2.10) into (2.9), we obtain:

$$\Delta''_n = \frac{W_t . e(B - \bar{a})}{\sum (a_i - \bar{a})^2/\delta_i} \tag{2.12}$$

From eq. (2.11), we obtain:

$$\Delta''_i = \frac{W_t . e . (a_i - \bar{a})}{\sum (a_i - \bar{a})^2/\delta_i} \tag{2.13}$$

and from eq. (2.9):

$$W''_i = W_t \cdot \frac{e(a_i - \bar{a})}{\delta_i . \sum (a_i - \bar{a})^2/\delta_i} \tag{2.14}$$

Combining eqs. (2.6) and (2.14), we obtain the final distribution of the wind force in the form:

$$W_i = W'_i + W''_i = W_t \cdot \left[\frac{1}{\delta_i . \sum 1/\delta_i} + \frac{e(a_i - \bar{a})}{\delta_i \sum (a_i - \bar{a})^2/\delta_i} \right] \tag{2.15}$$

Similarly, combining eqs. (2.12) and (2.13), we obtain:

$$\Delta_i = \Delta'_i + \Delta''_i = W_t \cdot \left[\frac{1}{\sum 1/\delta_i} + \frac{e(a_i - \bar{a})}{\sum (a_i - \bar{a})^2/\delta_i} \right] \tag{2.16}$$

Equations (2.15) and (2.16) can be used to obtain the coefficients for the distribution of wind forces and for the calculations of the actual deflections of frames. The procedure is illustrated by the following example.

Example 2.3

In Fig. 2.13, it is assumed that frames 1 to 6, when subjected to unit storey forces, deflect by 20, 40, 60, 120, 96, and 80 mm respectively.

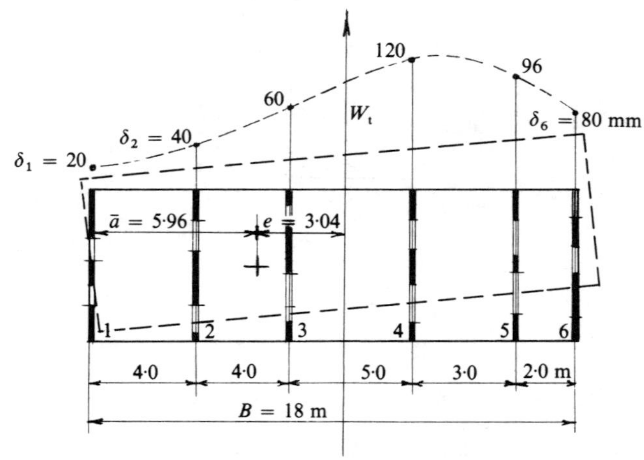

Fig. 2.13 Example of the analysis of distribution of lateral force for unsymmetrically loaded building

First, we calculate the value of $\sum 1/\delta_i = 29{\cdot}5/240$, and from eq. (2.6) we obtain:

$$W'_1 = 0{\cdot}4068\,W_t, \qquad W'_2 = 0{\cdot}2034\,W_t, \qquad W'_3 = 0{\cdot}1356\,W_t,$$
$$W'_4 = 0{\cdot}0678\,W_t, \qquad W'_5 = 0{\cdot}0847\,W_t, \qquad W'_6 = 0{\cdot}1017\,W_t.$$

From eq. (2.5), $\Delta' = 8{\cdot}13\,W_t$ (for $W_t = 6$ units, $\Delta' = 48{\cdot}8$ mm).

From eq. (2.8), $\bar{a} = 5{\cdot}96$ m and $e = 3{\cdot}04$ m. The value of $\sum (a_i - \bar{a})^2/\delta_i = 5{\cdot}2163$, and from eq. (2.14), we have:

$$W''_1 = -0{\cdot}1736\,W_t, \qquad W''_2 = -0{\cdot}0285\,W_t, \qquad W''_3 = +0{\cdot}0197\,W_t,$$
$$W''_4 = +0{\cdot}0340\,W_t, \qquad W''_5 = +0{\cdot}0608\,W_t, \qquad W''_6 = +0{\cdot}0875\,W_t.$$

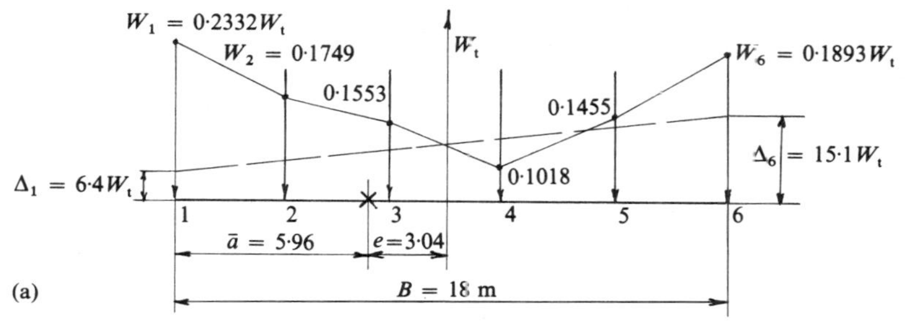

(b) Percentage distribution of total wind force

Based on translation only	40·7%	20·3%	13·5%	6·8	8·5	10·1%
Based on translation and rotation	23·3%	17·5%	15·5	10·2	14·5	19%
Frame No.	1	2	3	4	5	6

Fig. 2.14 Example of the analysis of distribution of lateral forces for eccentrically loaded building

Combining values W'_i and W''_i, we obtain the final distribution factors. These are as follows:

$$W_1 = 0{\cdot}2332\,W_t, \qquad W_2 = 0{\cdot}1749\,W_t, \qquad W_3 = 0{\cdot}1553\,W_t,$$
$$W_4 = 0{\cdot}1018\,W_t, \qquad W_5 = 0{\cdot}1455\,W_t, \qquad W_6 = 0{\cdot}1893\,W_t.$$

From eq. (2.16), we obtain:

$$\Delta_1 = 6{\cdot}4\,W_t, \qquad \Delta_6 = 15{\cdot}1\,W_t.$$

Figure 2.14(a) shows diagrammatically the distribution of the total force W_t to the individual frames and the table, Fig. 2.14(b), shows the comparison of the effect of translation only with the translation as well as rotation included in the analysis.

2.6 Preliminary Analysis of Shear Walls without the use of Computers

For a preliminary design, a method of analysis is needed which would furnish an approximate distribution of bending moments and deformations without setting out a matrix and without recourse to the computer. In the past, many approximate analyses were used, even for final design. The most commonly used were (a) the cantilever method, (b) the Portal Method, (c) Wilson–Maney Method, (d) Column Factor Method, (e) Oversway Method, etc. Most of these methods, with the exception of (d) and (e) are based on satisfying the conditions of equilibrium of forces only and do not generally satisfy the conditions of compatibility of deformations, and do not take into account the actual load–deflection characteristics of all members. These methods assume that beams and columns are bent into double curvature, with points of contraflexure being at the mid-lengths. These methods may give a satisfactory answer in the middle part of frames, but usually are in serious error in the lower part of the frame or shear wall and at points of discontinuities.

An approximate analysis can be made by the force–displacement method, satisfying the conditions of compatibility and equilibrium. It is assumed, however, that the stiffnesses of members remain repetitive throughout the height of the shear wall (although adjustments can be made to account for changes of sizes of members). No further assumptions or limitations need be imposed. The applied loads need not be uniformly distributed and can follow any given pattern.

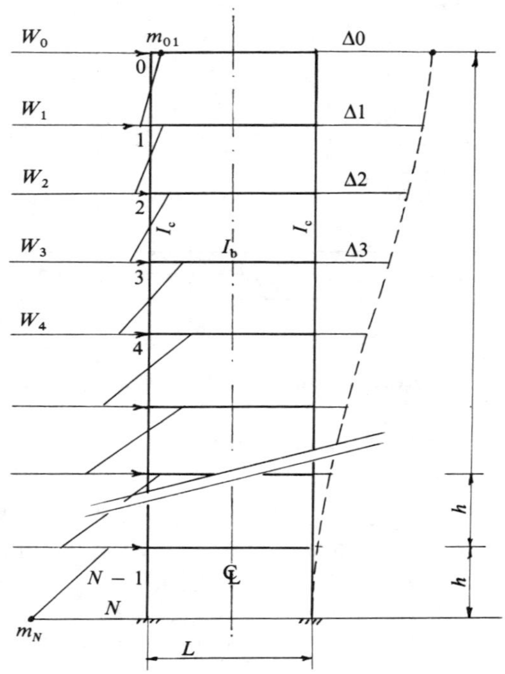

Fig. 2.15 Preliminary analysis of shear walls

Figure 2.15 shows a symmetrical shear wall of any number of floors loaded by any given system of lateral forces W_0, W_1, etc. The second moments of area of columns and beams are denoted I_c and I_b respectively. For the purpose of the development of the method, at first a full matrix is set out in its usual form for bending moments and displacements, as shown in Matrix 1.1 (Fig. 1.4). To facilitate presentation this matrix is set out for a given number of floors (ten in the example) using the frame shown in Fig. 2.15. The value of the parameter is:

$$r = \frac{L.I_c}{h.I_b} \tag{2.17}$$

and is referred to as the frame parameter. Introducing numerical values given in Fig. 1.5 the above matrix is solved for the purpose of comparing the results with the solution to be obtained later without the use of the matrix. For this frame the value of r is $=2{\cdot}5$, and the distribution of bending moments, forces, and deformations is as shown in Fig. 1.5 (see page 12).

Next, matrix 1.1 is reduced by eliminating all the deformations, and the resulting matrix is as shown in Table 2.4 in which

$$a = \frac{3}{r} + 1 \tag{2.18}$$

Table 2.4

$$\left(r = \frac{L.I_c}{h.I_b},\quad a = \frac{3}{r} + 1\right)$$

	0	1		2		3		4		5		6		7		8		9		10	
	0 1	1 0	1 2	2 1	2 3	3 2	3 4	4 3	4 5	5 4	5 6	6 5	6 7	7 6	7 8	8 7	8 9	9 8	9 10	10 01	
0–0′	1	1																			$V_{01}h$
1	$-a$	a	1																		0
1–1′			1	1																	$V_{12}h$
2		-1	$-a$	a	1																0
2–2′					F^e																$V_{23}h$
3																					0
3–3′							F^e														$V_{34}h$
4																					0
4–4′									F^e												$V_{45}h$
5																					0
5–5′											F^e										$V_{56}h$
6																					0
6–6′													F^e								$V_{67}h$
7																					0
7–7′															F^e						$V_{78}h$
8																					0
8–8′																	F^e				$V_{89}h$
9																					0
9–9′																			1	1	$V_{910}h$
10																		-1	$-a$	$a-1$	0

It is interesting to note that matrix 2.4, reduced to bending moments only, could be written directly without the use of matrix 1.1. The compatibility of deformations could be written at each floor level in the following way, e.g., at floor level 2, based on matrix 2.4, we have:

$$m_{21} + a.m_{23} - a.m_{32} - m_{34} = 0$$

or introducing the value for a, we obtain:

$$(m_{21} + m_{23})\frac{L}{6EI_b} + m_{23}.\frac{h}{2EI_c} - (m_{32} + m_{34})\frac{L}{6EI_b} - m_{32}.\frac{h}{2EI_c} = 0 \tag{2.19}$$

The equilibrium conditions remain in matrix 2.4 unaltered.

Equation (2.19) represents the combined condition of compatibility of deformations at the second floor level. Although it is written in an unfamiliar form it has a certain physical interpretation, which can be used in setting out the compatibility conditions without deflections. This not only reduces the size of the matrix, but makes it of a suitable form for a closed solution for bending moments at the extremities of the frame as is shown later. The deflections, if required, could then be calculated separately.

Referring to the matrix in Table 2.4, it should be noted that the last term along the diagonal (δ_{nn}) for any number of storeys is always equal to $(a-1)$ and not a as for the intermediate values along the diagonal.

Matrix 2.4 is diagonal and repetitive but is not diagonally symmetrical. However, it can be solved in a closed form for the bending moment m_{01} or for the moment at the encastre end m_{109}. The mathematical derivation is not given here but the moment at the top of the frame can be shown to be equal to

$$m_{01} = \frac{W.h}{2a.U_n - 1}.\sum_{1}^{n} s_n(a.U - 1) \tag{2.20}$$

in which the values of the parameters U are denoted by a, b, c, d, etc., at corresponding floor levels as shown in Fig. 2.16. These are referred to as floor parameters, the values of which depend on the dimensions of the shear wall and the sizes of members. The values of s are the shear coefficients in columns, e.g., for a ten-storey frame with uniform loading, these values are as shown in Fig. 2.17. A similar expression to eq. (2.20) can be developed for an unsymmetrical shear wall and for a shear wall with two rows of openings.

For tall shear walls, the unit values in eq. (2.20) become small and can be omitted and the equation can be reduced to

$$m_{01} = \frac{W.h}{2U}.\sum s.U \tag{2.21}$$

When the bending moment m_{01} is calculated from eq. (2.20) or eq. (2.21) all other moments in the shear wall can be calculated from the following conditions based on matrix 2.4:

$$\begin{aligned} m_{10} &= V_{01}.h - m_{01} \\ m_{12} &= a(m_{01} - m_{10}), \text{ etc.} \end{aligned} \tag{2.22}$$

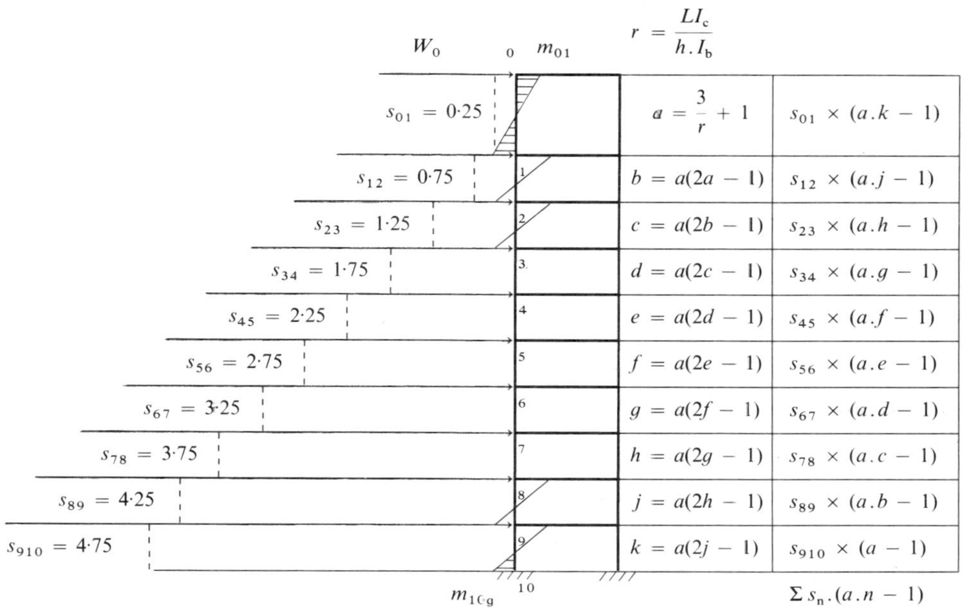

Fig. 2.16 Floor parameters *n*, in preliminary analysis of shear walls

Similarly, the general expression for the moment at the base is:

$$m_{n,n-1} = \frac{W.h.\sum s_n.U}{(a-1).(2U_{n-1}+1)+U_n} \tag{2.23}$$

Bending moments at other joints can be calculated from the eqs. (2.22) with the exception of the second last moment ($m_{n-1,n}$) which must be calculated, based on the matrix 2.4 from the relationship:

$$m_{n-1,n} = s_{n,n-1}.Wh - m_{n,n-1} \tag{2.24}$$

It should be noted that, although eqs. (2.20) and (2.23) are exact, the numerical values obtained will contain a percentage error due to the cumulative effect in numerical analysis.

The maximum deflection $\Delta 0$ can also be calculated, based on the distribution of bending moments now given. On the basis of matrix 1.1, it can be shown that:

$$\Delta 0 = \frac{W.h^3}{6EI_c} \times [m_{01} + (r+2)\sum(m_{10}+m_{21}+m_{32}+\cdots) +$$
$$+ (r-1)\sum(m_{01}+m_{12}+m_{34}+\cdots) + m_n] \tag{2.25}$$

Example 2.4

The procedure in the analysis is illustrated by the numerical example. The frame shown in Fig. 1.5 is used for comparison with the solutions obtained previously by the use of matrices shown on page 12 or 75. The frame parameter is as before:

$$r = \frac{L.I_c}{h.I_b} = 2{\cdot}5$$

and the floor parameters, based on the values shown in Fig. 1.5, are tabulated in Fig. 2.17, from which:

$$m_{01} = \frac{1{\cdot}084711}{4{\cdot}22244} W.h = 0{\cdot}2569\,Wh$$

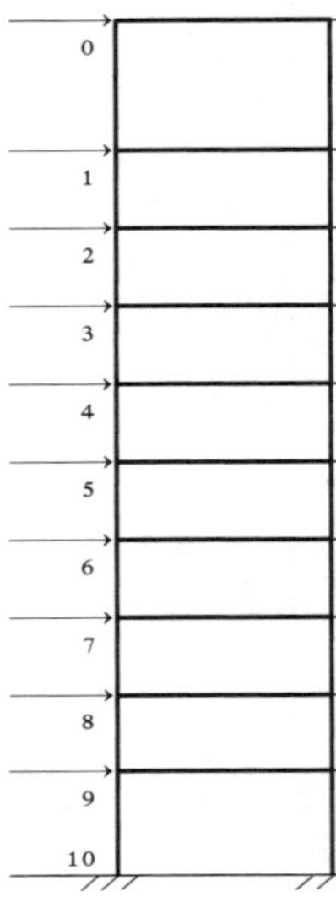

$a = \frac{3}{r} + 1 =$	2·2	$0{\cdot}25 \times (ak - 1) =$	527 805
$b = a(2a - 1) =$	7·5	$0{\cdot}75 \times (aj - 1) =$	359 868
$c = a(2b - 1) =$	30·7	$1{\cdot}25 \times (ah - 1) =$	136 315
$d = a(2 - 1) =$	132·9	$1{\cdot}75 \times (ag - 1) =$	43 374·9
$e = a(2d - 1) =$	582·6	$2{\cdot}25 \times (af - 1) =$	12 676·9
$f = a(2e - 1) =$	2561·0	$2{\cdot}75 \times (ae - 1) =$	3521·0
$g = a(2f - 1) =$	11 266·2	$3{\cdot}25 \times (ad - 1) =$	801·3
$h = a(2g - 1) =$	49 569·1	$3{\cdot}75 \times (ac - 1) =$	257·0
$j = a(2h - 1) =$	218 101·8	$4{\cdot}25 \times (ab - 1) =$	74·2
$k = a(2j - 1) =$	959 645·7	$4{\cdot}75 \times (a^2 - 1) =$	18·2
$2ak - 1 =$	4 222 440	$\Sigma =$	1 084 711

Fig. 2.17 Example of preliminary analysis of a shear wall

Alternatively, using eq. (2.21) we obtain:

$$\begin{aligned}\sum s.n &= 0{\cdot}25 \times 436\,321 + 0{\cdot}75 \times 99\,164 + 1{\cdot}25 \times 22\,537 + 1{\cdot}75 \times 5122 + \\ &\quad + 2{\cdot}25 \times 1164 + 2{\cdot}75 \times 264{\cdot}8 + 3{\cdot}25 \times 60{\cdot}4 + \\ &\quad + 3{\cdot}75 \times 13{\cdot}96 + 4{\cdot}25 \times 3{\cdot}4 + 4{\cdot}75 \times 2{\cdot}2 \\ &= 224\,210\end{aligned}$$

$$m_{01} = \frac{0{\cdot}224\,210}{2 \times 0{\cdot}436\,321} W.h = 0{\cdot}2569\,Wh$$

This value is 1·46 per cent higher than the exact value of 0·2532 Wh obtained from the matrix.

Based on eq. (2.23), bending moment at the foot of the frame is calculated as follows:

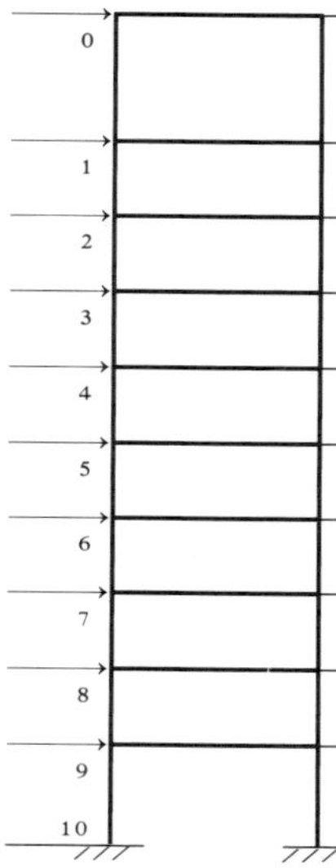

$a = \frac{3}{r} + 1 = 2{\cdot}2$	$0{\cdot}25a = 0{\cdot}55$
$b = a(2a - 1) = 7{\cdot}5$	$0{\cdot}75b = 5{\cdot}61$
$c = a(2b - 1) = 30{\cdot}7$	$1{\cdot}25c = 38{\cdot}39$
$d = a(2c - 1) = 132{\cdot}9$	$1{\cdot}75d = 232{\cdot}63$
$e = a(2d - 1) = 582{\cdot}6$	$2{\cdot}25e = 1311{\cdot}10$
$f = a(2e - 1) = 2561{\cdot}0$	$2{\cdot}75f = 7044{\cdot}7$
$g = a(2f - 1) = 11\,266{\cdot}2$	$3{\cdot}25g = 36\,625{\cdot}2$
$h = a(2g - 1) = 49\,596{\cdot}1$	$3{\cdot}75h = 185\,934{\cdot}7$
$j = a(2h - 1) = 218\,101{\cdot}8$	$4{\cdot}25j = 927\,185{\cdot}9$
$k = a(2h - 1) = 959\,645{\cdot}7$	$4{\cdot}75k = 4\,559\,562{\cdot}5$
	$\Sigma = 5\,717\,941$

from which

$$m_{10,9} = \frac{5\,717\,941\,.\,Wh}{1{\cdot}2(2 \times 218\,101{\cdot}8 + 1) + 959\,645{\cdot}7} = 3{\cdot}8543\,W.h$$

This value is 2·8 per cent higher when compared with 3·7491 Wh obtained from the matrix 1.1.

The maximum value of the deflection is calculated from eq. (2.25) as follows:

$$\Delta 0 = \frac{Wh^3}{6EI_c}[-0{\cdot}2532 + 4{\cdot}5(-0{\cdot}0032 + 0{\cdot}1859 + 0{\cdot}4214 + 0{\cdot}6681 + 0{\cdot}9182 + 1{\cdot}1720 + 1{\cdot}4387 + 1{\cdot}7582 + 2{\cdot}2971) + 1{\cdot}5(0{\cdot}5641 + 0{\cdot}8286 + 1{\cdot}0819 + 1{\cdot}3318 + 1{\cdot}5780 + 1{\cdot}8113 + 1{\cdot}9918 + 1{\cdot}9529 + 1{\cdot}0008) + 3{\cdot}7492]$$

$$= 62{\cdot}06\frac{Wh^3}{6EI_c} = 10{\cdot}34.\frac{Wh^3}{EI}$$

this is within 0·5 per cent of the value calculated previously (page 12).

A similar analysis can be made for an unsymmetrical frame on two or three columns.

2.7 Infilled Frames

When a frame is very flexible, brick or block infill can be used to increase its stiffness as shown in Fig. 2.18. Very flexible frames on their own have small resistance to lateral forces and show large deformations and rotations of joints. Similarly, block or brick walls, when on their own, fail mainly in shear under comparatively small loads after only a small displacement. Experiments and analysis show that the combined action of the two media can be very much greater than the sum of two separate, particularly in resisting lateral forces, providing lateral stability and reducing deformations.

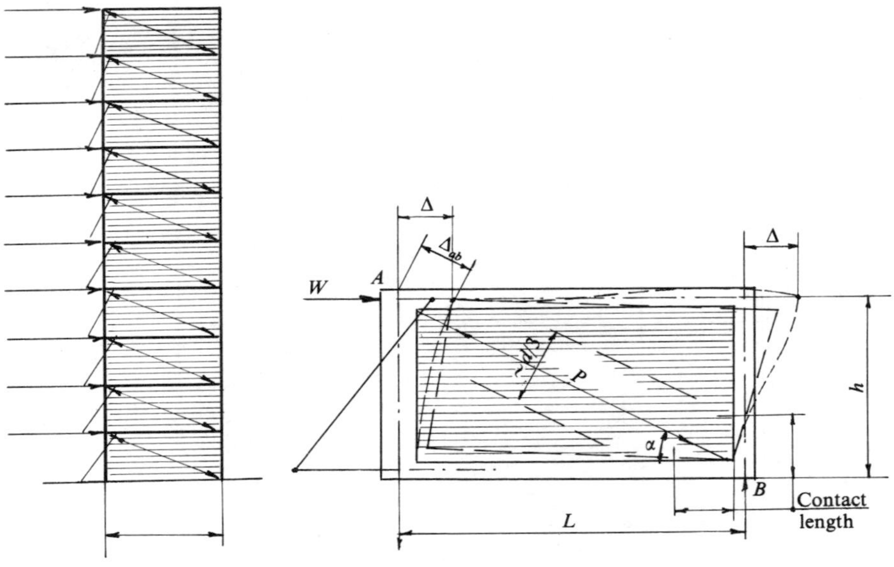

Fig. 2.18 Infilled frames

The great load-carrying capacity of infilled frames is due to the development of considerable axial forces as well as bending moments in the frame, and due to the confinement of the infill. The latter develops high diagonal forces but reduced shearing stresses and deformations.

For such combined action to develop, there should be a minimum of space left between the frame and infill. The effectiveness of the combined action depends greatly on the standard of workmanship and the materials used, but the increase in the stiffness is so great that even if only a small part of this combined action develops, the overall stiffness of the shear wall may still be sufficient for practical purposes.

It has been shown by Stafford Smith[14] and others[10,11,12] that the stiffness of an infilled panel depends on the area of contact between the frame and wall, and the quality of infill. The experimental values of the contact length is discussed in the above reference. M. Holmes[10] developed an expression for the change in length of the diagonal as well as the ultimate strength of an infilled panel. These are respectively:

$$\Delta_{ab} = \frac{(W - W\cos\alpha).h^3}{24EI_c}\left(1 + \frac{I_c}{I_b}\cot\alpha\right)\cdot\cos\alpha$$

$$W_{cr} = \frac{24EI_c.\varepsilon_c.d}{h^3(1 + I_c/I_b.\cot\alpha)} + A.f_c.\cos\alpha$$

in which: h, L, and d are the height, length, and the diagonal of the infill respectively, ε_c is strain in the infill material at the instant of maximum stress (it is suggested that this could be taken as 0·002 to 0·005 for weak and strong concrete respectively), f_c is the compression strength of the infilling material, A the cross-sectional area of the equivalent strut $= t \times d/3$, and t is the thickness of the wall.

The horizontal deflection at failure is, according to M. Holmes: $\Delta = \Delta ab . \cos \alpha = \varepsilon_c . d . \cos \alpha$.

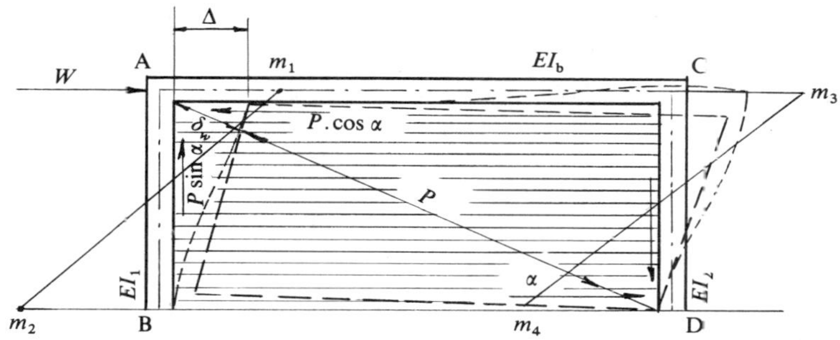

Fig. 2.19 Analysis of infilled frame by the force-displacement method

The force–displacement method can be used to analyse the distribution of bending moments in the frame and diagonal forces in the infilled panels, as well as deformations. In this method, the conditions of compatibility of deformations in the frame can be written as previously, but the conditions of equilibrium of forces must now be adjusted to account for the action of the infill, and one additional condition must be found to account for the additional unknown force in each panel.

For a portal frame in Fig. 2.19, the unknown moments and the force P in the assumed strut can be calculated from the following conditions:

A. $m_1(L/3EI_b + h/3EI_1) - m_2.h/6EI_1 - m_3.L/6EI_b = \Delta/h$

B. $m_2.h/3EI_1 - m_1.h/6EI_1 = \Delta/h$

C. $-m_1.L/6EI_b + m_3(L/3EI_b + h/3EI_2) - m_4.h/6EI_2 = \Delta/h$

D. $-m_3.h/6EI_2 + m_4.h/3EI_2 = \Delta/h$

S.C. $\dfrac{m_1 + m_2}{h} + \dfrac{m_3 + m_4}{h} + P.\cos \alpha = W$

A\D. $\Delta_{\text{frame}} = \delta_{\text{wall}}$

The last equation expresses the condition of compatibility of deformations between the frame and wall along the contact area across the diagonal AD. The value of Δ in the frame can be calculated (in terms of the unknown moments) comparatively accurately by the force–displacement method, as previously used. However, the value of the deformation of the wall along the diagonal can only be approximately estimated. This is a function of the relative stiffness of the frame and wall, length of contact, shrinkage of mortar and moisture expansion of bricks, quality of the materials and workmanship, etc.

The column formula:

$$\delta_w = P.d/A_w.E_w$$

may be used to estimate the value of the deformation. In this expression E_w is the modulus of elasticity of the wall, A_w is the cross-sectional area of the effective strut, and d is the length of the diagonal. For brick walls, the value of E_w varies between 7 and 28 kN/mm^2.

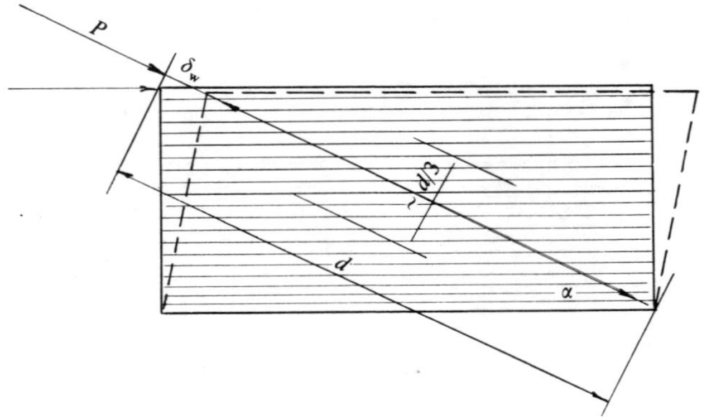

Fig. 2.20 Deformation of infilled panel

The shrinkage coefficient can be taken as for weak concrete, i.e., 0·2 to 0·5 per cent. The shear deformation of the wall can be taken as:

$$\delta_w = c.P.L/GA$$

or as shown by Timoshenko:[17]

$$\delta = P.L^3/3EI \times \left[1 + 0{\cdot}71\left(\frac{h}{L}\right)^2 - 0{\cdot}1\left(\frac{h}{L}\right)^3\right]$$

For rectangular walls, $c=6/5$.

When the value of the total deformation of the wall is estimated (in terms of the unknown diagonal force P), it can then be introduced in the matrix.

If a space is left between frame and infill, either from inaccurate workmanship or from shrinkage of mortar, then the infill will not come into operation until the gap is closed, as shown diagrammatically in Fig. 2.21. If λ_h and λ_v denote the magnitude of the spaces left, in horizontal and vertical directions respectively, then the corner A

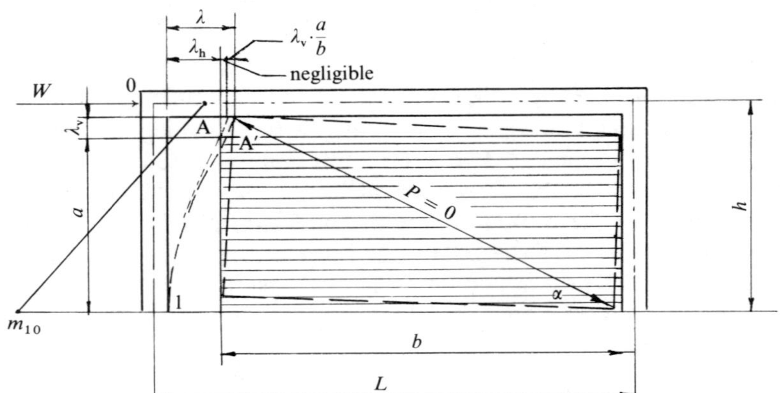

Fig. 2.21 Effect of spaces between frame and the infill

of the infill will move to A′ and the infill will still be unstressed while the frame will already develop bending moments. These moments must be added to those of the second stage when the infill becomes operative. It should be noted that the second stage analysis should be made for forces reduced by the forces of the first stage, i.e., the effectiveness of the infill with spaces is therefore reduced.

The magnitude of the movement of the frame λ can be taken as approximately equal to:

$$\lambda = \lambda_h + \lambda_v . \frac{a}{b}$$

The value of λ_h and λ_v must be either assumed or evaluated as indicated above.

Example 2.5

In Fig. 2.22, it is assumed that a portal frame of dimensions shown is infilled by the material of the modulus of elasticity $E_w = 14$ kN/mm², while the concrete frame has a modulus of elasticity of $E_c = 28$ kN/mm², i.e., $E_c/E_w = 2{\cdot}0$, and the thickness of the infill is 225 mm.

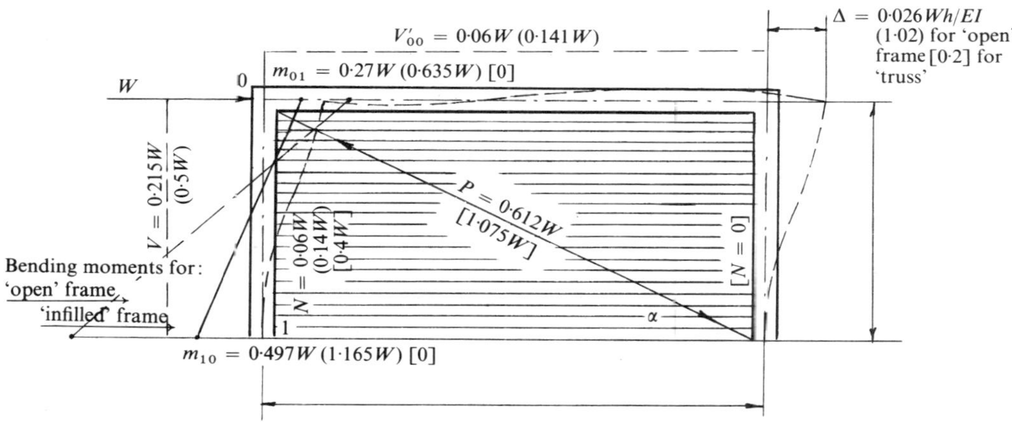

Fig. 2.22 Example of the analysis of infilled portal frame

If at first we assume that no gap is left between the frame and the infill, the required value of δ_w is calculated in terms of the unknown force P as follows:

$$\frac{\delta_w}{\cos\alpha} \cdot \frac{E_c . I_c}{h} = \frac{P . d}{E_w . A_w . \cos\alpha} \cdot \frac{E_c . I_c}{h} = 0{\cdot}0428 P$$

The matrix becomes now:

$$\begin{bmatrix} 2{\cdot}7 & -0{\cdot}6 & 0 & -1 \\ -0{\cdot}6 & 1{\cdot}2 & 0 & -1 \\ 1 & 1 & 1{\cdot}695 & 0 \\ 0 & 0 & -0{\cdot}0428 & 1 \end{bmatrix} \begin{Bmatrix} m_{01} \\ m_{10} \\ P \\ \Delta \end{Bmatrix} = \begin{Bmatrix} 0 \\ 0 \\ 1{\cdot}8W \\ 0 \end{Bmatrix}$$

from which $m_{01} = 0{\cdot}27W$, $m_{10} = 0{\cdot}497W$, $P = 0{\cdot}612W$, and $\Delta = 0{\cdot}026Wh/EI$.

Figure 2.22 shows the distribution of bending moments, forces, and deformations, together with the similar values (shown in parentheses), related to the frame without infill. Figures in square brackets show forces and deformations obtained from the system treated as a pin-connected truss.

This analysis shows that the stresses and deformations of the combined system of the frame and infill are considerably reduced when compared with those of the frame alone or even of the truss.

Example 2.6

In this example, the effect of spaces left between the frame and infill is evaluated. It is assumed that the spaces are: $\lambda_h = \lambda_v = 1$ mm so that $\lambda = 1{\cdot}42$ mm, i.e., the frame can move by this amount before it comes into contact with the infill, and the bending moments calculated in this stage must be added to those previously calculated. The required value of $\lambda . E_c I_c / h$ is equal to 60·0.

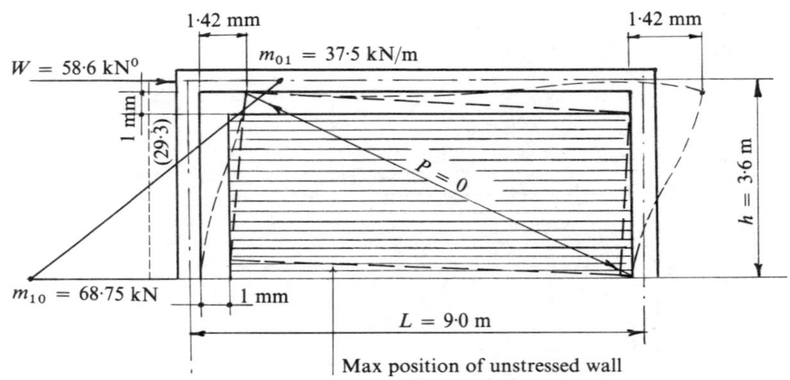

Fig. 2.23 Analysis of the effect of spaces between the frame and infill

The compatibility and equilibrium conditions result in the following matrix for the evaluation of moments and force W:

$$\begin{bmatrix} 2{\cdot}7 & -0{\cdot}6 & 0 \\ -0{\cdot}6 & 1{\cdot}2 & 0 \\ 1 & 1 & -1{\cdot}8 \end{bmatrix} \begin{Bmatrix} m_{01} \\ m_{10} \\ W \end{Bmatrix} = \begin{Bmatrix} 60{\cdot}0 \\ 60{\cdot}0 \\ 0 \end{Bmatrix}$$

from which $m_{01} = 36{\cdot}82$ kN/m, $m_{10} = 67{\cdot}5$ kN/m, and $W = 58{\cdot}6$ kN (the value of P is, of course, equal to zero).

Example 2.7

In this example, the frame previously analysed as an open frame (see Fig. 1.7), is now assumed to be infilled by material of modulus of elasticity $E_w = 14$ kN/mm^2, while the modulus of the concrete frame is assumed to be $E_c = 28$ kN/mm^2. If no spaces exist between the frame and the infill, then the force–displacement matrix for the evaluation of moments in the frame, diagonal forces in the infill, and deformations

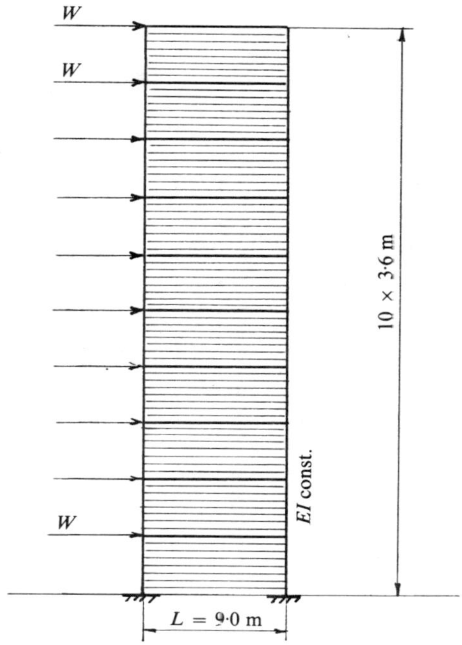

Fig. 2.24 Example of the analysis of tall infilled shear wall

can be set out in a similar way to Example 2.5. The overall matrix is shown in Table 2.5.

Figure 2.25 shows the resulting distribution of moments, forces, and deformations in the system. To facilitate the comparison, similar data are shown for the frame without infill, as well as for the system treated as a pin-connected truss. It can be seen again that the introduction of the infill reduces bending moments and deflection in the frame, although direct stresses in the frame are increased when compared with an open frame. The assumption of the 'pin-connected' truss results in absence of moments but in the increase of direct forces. The effect of axial forces can be included in the analysis, but this is discussed later in the text.

Example 2.8

In this example, the effect of spaces between the frame and the infill is considered on the initial bending moments and deformations, prior to the frame coming into contact with the infill. Assuming that all spaces are equal, and that $\lambda_h = \lambda_v = 1$ mm, the compatibility equations for the frame can be set out in the usual form, e.g., at the point 0, in Fig. 2.26 we can write:

$$-m_{10} \cdot \frac{h}{6E_c I_c} + m_{01}\left(\frac{h}{3E_c I_c} + \frac{L}{6E_c I_b}\right) = \frac{\lambda_0 - \lambda_1}{h}$$

in which:

$$(\lambda_0 - \lambda_1) . EI/h = 60{\cdot}0$$

The equilibrium conditions are related to the unknown forces W.

Table 2.5 Force–displacement matrix for the analysis of infilled frame in Fig. 2.25

	0			1				2				3				4			
	0 1	P_0	Δ0	1 0	1 2	P_1	Δ1	2 1	2 3	P_2	Δ2	3 2	3 4	P_3	Δ3	4 3	4 5	P_4	Δ4
1 0 0′	2·7	0	−1	−0·6	0	0	+1												
1 / 0	0	−0·0428	+1	0	0	0	−1												
0–1	+1	1·6711	0	+1	0	0	0												
2 1 1′	0	0	0	1·5	2·7	0	−1	−0·6	0	0	1								
0 1 2	−0·6	0	−1	1·2	−1·2	0	2	0·6	0	0	−1								
2 / 1	0	0	0	0	0	−0·0428	1	0	0	0	−1								
1–2	0	0	0	0	1	1·6711	0	1	0	0	0								
3 2 2′																			
1 2 3										F^e									
3 / 2																			
2–3																			
4 3 3′																			
2 3 4														F^e					
4 / 3																			
3–4																			
5 4 4′																			
3 4 5																		F^e	
5 / 4																			
4–5																			
6 5 5′																			
4 5 6																			
6 / 5																			
5–6																			
7 6 6′																			
5 6 7																			
7 / 6																			
6–7																			
8 7 7′																			
6 7 8																			
8 / 7																			
7–8																			
9 8 8′																			
7 8 9																			
9 / 8																			
8–9																			
10 9 9′																			
8 9 10																			
10 / 9																			
9–10																			
10 9																			

5				6				7				8				9				10	
5 4	5 6	P_5	Δ5	6 5	6 7	P_7	Δ7	7 6	7 8	P_8	Δ8	8 7	8 9	P_8	Δ8	9 8	9 10	P_9	Δ9	10 9	
																					0
																					0
																					1·8W
																					0
																					0
																					0
																					3·6W
																					0
																					0
																					0
																					5·4W
																					0
																					0
																					0
																					7·2W
																					0
																					0
																					0
																					9·0W
																					0
		F^e																			0
																					0
																					10·8W
																					0
						F^e															0
																					0
																					12·6W
																					0
										F^e											0
																					0
																					14·4W
																					0
														F^e							0
																					0
																					16·2W
													0	0	0	1·5	2·7	0	−1	−0·6	0
													−0·6	0	−1	1·2	−1·2	0	2	0·6	0
													0	0	0	0	0	−0·0428	1	0	0
													0	0	0	0	1	1·6711	0	1	18·0W
																	−0·6	0	−1	1·2	0

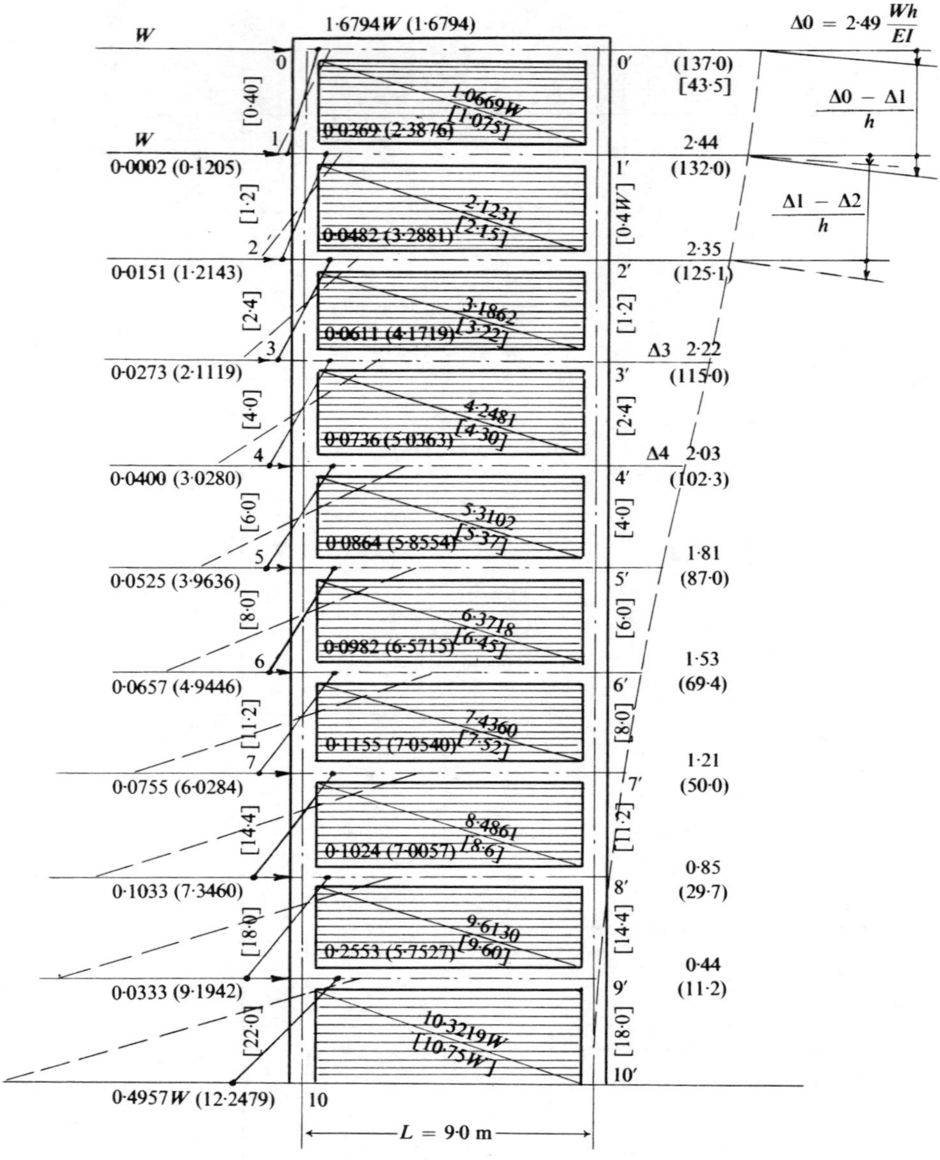

Fig. 2.25 Example of the analysis of infilled frame

The element force-displacement matrix is now as follows:

$$\begin{bmatrix} 0 & 0 & 1{\cdot}5 & 2{\cdot}7 & 0 & -0{\cdot}6 \\ -0{\cdot}6 & 0 & 1{\cdot}2 & -1{\cdot}2 & 0 & 0{\cdot}6 \\ -1 & 0 & -1 & 1 & -1{\cdot}8 & 1 \end{bmatrix} \begin{Bmatrix} m_{01} \\ W_0 \\ m_{10} \\ m_{12} \\ W_1 \\ m_{21} \end{Bmatrix} = \begin{Bmatrix} 60 \\ 0 \\ 0 \end{Bmatrix}$$

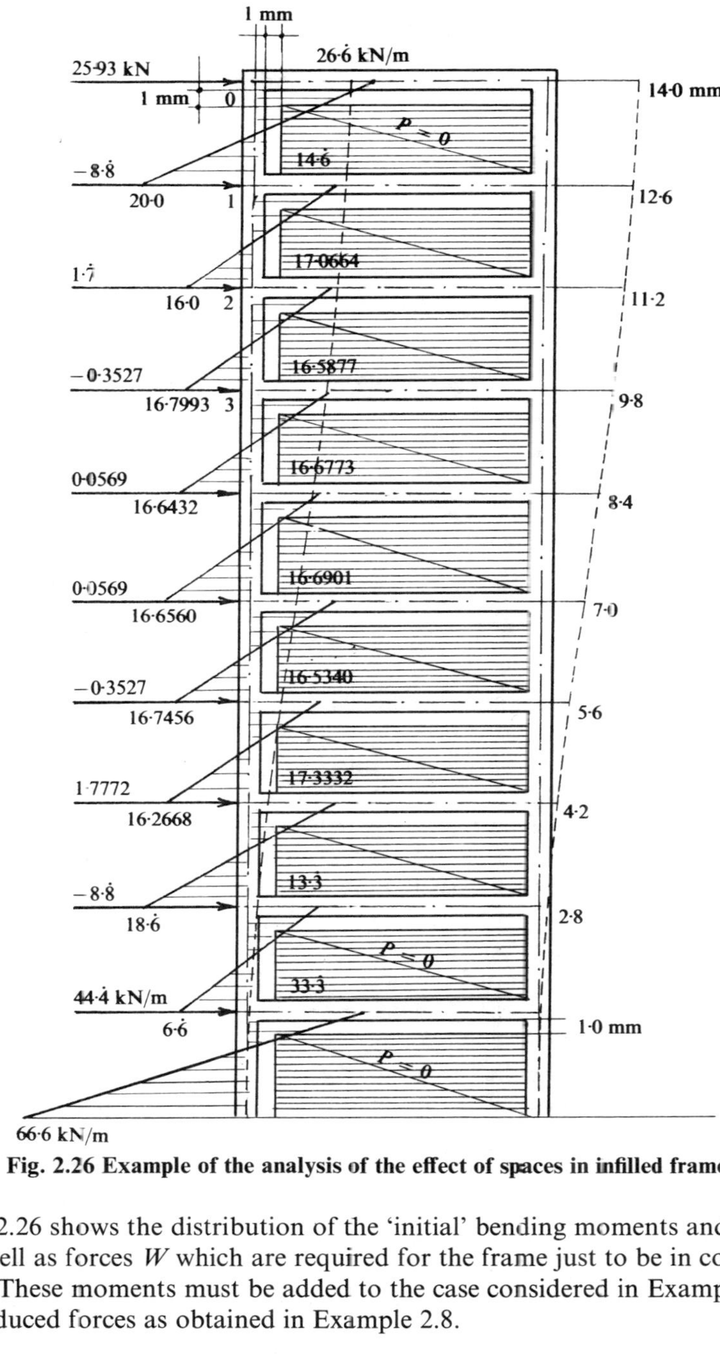

Fig. 2.26 Example of the analysis of the effect of spaces in infilled frame

Figure 2.26 shows the distribution of the 'initial' bending moments and deformations as well as forces W which are required for the frame just to be in contact with the infill. These moments must be added to the case considered in Example 2.7, but for the reduced forces as obtained in Example 2.8.

2.8 Effect of Perforations in the Analysis of Infilled Frames

Infilled panels are often perforated for windows and doors and, for this reason, such infills are considered to be of little structural value. However, if the opening is small and a parapet beam or infill above the door lintel is left continuous and tightly fitted between columns, then such perforated infills may still contribute to the stiffness of the system. It is, of course, an advantage if reinforcement is used in the infill, at least along the outer boundary.

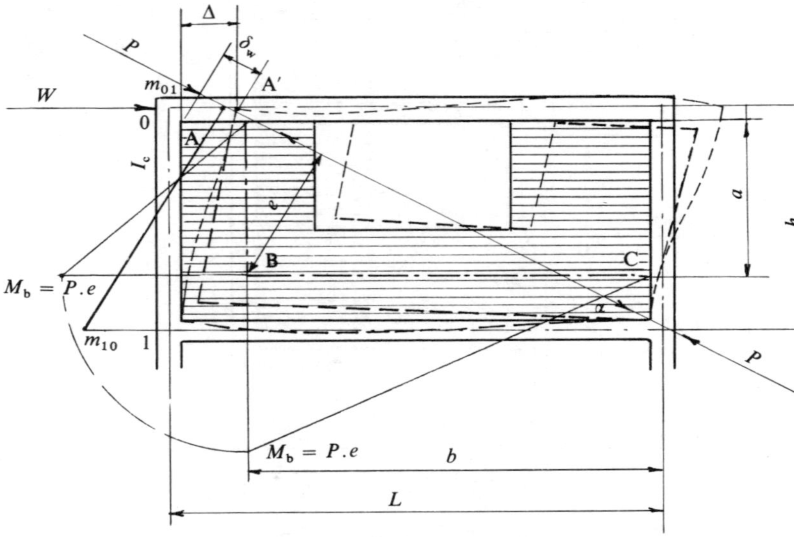

Fig. 2.27 Effect of perforations in the infilled panels

It is possible to analyse such frames by introducing into the matrix the deformation of the corner point (point A in Fig. 2.27) along the diagonal. This deformation is difficult to evaluate precisely and may be mainly of the shear type, depending on the relative sizes of the infill, its shear modulus, and the performance of the joints. In addition to the diagonal force P, such infills are subjected to bending $M_b = P.e$ and shear, and if not reinforced may crack or joints may shear off. Frequent cracks above door lintels and in window parapets may be due to this effect rather than to the frequently suspected effect of differential settlement or thermal changes.

The analysis of such perforated infilled frames can be made by setting the conditions of the equilibrium and compatibility of deformations of the frame as well as along the diagonal between the frame and the infill.

First, however, the deformation of the infill Δ along the diagonal must be determined. This can be conveniently done by the unit load method or by the force–displacement method. Using the latter approach, an L-shaped element ABC in Fig. 2.28, continuous at point B, is assumed to be subjected to the force P acting along the diagonal AC. At release at point B the discontinuity is Δ/a and this is restored by applying the bending moment $M = P.e$, and we can write that:

$$(P.e)\left(\frac{a}{3EI_1} + \frac{b}{3EI_2}\right) = \frac{\Delta}{a}$$

from which

$$\Delta = P\frac{e.a}{3E_w I_{w1}}(a + nb) \tag{2.26}$$

where $n = I_1/I_2$.

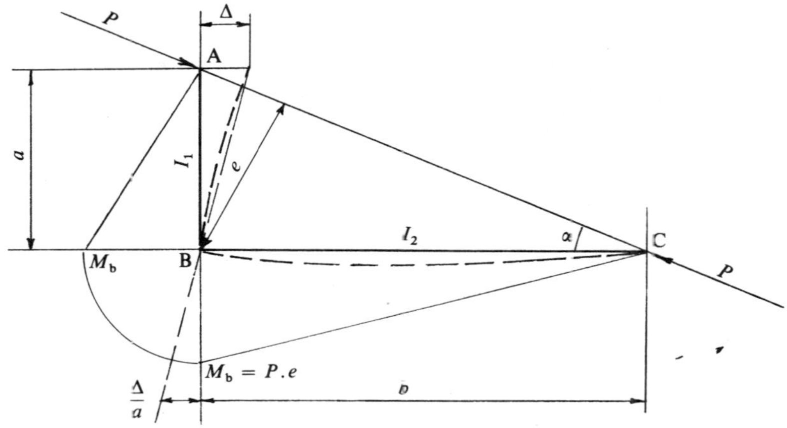

Fig. 2.28 Analysis of flexural deformations of infill

Referring to Fig. 2.27, the matrix can now be set out expressing the conditions of the compatibility of deformations in the frame at points 0 and 1, and along the diagonal 0–1′ between the frame and infill as well as the condition of equilibrium of forces. These are as follows:

$$
\begin{aligned}
&0. \quad -m_{10}\cdot\frac{h}{6E_cI_c} + m_{01}\cdot\left(\frac{h}{3E_cI_c} + \frac{L}{3E_cI_b}\right) = \frac{\Delta}{h} \\
&1. \quad m_{10}\cdot\frac{h}{3E_cI_c} - m_{01}\cdot\frac{h}{6E_cI_c} = \frac{\Delta}{h} \\
&0\text{–}1. \quad 2\frac{m_{01} + m_{10}}{h} + P.\cos\alpha = W \\
&0\searrow 1'. \quad \Delta.\cos\alpha = \Delta w
\end{aligned}
\tag{2.27}
$$

from which the statically indeterminate values can be obtained.

The following example is used to illustrate the procedure.

Example 2.9

In this example, the same frame is used as analysed previously (see Example 2.5), but it is now assumed that an opening is introduced of size 2 × 3 m and that only the flexural stiffness of the infill is considered. It is assumed that $a=2{\cdot}6$ m, $b=7$ m, $e=2{\cdot}45$ m, $I_w=0{\cdot}39$ m^4, $I_w=0{\cdot}041$ m^4, $(n=9{\cdot}5)$, $I_c/I_w=0{\cdot}014$ and $E_c/E_w=2{\cdot}0$. The following values are required in the matrix:

$$\frac{h}{2}.\cos\alpha = 1{\cdot}6956$$

and

$$\Delta_w\cdot\frac{E_c.I_c}{h.\cos\alpha} = P\cdot\frac{e.a.(a+nb)}{3h.\cos\alpha}\cdot\frac{E_c}{E_w}\cdot\frac{I_c}{I_1} = 1{\cdot}225P$$

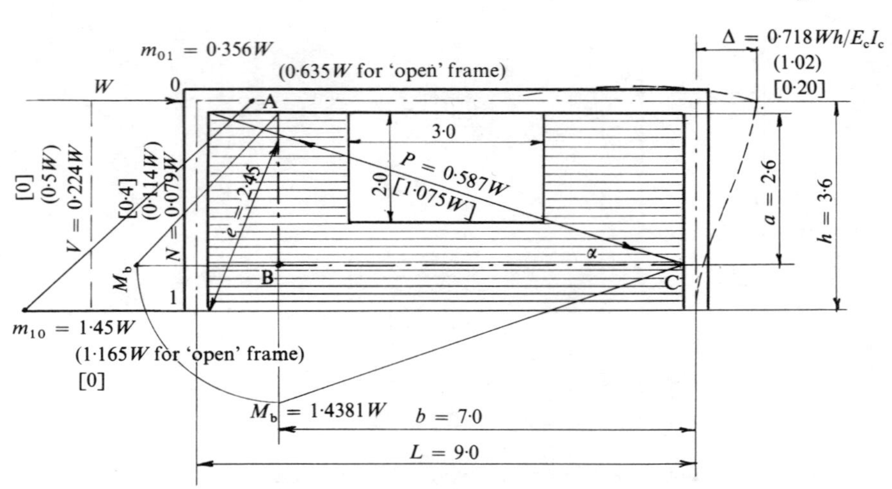

Fig. 2.29 Example of the analysis of infilled portal frame with window opening

From eqs. (2.27), we obtain the following matrix:

$$\begin{bmatrix} 2{\cdot}7 & -0{\cdot}6 & 0 & -1 \\ -0{\cdot}6 & 1{\cdot}2 & 0 & -1 \\ 1 & 1 & 1{\cdot}6956 & 0 \\ 0 & 0 & 1{\cdot}2250 & -1 \end{bmatrix} \begin{Bmatrix} m_{01} \\ m_{10} \\ P \\ \Delta \end{Bmatrix} = \begin{Bmatrix} 0 \\ 0 \\ 1{\cdot}8W \\ 0 \end{Bmatrix}$$

from which $m_{01} = 0{\cdot}356W$, $m_{10} = 0{\cdot}45W$, $P = 0{\cdot}587W$, and $\Delta = 0{\cdot}718\,Wh/EI_c$. These values are shown in Fig. 2.29, together with the values in parentheses for the frame without infill, and values in square brackets related to the system treated as a pin-connected truss.

This example shows that even a perforated infill can contribute to the overall stiffness of the frame even if only the bending deformations of the infill are considered. However, a large moment and shear force may develop in the infill and may cause cracks. Reinforcement could be provided along the outer periphery of the infill in the mortar joint for this purpose.

2.9 Analysis of Frames with Parapet Infills

Parapet infills are not usually considered as contributing much to the overall stiffness of frames but, if required, their effect on the performance of frames can be approximately evaluated as follows.

When a frame (a portal or a multi-storey) is subjected to the effect of lateral forces as in Fig. 2.30, then a thrust P in the infill may still materialize, provided that the space between the frame and the infill is not large, and that the infill material does not become crushed. This thrust may act horizontally or may be inclined, depending on the distribution of pressure along the length of contact, i.e., whether it is uniformly distributed, triangular, or curvilinear. This, however, is a matter for research and is difficult to ascertain analytically only. The effect of the form of contact pressure can be expected to be small on the overall performance of the frame, although it may be significant for the evaluation of local stresses in the infill.

Considering the frame in Fig. 2.30, the statically indeterminate values, when using the force–displacement method, are: bending moments at the points 0 and 1, deflection Δ, and the force P. These values can be evaluated by setting out the conditions of compatibility of deformations at the points 0 and 1 and along the diagonal AB and using the condition of equilibrium of forces.

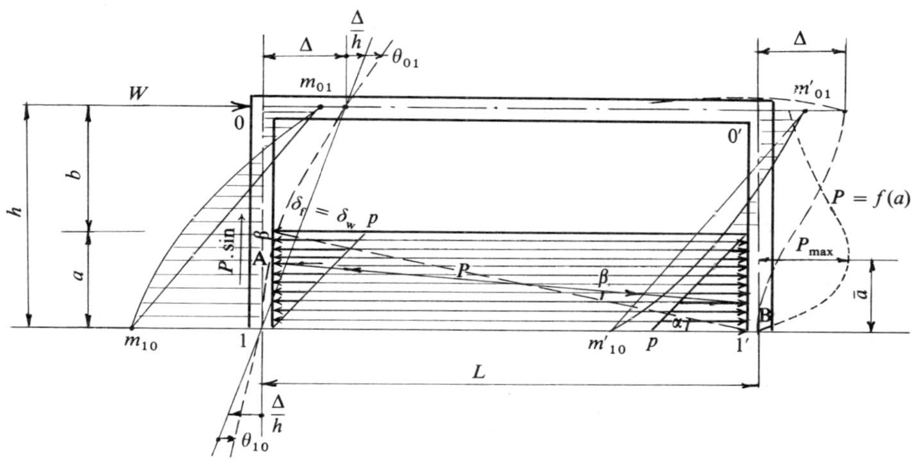

Fig. 2.30 Effect of parapet infills on frames

It should be noted that, at release, both columns are subjected to end rotations θ_{01} and θ_{10} from the pressure exerted by the wall. The latter can be evaluated separately in terms of the unknown force P (or pressure p) as follows:

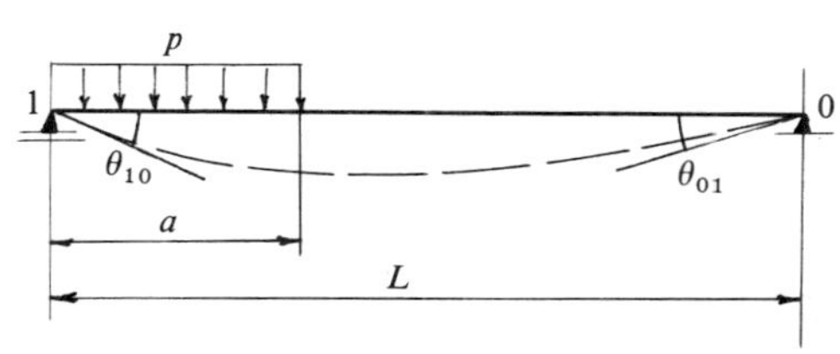

Fig. 2.31

(a) If the pressure is assumed to be uniformly distributed along the length of contact with the frame, as shown in Fig. 2.31, then it can be shown that:

$$\theta_{10} = \frac{p.a^2}{24L.EI}(2L - a)^2 = P\cdot\frac{a}{24L.EI}(2L - a)^2 \tag{2.28}$$

$$\theta_{01} = \frac{p.a^2}{24L.EI}(2L^2 - a^2) = P\cdot\frac{a}{24L.EI}(2L^2 - a^2) \tag{2.29}$$

(b) If the contact pressure is assumed to be triangular, as shown in Fig. 2.32, it can be shown, e.g., by the force–displacement method, that:

$$\theta_{01} = \frac{p.a^3}{45EI}\left(1 + \frac{4b}{L} + \frac{5b^2}{2L.a}\right) \tag{2.30}$$

$$\theta_{10} = \frac{7}{8}\cdot\frac{p.a^3}{45EI}\cdot\left[1 + \frac{b}{a}\left(1 + \frac{3b}{7L}\right)\right] \tag{2.31}$$

The derivation of these values is omitted but it can be made by introducing a release at point 2, and introducing also an additional unknown deflection λ, as shown in Fig. 2.32.

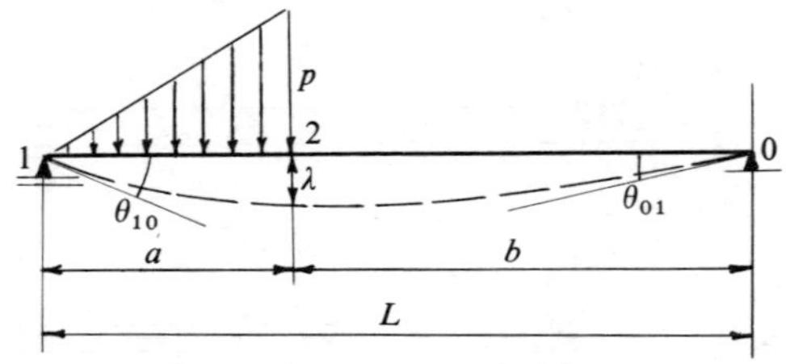

Fig. 2.32

It should be noted that column 0′1′ in Fig. 2.30 is subjected to differently orientated triangular pressure than column 01 and eqs. (2.30) and (2.31) should be used for column 01. The magnitude of pressure in column 0′1′ may also be different to that of column 01 and, in addition, in multi-storey frames, beams are also subjected to different pressures at the ends. For these reasons, bending moments in both columns may differ even if the frame is symmetrical.

The force–displacement matrix for the frame in Fig. 2.30 can now be set out in general terms as follows:

$$\begin{aligned}
&0. && -m_{10}\cdot\frac{h}{6EI_c} + m_{01}\cdot\left(\frac{h}{3EI_c} + \frac{L}{6EI_b}\right) = \frac{\Delta}{h} + \theta_{01} \\
&1. && m_{10}\cdot\frac{h}{3EI_c} - m_{01}\cdot\frac{h}{6EI_c} = \frac{\Delta}{h} - \theta_{10} \\
&0\text{–}1. && 2\cdot\frac{m_{01} + m_{10}}{h} + P\cdot\frac{a/2}{h} = W \\
&0\diagdown 1'. && \delta_f = \delta_w
\end{aligned} \tag{2.32}$$

The first three conditions are of the usual form, except for the values of θ as discussed above. The last equation expresses the compatibility of deformations of the frame and the infill along the line of the assumed strut. These are:

$$\delta_f = \Delta\cdot\frac{a}{h.\cos\alpha}$$

$$\delta_w = \frac{P.L}{E_w.A_w} \tag{2.33}$$

An additional difficulty arises: while for solid walls the expression (2.33) can be used directly, it is usual practice that the external wall is of cavity construction, and often consists of two leaves of different quality materials. It is necessary, therefore, to evaluate the deflection δ_w of the composite wall as well as the distribution of pressure in each leaf.

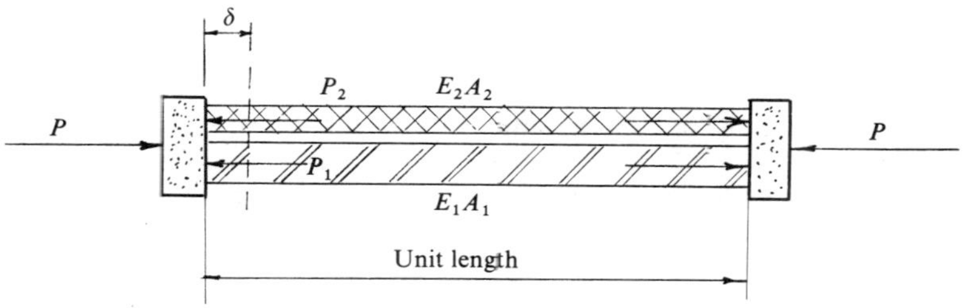

Fig. 2.33

Figure 2.33 shows an idealized case in which a force P is applied by means of concrete members to the composite wall of moduli of elasticity E_1 and E_2. If the axial deformation δ is assumed to be the same in both leaves (without bending in the horizontal plane), we can write that:

$$\frac{P_1}{E_1A_1} = \frac{P_2}{E_2A_2}$$

from which:

$$P_1 = P\cdot\frac{k}{1+k} \tag{2.34}$$

$$P_2 = P\cdot\frac{1}{1+k} \tag{2.35}$$

and

$$k = \frac{E_1 . A_1}{E_2 . A_2} \tag{2.36}$$

Other expressions can be derived based on different assumptions.

Using eqs. (2.34) and (2.35), we can now express the deformation of the composite wall as:

$$\delta_w = \frac{P.L}{E_1 . A_1}\cdot\frac{k}{1+k} \tag{2.37}$$

so that the last equation in the matrix 2.1 becomes:

$$\Delta\cdot\frac{a}{h.\cos\alpha} = \frac{P.L}{E_1 . A_1}\cdot\frac{k}{1+k}$$

or in the form consistent with the remaining equations:

$$\Delta\cdot\frac{E_c . I_c}{h} - P\cdot\frac{L}{a}\cdot\frac{E_c}{E_1}\cdot\frac{I_c}{A_1}\cdot\frac{k}{1+k}\cdot\cos\alpha = 0 \tag{2.38}$$

Using the relationships (2.28) to (2.38), the matrix can be set out and the statically indeterminate values obtained.

Example 2.10

In this example, the infill is assumed to be solid, 225 mm thick as in Example 2.5, so that comparison can be made. It is assumed that $E_c = 28$ kN/mm^2 and $E_w = 14$ kN/mm^2. The following value is also required in the matrix:

$$P \cdot \frac{L}{a} \cdot \frac{E_c}{E_w} \cdot \frac{I_c}{A_w} \cdot \cos\alpha = 0{\cdot}1880P$$

The force–displacement matrix becomes now:

$$\begin{bmatrix} 2{\cdot}7 & -0{\cdot}6 & -0{\cdot}1645 & -1 \\ -0{\cdot}6 & 1{\cdot}2 & 0{\cdot}1705 & -1 \\ 1 & 1 & 0{\cdot}3750 & 0 \\ 0 & 0 & 0{\cdot}1880 & -1 \end{bmatrix} \begin{Bmatrix} m_{01} \\ m_{10} \\ P \\ \Delta \end{Bmatrix} = \begin{Bmatrix} \mathrm{V} \\ 0 \\ 1{\cdot}8W \\ 0 \end{Bmatrix}$$

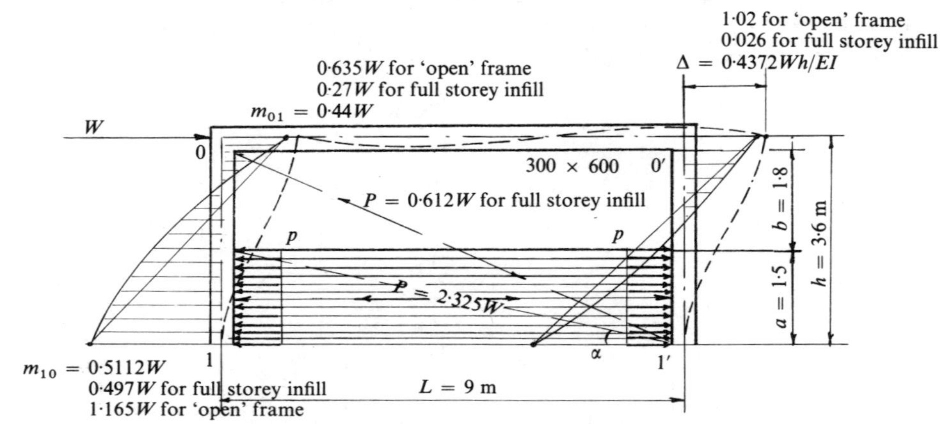

Fig. 2.34 Example of the analysis of a frame with parapet infill

Figure 2.34 shows the resulting distribution of bending moments, forces, and deflections for the frame with parapet infill, for the frame with full storey infill and for 'open' frame without infill. It can be seen that the presence of the parapet infill has considerable effect on the stiffness of the frame and reduces overall bending moments and deflections, provided that the infill does not crack or become separated from the frame by shrinkage or temperature changes. In this respect, it should be noted that the force P is now increased nearly four times when compared with the full storey infill. It can be expected that the force P is at its minimum for $a = h$, with zero value for $a = 0$ and reaches a maximum value for an intermediate depth of the infill $\bar{a}$ as shown diagrammatically in Fig. 2.30. This value is difficult to obtain in explicit form, although theoretically it could be obtained from eq. 2.32 by setting the condition $\mathrm{d}P/\mathrm{d}a = 0$.

No direct comparison can be made with the frame in Fig. 2.29, because the latter was analysed including the bending deformations of the infill only, and no direct stresses in the parapet were considered.

A similar analysis can be made for multi-storey frames with parapet infills including even the effect of spaces.

It should be noted, based on the above discussion, that the case of a frame with a parapet infill differs basically from that of a frame in which the parapet forms an integral part of the frame. While the latter can be analysed 'exactly' within the elastic range, an exact analysis of the parapet infilled frame is difficult because of the uncertainty of many parameters such as: the spaces left at interface, moduli of elasticity of composite wall, effect of shrinkage of mortar, moisture expansion of bricks, temperature changes, the form of pressure distribution at the interface, etc. However, an approximate solution can be obtained as discussed above, or similar solution, depending on the assumptions made.

Example 2.11

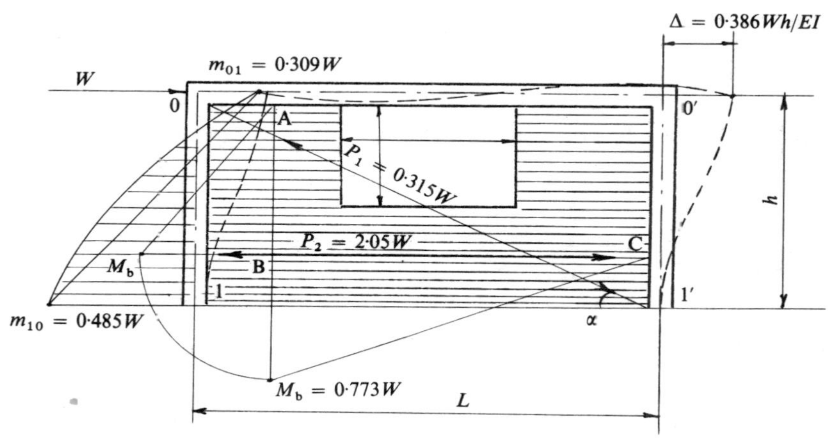

Fig. 2.35 Example of the analysis of a frame assuming double strut action

The infilled frame in Fig. 2.35 is identical with the frame in Example 2.9 (Fig. 2.29), but, in this case, both bending and direct force resistance of the infill are included in the analysis. The matrix now contains five unknown values, two of which are forces in the infill: one assumed to be positioned diagonally between points 0 and 1′ and the second horizontally at the centre of the parapet. Based on these assumptions, the matrix becomes:

0. $$-m_{10}\cdot\frac{h}{6E_cI_c} + m_{01}\left(\frac{h}{3E_cI_c} + \frac{L}{3E_cI_b}\right) = \frac{\Delta}{h}$$

1. $$m_{10}\cdot\frac{h}{3E_cI_c} - m_{01}\cdot\frac{h}{6E_cI_c} = \frac{\Delta}{h}$$

0–1. $$2\cdot\frac{m_{01}+m_{10}}{h} + P_1\cos\alpha + P_2\cdot\frac{a}{2h} = W$$

0∖1′. $$\Delta.\cos\alpha - P_1\cdot\frac{e.a}{3E_w.A_{w1}}\cdot(a+nb) = 0$$

2∖1′. $$\Delta\frac{a}{h}\cos\beta - P_2\frac{L}{E_wA_{w2}} = 0$$

and, introducing numerical values, we obtain:

$$\begin{bmatrix} 2{\cdot}7 & -0{\cdot}6 & 0 & -0{\cdot}1645 & -1 \\ -0{\cdot}6 & 1{\cdot}2 & 0 & 0{\cdot}1705 & -1 \\ 1 & 1 & 1{\cdot}6956 & 0{\cdot}3750 & 0 \\ 0 & 0 & 1{\cdot}2250 & 0 & -1 \\ 0 & 0 & 0 & 0{\cdot}1880 & -1 \end{bmatrix} \begin{Bmatrix} m_{01} \\ m_{10} \\ P_1 \\ P_2 \\ \Delta \end{Bmatrix} = \begin{Bmatrix} 0 \\ 0 \\ 1{\cdot}8W \\ 0 \\ 0 \end{Bmatrix}$$

The resulting distribution of bending moments and forces is shown in Fig. 2.35. It can be seen that bending moments in the frame did not change appreciably when compared with the previous analyses but the forces in the infill have assumed smaller values, deflection has decreased as well as the moment (and therefore shear force), in the infill.

2.10 Effect of Contact Pressure in the Analysis of Infilled Frames

An improvement in the above analyses can be made if we consider, as confirmed by experiments,[14] that assumed struts exert pressure on a frame along a given length of contact λL. This contact length can be taken based on, e.g., Dr Stafford-Smith's research.[13,14] The problem is complex but this effect can be included in the analysis if we assume that the pressure distribution is linear along the contact length. Dead weight of the panel can also be included as well as bending and shearing deformations, if required. The following suggested procedure could be adopted based on the force–displacement method.

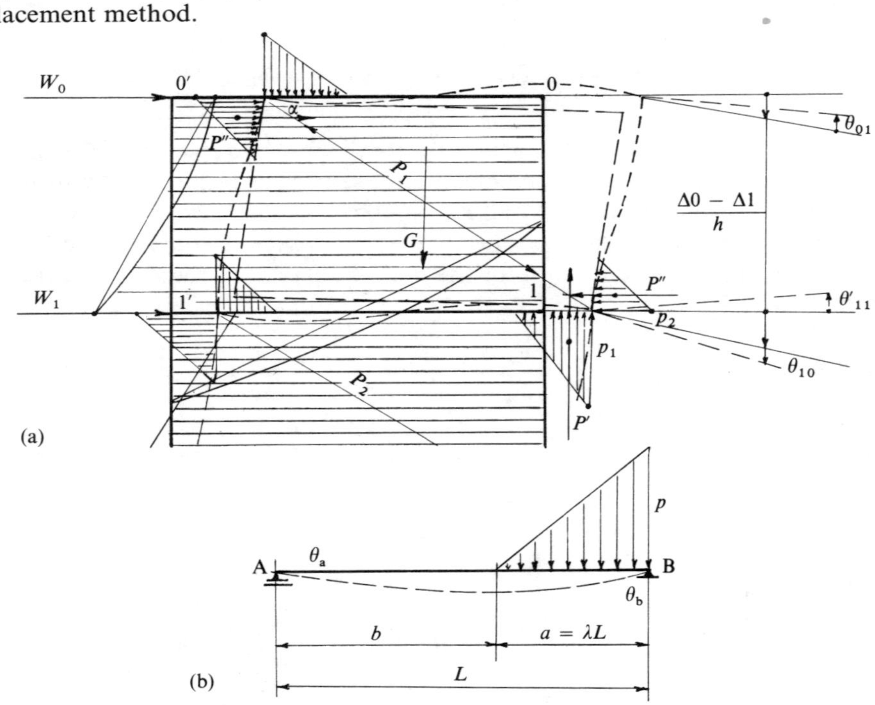

Fig. 2.36 Effect of contact pressure on the analysis of infilled frames

It can be shown that, for any given pressure, p, at the point B, in Fig. 2.36(b), extending linearly over the length $a = \lambda L$, the angular deformations are:

$$\theta_b = \frac{p.a^3}{45EI} + \frac{p.a^2.b}{6LEI}\left(\frac{b}{3} + \frac{a}{20}\right)$$

$$\theta_a = \frac{7}{8}\cdot\frac{pa^3}{45EI} + \frac{p.a^2.b}{18EI} \tag{2.39}$$

The values of θ can be introduced in the matrix, but first the values of p_1, p_2, etc., must be expressed in terms of the unknown diagonal forces, P. From Fig. 2.36(a) we can write that:

$$P' = P.\sin\alpha + G$$

and

$$P'' = P.\cos\alpha + G\cdot\frac{l}{2(h - 2\lambda h/3)} \tag{2.40}$$

in which G is the weight of the infilled panel, now assumed to take bearing on the frame along the contact length λL at the top and bottom corners of the frame as shown in Fig. 2.36(a).

From eq. (2.40), we obtain:

$$p_1 = \frac{2}{\lambda L}(P.\sin\alpha + G)$$

$$p_2 = \frac{2}{\lambda h}\left(P.\cos\alpha - G\cdot\frac{l}{2(h - 2\lambda h/3)}\right) \tag{2.41}$$

These values can now be introduced into eq. (2.39), thus expressing the values of θ in terms of P. Similar values can be obtained for the opposite corner in each panel and then introduced in the main force–displacement matrix.

It should be pointed out, however, that strictly the contact length depends on the magnitude of the applied loads, as well as on the relative stiffnesses of the frame and the infill, and is, therefore, variable and indeterminate.

2.11 Stress Concentrations in Frames and Shear Walls

In linearly elastic analysis of frames and shear walls, beams and columns are often assumed to be linear and their connections are assumed to be at one point only (Fig. 2.37(a)). However, in reality, all structural members have depth and width as well as length, and the actual intersection takes place at a common block at which stress interaction occurs. The stress distribution in this area is referred to as the 'stress concentration' problem. Stresses in this area are usually very high and can affect overall distribution of bending moments and deflections of a structure. Cracks often appear in the immediate vicinity of this area and have further effect on a structure. The mechanism by which moments in the column change rapidly above and below each junction is not yet fully resolved, although in homogeneous materials photoelastic models can give certain indications of stress distribution.

There are other causes of stress concentration in beams and columns, e.g., due to local holes and notches, cracks, fabrication and welding stresses, etc. Some of these

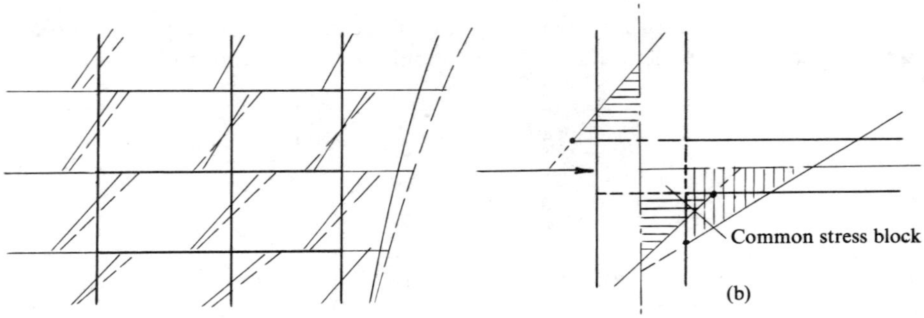

(a) Re-distribution of bending moments and deflections

Fig. 2.37 Stress concentrations in shear walls

problems have received considerable attention in the literature, but are not discussed here. Reference can be made to the bibliography. The effect of local stress concentrations due to the applied loads is seldom included in the analysis because, on the basis of St Venant's principle, it is generally assumed that the stress distribution at sections remote from an applied force is independent of the actual distribution of stresses at this point and depends only on its magnitude. Although local stress concentrations are not taken into account in the design, it is important that they should not cause local damage or reduce the load-carrying capacity of a member. At present, there is no exact analysis available of stress concentrations as occurs at joints or members in tall buildings. However, some aspects of stress concentration in the 'common block' will be briefly discussed.

It is expected that the stresses in the common block can have effect on (a) local redistribution of moments and deformations and (b) on overall response of a structure. It is usually assumed that the preliminary linear analysis will furnish approximate values of moments, shear and axial forces acting on a junction as indicated in Fig. 2.37. These forces can now be used for more detailed analysis of local elastic deformations of members at the joint, and these in turn can be used for setting out a more refined matrix which will include these effects.

Hahn and Walter used the preliminary values of bending moments, shear and axial forces to determine the stress at any point P within the 'common block' area (Fig. 2.38). These authors use only the conditions of equilibrium, and bending stresses are assumed to be linear and shear stresses parabolic. Based on these assumptions, the above authors suggest approximate formulae for the stresses σ_x, σ_y, and τ_{xy} in this common block area.

Dr Bayly,[15] using Fourier series in conjunction with the Airy's stress function:

$$\frac{\partial^4 \phi}{\partial x^4} + 2\frac{\partial^4 \phi}{\partial x^2 \partial y^2} + \frac{\partial^4 \phi}{\partial y^4} = 0$$

obtained a particular solution in the form:

$$\phi(x, y) = (K_1 . \cos \lambda y + K_2 . \sin \lambda y)(A . \cosh \lambda y + B . \sinh \lambda y + C . y . \cosh \lambda y + \\ + D . y . \sinh \lambda y)$$

where $\lambda = n\pi/L$.

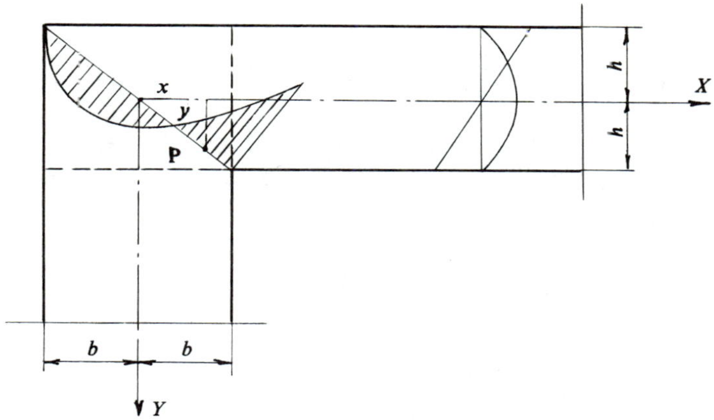

Fig. 2.38

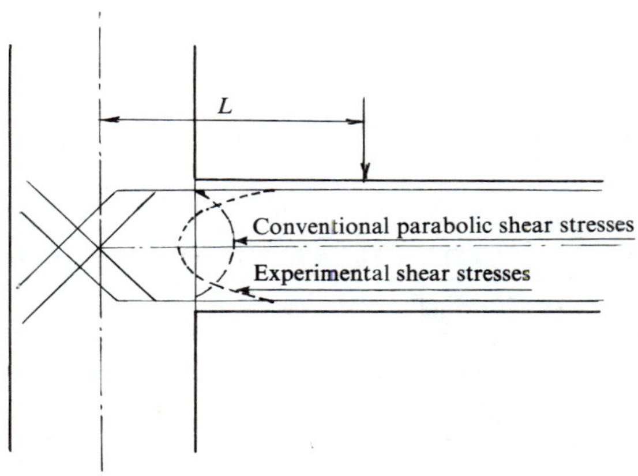

Fig. 2.39

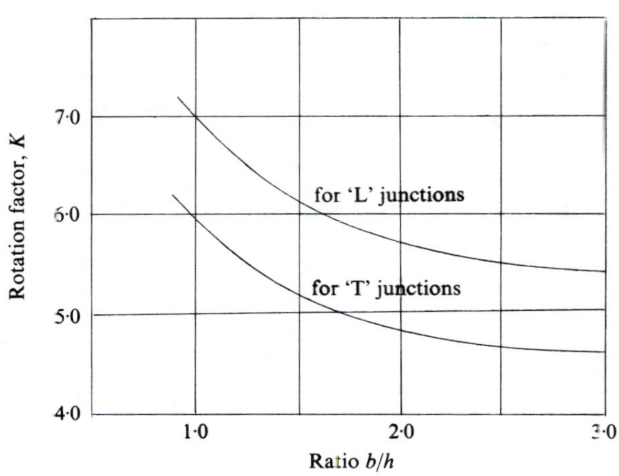

Fig. 2.40 Joint rotation factors

This research has shown that (a) the bending stresses at the junction are approximately linear when away from the actual corners, (b) that the shear stresses depart considerably from classical parabolic shape, and even shear stresses of the opposite sign have been frequently observed, (c) for maximum strength of a joint (of up to 80 per cent of ultimate), the most efficient arrangement of reinforcement is as shown in Fig. 2.39, and (d) that the joint rotation can be estimated by means of the equation:

$$\phi = K \cdot \frac{M}{4Eh^2} \tag{2.42}$$

in which the coefficient K is shown in Fig. 2.40. M is the applied bending moment and h is the half-depth of the beam. The factor K depends on the ratio of the depth of beam to column, as shown in Fig. 2.40.

These joint rotation factors can now be used in setting out of the matrix. Analysis shows that in some cases this effect can be significant.

References

1. R. Rosman, 'Stress Analysis of Horizontally Loaded Shear Walls of Tall Buildings', *Bauingenieur*, **35** (1960), (Part 1); **37** (1962), (Parts II and III).
2. A. Coull and J. R. Choudhury, 'Stresses and Deflections in Coupled Shear Walls', *JACI*, **64** (1967).
3. A. Coull and R. D. Puri, 'Analysis of Pierced Shear Walls', *Journal of Structural Division Proc. ASCE*, Jan. 1968.
4. H. Beck, 'Contribution to the Analysis of Coupled Shear Wall', *J. ACI*, **59** (1962).
5. I. A. Macleod, *Analysis of Shear Wall Structure*, Ph.D. Thesis, Glasgow University (1966).
6. I. A. Macleod, 'Lateral Stiffness of Shear Walls with Openings', *Symposium on Tall Buildings*, Southampton (April, 1966).
7. R. Rosman, 'Approximate Analysis of Shear Walls subjected to Lateral Loads', *J. ACI*, **61**, (1964).
8. R. K. Livesley, *Matrix Methods of Structural Analysis*, Pergamon Press, Oxford (1969).
9. I. A. Macleod, 'New Rectangular Finite Element for Shear Wall Analysis', *Journal of the Structural Division of ASCE*, **95** (1969).
10. M. Holmes, 'Steel frames with brickwork and concrete infilling', *Proc. Instn. civ. Engrs*, **19**, 473–475 (1966).
11. R. H. Wood, 'The stability of tall buildings', *Proc. Instn civ. Engrs*, **11**, 69–102 (1958).
12. R. H. Wood, *Studies in composite construction—1. The composite action of brick panel walls supported on concrete beams.* National Building Studies, Research Paper No. 13, H.M.S.O., London (1952).
13. B. Stafford Smith, 'Lateral stiffness of infilled frames', *Proc. ASCE*, **88**, No. S.T.6, 183–99 (1962).
14. B. Stafford Smith, *Infilled Frames*, Ph.D. Thesis, University of Bristol (1963).
15. D. J. Pearce and D. D. Matthews, *Shear Walls, an appraisal of their design in box-frame structures*, DoE Directorate of Civil Engineering Development (1972).
16. D. Bayly, *Stresses concentrations in shear Walls*. Ph.D. Thesis, The City University (1969).
17. S. Timoshenko, 'Strength of Materials', part 1, p. 270, 2nd Ed. (1946).
18. M. Smolira, 'Analysis of infilled shear walls', *Proc. Instn civ. Engrs*, **55**, Dec., 896–912 (1973).

3 Principles of Three-dimensional Analysis of Tall Buildings

Most engineering structures are three-dimensional and possess a certain degree of continuity of members meeting at joints in all three directions. In such structures, columns are subjected to bending moments in two directions as well as torsion. Similarly, beams develop bending moments and torsion. These torsional moments alone can be very high, particularly in columns. However, in nearly all cases, structures are analysed as two-dimensional units separately, and this appears to be adequate for the majority of practical work. In some cases, it may be advisable or even necessary to consider three-dimensional action of a building as a whole.

Three-dimensional action develops mainly because of the high degree of continuity at joints, and is also the result of lack of symmetry in a structural system, eccentricity of lateral or gravity forces, differential settlement of foundations, temperature changes, etc.

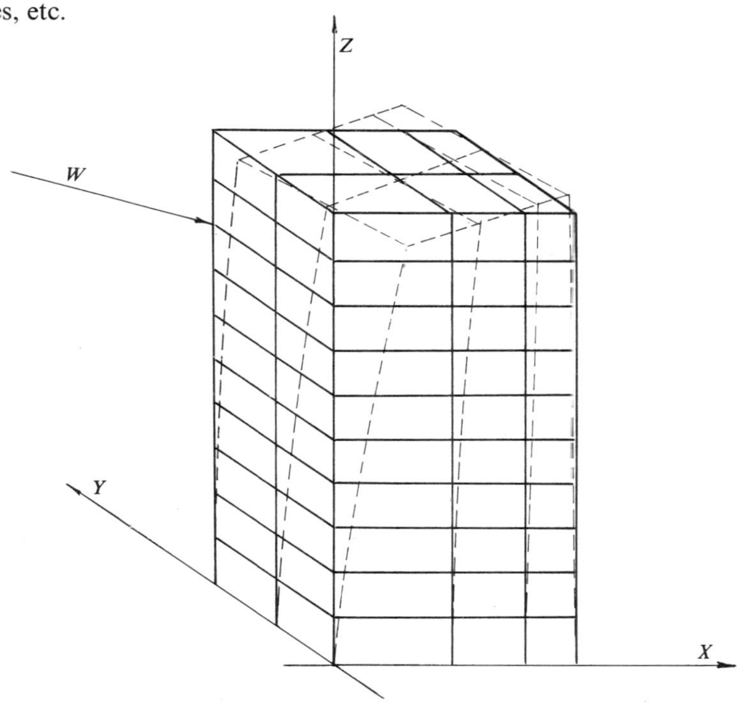

Fig. 3.1 Analysis of three-dimensional frames

It appears that, till now, little work has been done in this field, presumably because of the large computer storage required. However, both the stiffness and the force–displacement methods can be used in setting out a matrix for a structure as a whole. In the following, only the force–displacement concept will be used.

In this method, a plane joint with four members has three indeterminate bending moments and one translation (common to all joints at a particular floor level). A three-dimensional joint, such as shown in Fig. 3.2, has four unknown moments in each plane XZ and YZ, one torsional bending moment, two translations, and one rotation. However, because of the rigidity of the floor system, deformations of all points in the floor system are related by geometrical requirements.

A typical three-dimensional joint, such as in Fig. 3.2, has, therefore, ten unknown moments, two translational deformations, and one rotation. Using the force–displacement concept, a matrix can be set out expressing conditions of compatibility of deformations in three directions and equilibrium of forces in the directions X and Y. First, however, the position of the centre of rotation of each floor must be determined.

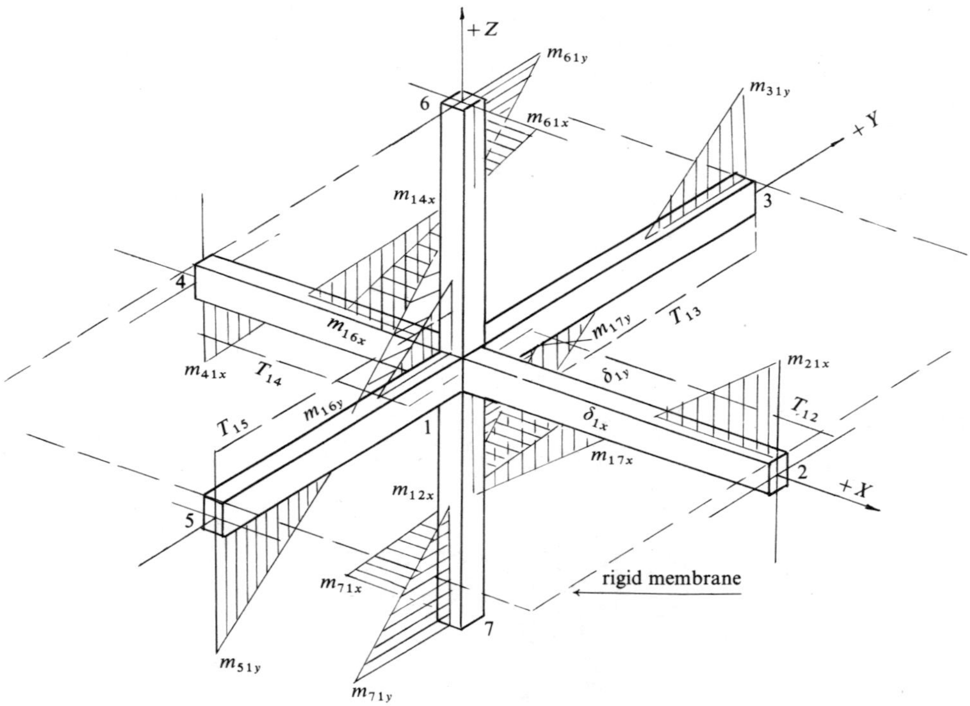

Fig. 3.2 A typical three-dimensional joint (with floor slab removed)

The centre of rotation of a floor system (point O in Fig. 3.3) is a generalized centre of shear and is defined as a point in the plane of the floor with respect to which all points of the floor rotate, with the exception of this point only. Such point is, therefore, subjected only to translations, Δ_{0x} and Δ_{0y}, while all other points rotate with respect to this point by a constant angle ϕ_0. If the floor systems were flexible in their own planes, then each joint would rotate by different amounts. However, actual floors in tall buildings are nearly always rigid in their own planes. Such floors, under the effect of eccentric forces, will first be subjected to translations Δ_{0x} and Δ_{0y}, and then to rotation ϕ_0 (Fig. 3.3).

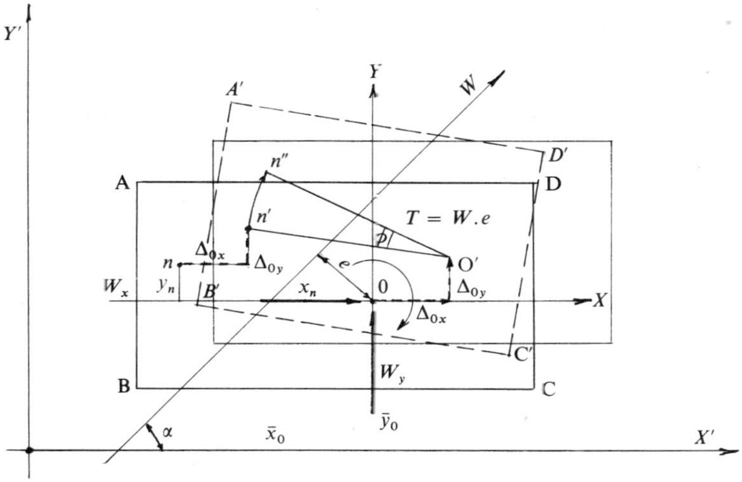

Fig. 3.3 Centre of rotation of a rigid floor system

The final position of any point, such as point n in Fig. 3.3, in a rigid floor system, can be defined by its coordinates in the undeflected, original position of this point, x_n and y_n, and the values of Δ_{0x}, Δ_{0y}, and ϕ_0. These deformations, as well as bending moments in the directions X and Y, represent statically indeterminate values. The position of the centre of rotation of each floor system is therefore fundamental to the theory of three-dimensional analysis.

The position of the centre of rotation is a function of the geometry of the system, stiffnesses of members, quality of materials, end conditions, stress–strain relationship of members, and depends even on the formation of cracks.

In tall buildings, end conditions are generally such that both the translations and rotations take place, which are unknown until the frame is fully analysed and designed. These are indeterminate but, as a first approximation, it could be assumed that, e.g., far ends of supporting systems are fixed, hinged, or as derived from two-dimensional analysis.

In Fig. 3.3, the points ABCD represent a rigid floor system in Cartesian coordinates $X'Y'$. Points 1, 2, 3, . . . , n represent centres of columns and walls. W is the storey wind force and point O is the centre of rotation. The force W is now resolved at this point into:

$$W_x = W.\cos\alpha$$
$$W_y = W.\sin\alpha$$
$$M = W.e$$

From the conditions of static equilibrium of moments of X-axis, we can write:

$$W_x.\bar{y}_0 = \sum_{1}^{n} V_{nx}.y_n$$

and introducing the value for V_n from:

$$V_n = 12EI_{nx}.\delta_{nx}/h^3$$

we obtain:

$$\bar{y}_0 = \sum_1^n \frac{I_{nx}}{h^3} \cdot \frac{y_n}{\sum (I_{nx}/h^3)}$$

Similarly, $$\bar{x}_0 = \sum_1^n \frac{I_{ny}}{h^3} \cdot \frac{x_n}{\sum (I_{ny}/h^3)}$$

For equilibrium of torsional moments, we can write:

$$W.e = \sum T_n + \sum V_x . e_x + \sum V_y . e_y \tag{3.1}$$

in which, by St Venant:

$$T_n = K.G.\phi \tag{3.2}$$

For rectangular concrete sections, $\mu = 0{\cdot}15$ to $0{\cdot}20$, and $G = E/2(1+\mu)$.

Taking $\mu = 0{\cdot}2, \quad G = E/2{\cdot}4$

and introducing this value into eq. (3.2), we obtain an approximate value of rotation:

$$\phi_n = \frac{W.e}{4{\cdot}8E \sum I_n . \beta_n . (b/d)^2} \tag{3.3}$$

The coefficients β are functions of the ratio d/b and are given by Timoshenko (see ref. 17, page 102). These are as follows:

d/b	1·0	1·5	2·0	2·5	3·0	4·0	6·0	10·0	∞
β	0·141	0·196	0·229	0·249	0·263	0·281	0·299	0·333	1/3

The value of ϕ is, strictly, statically indeterminate and must enter as such in the matrix based on the condition expressed by eq. (3.1).

It is now necessary to express the position of each point in the floor system by its coordinates *xn* and *yn*, and by statically indeterminate deflections of the point of rotation, i.e., Δ_{0x}, Δ_{0y}, and ϕ_0.

From Fig. 3.4, we can write:

$$\begin{aligned} \delta_x &= x_a - x'_a = R.\cos\alpha - R.\cos(\alpha + \phi) \\ &= x_a(1 - \cos\phi) + y_a . \sin\phi \end{aligned} \tag{3.4}$$

Similarly:

$$\begin{aligned} \delta_y &= y'_a - y_a = R.\sin(\alpha + \phi) - R.\sin\alpha \\ &= x_a . \sin\phi - y_a .(1 - \cos\alpha) \end{aligned} \tag{3.5}$$

If the rotation of the floor system ϕ_0 is small, then eqs. (3.4) and (3.5) can be simplified by assuming that:

$$\sin\phi \rightarrow \phi \quad \text{and} \quad \cos\phi \rightarrow 1$$

so that:

$$\begin{aligned} \delta_{nx} &= y_a . \phi_0 + \Delta x_0 \\ \delta_{ny} &= x_a . \phi_0 + \Delta y_0 \end{aligned} \tag{3.6}$$

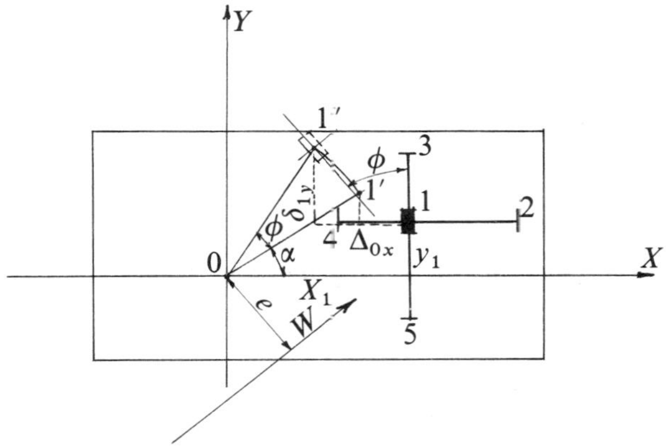

Fig. 3.4

Figure 3.4 shows a plan of a typical three-dimensional junction, marked as number 1, with far ends marked 2 to 5. (Points 6 and 7 are on the Z axis.) The point 1′ denotes the position of the junction after translations Δ_{0x} and Δ_{0y} have taken place, these being common to all points in the particular floor system. Point 1″ is the final position of the junction 1 after rotation ϕ_0 has taken place. The values of δ_{1x} and δ_{1y} are the translations of the point 1 from its original position defined by its coordinates x_1 and y_1. These deflections do not directly represent unknown values but are related to the deflections of the centre of rotation 0, as shown above.

Figures 3.5(a) and (b) show the distribution of bending moments, torsions, and deflections drawn separately in the planes ZX and ZY. It can be seen from these diagrams that in each vertical plane passing through junction 1 there are four unknown bending moments forming a system in equilibrium with torsional moments at right angles to each plane. Therefore, five independent compatibility conditions must be found in each plane. These can be taken as follows: three flexural conditions 214, 617, and 217, one flexure–torsion condition $F17$–$T13$, and one condition of $M_x=0$. Other arrangements of the first four conditions can, of course, be made.

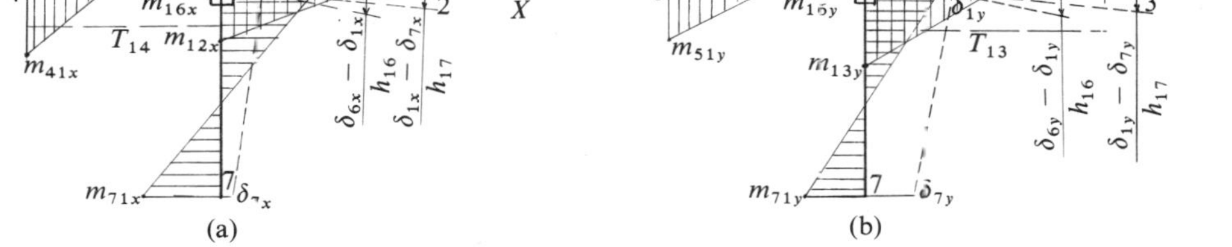

Fig. 3.5

Table 3.1 Element force–displacement matrix for the analysis of three-dimensional frames

Joint	0		1										
$m.T_{or}$ Δ	Δ_{0x}	Δ_{0y}	m_{12x}	m_{13y}	m_{14x}	m_{15y}	m_{16x}	m_{16y}	m_{17x}	m_{17y}	T_{12}	T_{13}	T_{14}
$F.214$	0	0	$-\dfrac{l_{12}}{3EI_{12x}}$	0	$\dfrac{l_{14}}{3EI_{14z}}$	0	0	0	0	0	0	0	0
$F.617_x$	$-\left(\dfrac{1}{h_{16}}+\dfrac{1}{h_{17}}\right)$	0	0	0	0	0	$\dfrac{h_{16}}{3EI_{16x}}$	0	$-\dfrac{h_{17}}{3EI_{17x}}$	0	0	0	0
$F.217$	$-\dfrac{1}{h_{17}}$	0	$\dfrac{l_{12}}{3EI_{12z}}$	0	0	0	0	0	$\dfrac{h_{17}}{3EI_{17x}}$	0	0	0	0
$F.17x-$ $-T.13y$	$-\dfrac{1}{h_{17}}$	0	0	0	0	0	0	0	$\dfrac{h_{17}}{3EI_{17x}}$	0	0	$\phi_{13}.l_{13}$	0
$\sum M_x$	0	0	+1	0	+1	0	−1	0	−1	0	0	−1	0
$F.315$	0	0	0	$\dfrac{l_{13}}{3EI_{13z}}$	0	$-\dfrac{l_{15}}{3EI_{15z}}$	0	0	0	0	0	0	0
$F.617_y$	0	$\dfrac{1}{h_{17}}+\dfrac{1}{h_{16}}$	0	0	0	0	0	$\dfrac{h_{16}}{3EI_{16y}}$	0	$-\dfrac{h_{17}}{3EI_{17y}}$	0	0	0
$F.317$	0	$-\dfrac{1}{h_{17}}$	0	$\dfrac{l_{13}}{3EI_{13z}}$	0	0	0	0	0	$\dfrac{h_{17}}{3EI_{17y}}$	0	0	0
$F.17y-$ $-T.12x$	0	$-\dfrac{1}{h_{17}}$	0	0	0	0	0	0	0	$\dfrac{h_{17}}{3EI_{17y}}$	$\phi_{12}.l_{12}$	0	0
$\sum M_y$	0	0	0	1	0	1	0	−1	0	−1	−1	0	−1

2	3	4	5	6				7				
$_{21x}$	m_{31y}	m_{41x}	m_{51y}	m_{61x}	m_{61y}	$\Delta 6x$	$\Delta 6y$	m_{71x}	m_{71y}	$\Delta 7x$	$\Delta 7y$	
$\frac{{}_{12}}{I_{12z}}$	0	$-\frac{l_{14}}{6EI_{14z}}$	0	0	0	0	0	0	0	0	0	0
0	0	0	0	$-\frac{h_{16}}{6EI_{16x}}$	0	$\frac{y_6}{h_{16}}$	0	$\frac{h_{17}}{6EI_{17x}}$	0	$\frac{y_7}{h_{17}}$	0	$-\frac{y_6}{h_{16}}\phi_6 + y_1\left(\frac{1}{h_{16}} + \frac{1}{h_{17}}\right)\phi_0 - \frac{y_7}{h_{17}}\cdot\phi_7$
$\frac{l_{12}}{EI_{12z}}$	0	0	0	0	0	0	0	$-\frac{h_{17}}{6EI_{17x}}$	0	$\frac{1}{h_{17}}$	0	$-\frac{y_1}{h_{17}}\phi_0 + \frac{y_7}{h_{17}}\cdot\phi_7$
0	0	0	0	0	0	0	0	$-\frac{h_{17}}{6EI_{17x}}$	0	$\frac{1}{h_{17}}$	0	$-\frac{y_1}{h_{17}}\phi_0 + \frac{y_7}{h_{17}}\phi_7$
0	0	0	0	0	0	0	0	0	0	0	0	0
0	$-\frac{l_{13}}{6EI_{13z}}$	0	$\frac{l_{15}}{6EI_{15z}}$	0	0	0	0	0	0	0	0	0
0	0	0	0	0	$-\frac{h_{16}}{6EI_{16y}}$	0	$-\frac{1}{h_{16}}$	0	$\frac{h_{17}}{6EI_{17y}}$	0	$-\frac{1}{h_{17}}$	$-x_1\left(\frac{1}{h_{16}} + \frac{1}{h_{17}}\right)\phi_0 + \frac{x_6}{h_{16}}\phi_6 + \frac{x_7}{h_{17}}\phi_7$
0	$-\frac{l_{13}}{6EI_{13z}}$	0	0	0	0	0	0	0	$-\frac{h_{17}}{6EI_{17y}}$	0	$\frac{1}{h_{17}}$	$-\frac{x_1}{h_{17}}\phi_0 + \frac{x_7}{h_{17}}\phi_7$
0	0	0	0	0	0	0	0	0	$-\frac{h_{17}}{6EI_{17y}}$	0	$\frac{1}{h_{17}}$	$-\frac{x_1}{h_{17}}\phi_0 + \frac{x_7}{h_{17}}\cdot\phi_7$
0	0	0	0	0	0	0	0	0	0	0	0	0

The last two conditions, corresponding to the unknown deformations Δ_{0x} and Δ_{0y}, are set out using the condition of equilibrium of forces in the directions X and Y. These equations can be written based on Fig. 3.5 as follows:

F.214	$-m_{41x}\cdot\frac{l_{14}}{6EI_{14z}}+m_{14x}\cdot\frac{l_{14}}{3EI_{14z}}-m_{12x}\cdot\frac{l_{12}}{3EI_{12z}}+m_{21x}\cdot\frac{l_{12}}{6EI_{12z}}$	0
F.617x	$-m_{61x}\frac{h_{16}}{6EI_{16x}}+m_{16x}\frac{h_{16}}{3EI_{16x}}-m_{17x}\cdot\frac{h_{17}}{3EI_{17x}}+m_{71x}\cdot\frac{h_{17}}{6EI_{17x}}$	$\frac{\delta_{6x}-\delta_{1x}}{h_{16}}-\frac{\delta_{1x}-\delta_{7x}}{h_{17}}$
F.217	$-m_{21x}\frac{l_{12}}{6EI_{12z}}+m_{12x}\frac{l_{12}}{3EI_{12z}}+m_{17x}\cdot\frac{h_{17}}{3EI_{17x}}-m_{71x}\frac{h_{17}}{6EI_{17x}}$	$\frac{\delta_{1x}-\delta_{7x}}{h_{17}}$
F.17x–T.13y	$m_{17x}\frac{h_{17}}{3EI_{17x}}-m_{71x}\frac{h_{17}}{6EI_{17x}}+T_{13}.\phi_{13}.l_{13}$	$\frac{\delta_{1x}-\delta_{7x}}{h_{17}}$
$\sum Mx$	$m_{12x}+m_{14x}-m_{16x}-m_{17x}-T_{13}-T_{15}$	0
F.315	$-m_{31y}\frac{l_{13}}{6EI_{13z}}+m_{13y}\frac{l_{13}}{3EI_{13z}}-m_{15y}\frac{l_{15}}{3EI_{13z}}+m_{51y}\frac{l_{15}}{6EI_{15z}}$	0
F.617y	$-m_{61y}\frac{h_{16}}{6EI_{16y}}+m_{16y}\frac{h_{16}}{3EI_{16y}}-m_{17y}\frac{h_{17}}{3EI_{17y}}+m_{71y}\frac{h_{17}}{6EI_{17y}}$	$\frac{\delta_{6y}-\delta_{1y}}{h_{16}}-\frac{\delta_{1y}-\delta_{7y}}{h_{17}}$
F.317	$-m_{31y}\frac{l_{13}}{6EI_{13z}}+m_{13y}\frac{l_{13}}{3EI_{13z}}+m_{17y}\frac{h_{17}}{3EI_{17y}}-m_{71y}\frac{h_{17}}{6EI_{17y}}$	$\frac{\delta_{17y}-\delta_{71y}}{h_{17}}$
F.17y–T.12x	$m_{17y}\frac{h_{17}}{3EI_{17y}}-m_{71y}\frac{h_{17}}{6EI_{17y}}+T_{12}.\phi_{12}.l_2$	$\frac{\delta_{17y}-\delta_{71y}}{h_{17}}$
$\sum My$	$m_{13y}+m_{15y}-m_{16y}-m_{17y}-T_{12}-T_{14}$	0

The conditions of equilibrium of forces are:

$$\sum Vx = Wx$$
$$\sum Vy = Wy$$

The values of the shear forces Vx and Vy can now be expressed in terms of the unknown bending moments. These equations, however, must be written for the whole floor, i.e., for all columns or shear walls below and above the particular floor system.

Substituting values for deflections (eq. (3.3)) in terms of respective translations and rotations of each floor, the force–displacement matrix for a three-dimensional junction can be written in a general form as shown in Table 3.1.

The main matrix can now be set out using this element matrix. However, this needs to be programmed for repetitive generation of such element matrices for all junctions. For this purpose, it is an advantage to consider all the junctions as typical, and subsequently to zero-ize all forces and deformations in the non-existent members, in a similar way to that shown in Table 2.1.

4 Analysis of Perforated Cores

4.1 Introduction

The structural function of lift shafts in tall buildings is to resist horizontal forces from wind and earthquakes, and to provide lateral stability. In addition, it is an advantage for cores to be designed to carry large gravity forces.

Cores in tall buildings are always perforated for door openings leading to lifts, stairs, etc., and are stiffened by floor diaphragms which prevent distortion of cross-sections taking place. Such cores, when subjected to gravity and lateral forces, develop three-dimensional stress distribution in bending, torsion, shear, and direct stresses. In addition, warping and shear lag stresses take place.

An exact analysis of perforated cores is complex but, fortunately, is seldom necessary for practical purposes. Finite elements method can be used but requires a considerable amount of work and large storage of computers. However, approximate analysis can be made based on various simplifying assumptions. In the following analysis, it will be assumed that:

(a) linear relationship between stresses and strains applies,
(b) method of superposition is valid,
(c) materials are homogeneous, isotropic, and monolithic,

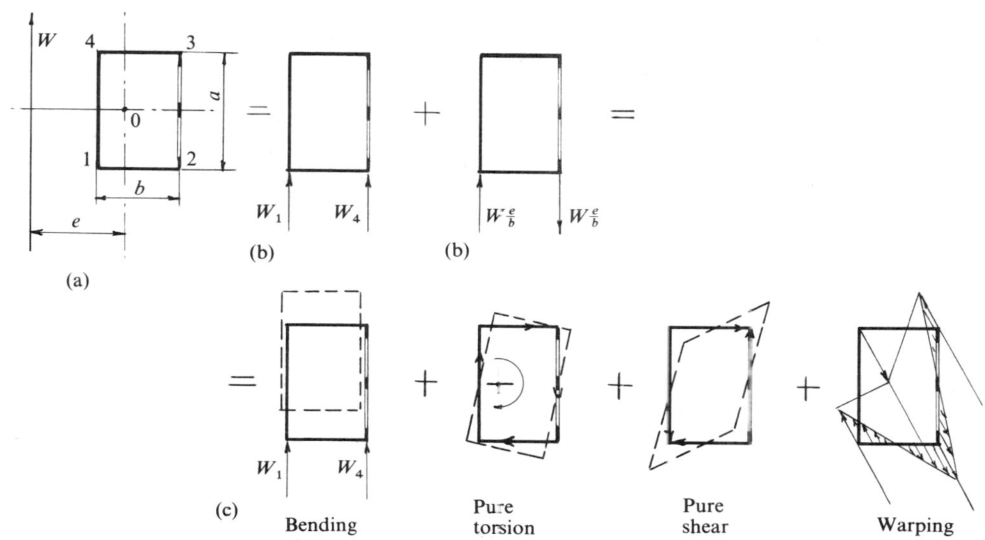

Fig. 4.1 Cores subjected to eccentric lateral forces

(d) deformations are small, so that the equilibrium is satisfied in undeformed configuration, and that instability effects are not appreciable,
(e) thicknesses of walls are small in relation to their width, and widths are small in relation to the length, and
(f) the effect of perforations can be considered separately.

Figure 4.1 shows diagrammatically a perforated core subjected to the effect of lateral forces W, acting at an eccentricity e. These could vary at each floor level. Such a system can be resolved into a system of forces acting symmetrically and anti-symmetrically, as shown in Fig. 4.1(c), in which the first part is referred to as the 'engineering bending'. It can be shown that the unsymmetrical system of forces can be further resolved into pure torsion and pure shear as shown. In addition, warping stresses take place in the longitudinal direction, due to the effect of unsymmetrical loads. Warping stresses are usually sufficiently high to be considered.

The point 0 in Fig. 4.1(a) is the centre of rotation of the core or its generalized shear centre, i.e., when the force W is transferred to this point, a structure is subjected to translations only. The force W is now resolved in the directions of walls parallel to the direction of this force (W_1 and W_4 in Fig. 4.1(c)). These forces produce bending in walls, whether solid or perforated. The exact analysis must include the shear transfer at the corners and the shear lag stresses in walls perpendicular to the direction of lateral forces.

Torsional stresses can be very high and, because of perforations, high bending stresses can develop in beams and columns. These will be discussed separately.

The effect of pure shear in cores is negligible, because of the stiffening action of floor diaphragms, which effectively prevent distortions of cross-sections taking place, and need not be considered.

4.2 Pure Torsion in Thin Tubes

The analysis of torsional stresses and deformations in thin tubes is usually based on St Venant's theory. The tube in Fig. 4.2 is assumed to be fixed at one end, but free to distort at the other end, when subjected to a torsional moment T. This torque is resisted by the shearing stresses $\tau = q/t$ distributed along the circumference of the tube which, for equilibrium, must also exist along the length of the tube. The shear forces q along AB must be equal but opposite to the shear along section CD and are equal to:

$$q = \tau . t$$

in which q is the shear force per unit length, but for total thickness of the tube, and is constant throughout the tube. The distribution of these shear forces, q, is called the shear flow.

A small element ds, in Fig. 4.2, is subjected to a force $q.\mathrm{d}s$. Taking moment of this force with respect to any point 0, we can write for equilibrium that:

$$T = \int_s (q.\mathrm{d}s).r = q \int_s r\,\mathrm{d}s \tag{4.1}$$

The integral $\int_s r\,\mathrm{d}s = 2A$, i.e., is equal to twice the area enclosed by the centre line of

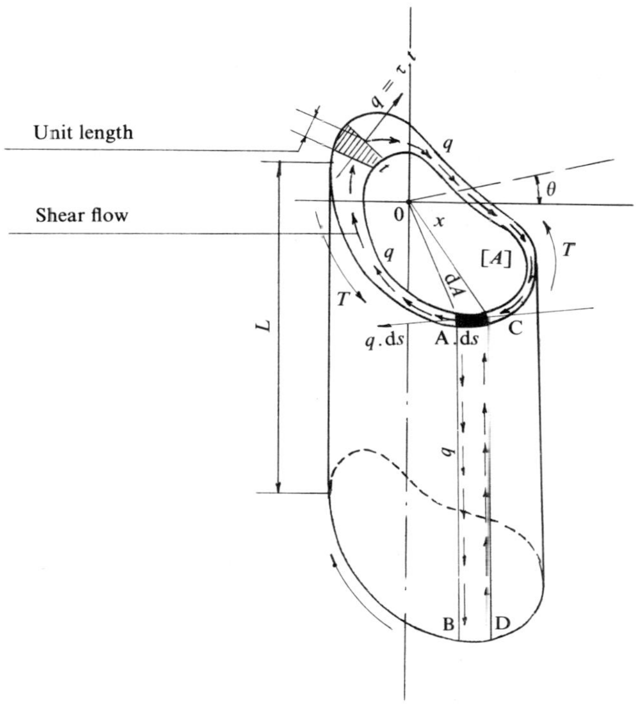

Fig. 4.2 Torsion of thin tubes

the tube; e.g., for circular tube:

$$\int_s r \, ds = \int_0^{2\pi} r(r \, a\phi) = 2r^2\pi$$

From eq. (4.1):

$$T = 2A.q$$

and the shearing stress at any point is therefore:

$$\tau = \frac{q}{t} = \frac{T}{2At} \tag{4.2}$$

To find the angle of twist of the tube, we consider the strain energy stored in the tube and equate it to the work done by the torque T. When a material is subjected to shearing stress τ, the strain energy stored per unit volume of the material is:

$$\frac{\tau^2}{2G} = \tfrac{1}{2}G\gamma^2$$

where G is the shear modulus of the material.

The strain energy stored in a longitudinal element of length L, width ds, and thickness t, is:

$$\frac{\tau^2}{2G} \cdot L.t.ds$$

and the total strain energy stored in the tube is therefore:

$$U = \int_s \frac{\tau^2}{2G} \cdot L.t.\mathrm{d}s$$

Since $q = \tau.t$ is constant, we can write:

$$U = \int_s \frac{\tau^2}{2G} \cdot L.t^2 \cdot \frac{\mathrm{d}s}{t} = \frac{q^2 L}{2G} \int_s \frac{\mathrm{d}s}{t} \tag{4.3}$$

The work done by the torque T is:

$$W = \tfrac{1}{2}T.\theta \tag{4.4}$$

Equalizing $U = W$, we obtain:

$$\theta = \frac{q^2 L}{GT} \int_s \frac{\mathrm{d}s}{t} \tag{4.5}$$

Substituting $q = T/2A$, eq. (4.5) may be written in the well-known form:

$$\theta = \frac{T.L}{4A^2.G} \int_s \frac{\mathrm{d}s}{t} \tag{4.6}$$

For a tube of uniform thickness, this equation becomes:

$$\theta = \frac{T.L}{4A^2.G} \cdot \frac{s}{t}$$

Equation (4.6) can be written in the form:

$$\theta = \frac{T.L}{G.J}$$

where

$$J = \frac{4A^2}{\int_s \mathrm{d}s/t}$$

J is the torsion constant for the section. For circular sections, J is equal to the polar second moment of area, but this is not true for other sections.

However, in addition to the above effect, torsion of non-circular members gives rise to warping stresses and deformations, which are discussed separately.

Example 4.1

The lift shaft shown in Fig. 4.3(a) and (b) is assumed to be subjected to the lateral forces $W = 50$ kN per floor, acting at an eccentricity $e = 3$ m. Assuming that $E = 28$ kN/mm^2, $\mu = 0{\cdot}18$, and $G = E/2(1+\mu) = 12$ kN/mm^2, we have:

$$T = W.e = 150 \text{ kN/m per floor}$$
$$A = 8{\cdot}0 \text{ m}^2$$

and the shear flow $q = T/2A = 9{\cdot}4$ kN per metre run.

Shear stresses in 200 mm walls are:

$$\tau = 0{\cdot}047 \text{ N/mm}^2 \text{ per floor}$$

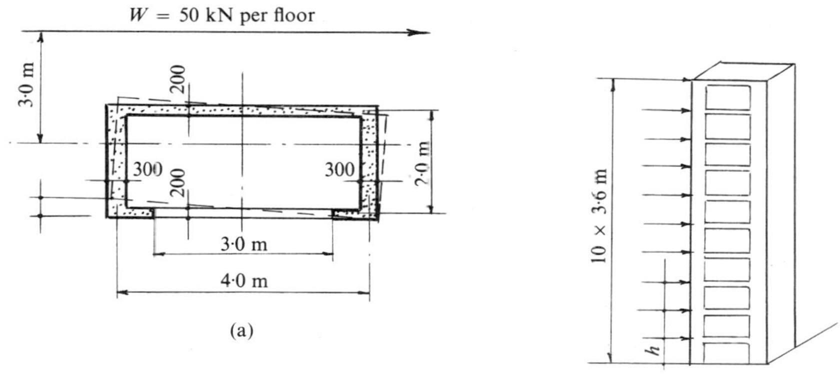

Fig. 4.3 Example

and in 300 mm walls:

$$\tau = 0{\cdot}0314 \text{ N/mm}^2 \text{ per floor}$$

$$\int_s \frac{ds}{t} = 53{\cdot}34$$

and maximum twist is:

$$\theta = \frac{T.h}{4A^2G}\int \frac{ds}{t} = 0{\cdot}000\ 021 \text{ rad per floor}$$

The maximum torsional deflection at the top of a ten-storey lift shaft is:

$$\delta = \sum_1 \theta \cdot \frac{l}{2} = 8{\cdot}3 \text{ mm}$$

The presence of openings affects the distribution of torsional stresses and deformations but the exact analysis is complicated. An approximate estimate can be made if

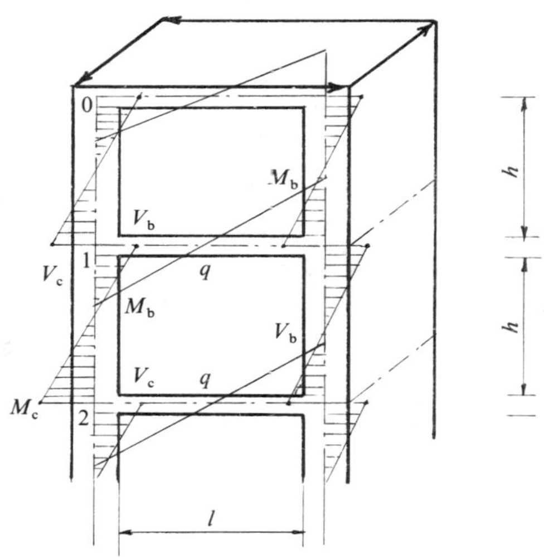

Fig. 4.4 Moments around perforations due to torsion only

we consider that columns and beams in the vicinity of the opening transfer the torsional shear flow. We could then write approximately, that the total shear forces to which columns and beams are subjected are:

$$V_c = q.l \quad \text{and} \quad V_b = q.h$$

Bending moments could then be calculated from:

$$M_c = M_b = V_c \cdot \frac{h}{2} = V_b \cdot \frac{l}{2} = \frac{1}{2} \cdot qlh$$

Such analysis can be done at each floor level for gradually increased torsional moments, and thus for the shear flow q.

In the previous example, we have:

$$V_c = q.l = 9{\cdot}4 \times 3{\cdot}0 = 29{\cdot}2 \text{ kN}$$
$$V_b = 9{\cdot}4 \times 2{\cdot}6 = 24{\cdot}4 \text{ kN}$$

and the bending moments are:

$$M_b = M_c = 36{\cdot}5 \text{ kN/m per floor.}$$

These forces and moments will, of course, be progressively increased in each floor from the top down.

An alternative approximate method of determining bending moments in a perforated core, resulting from torsion only, can be made as follows: First, torsional deformations of points 1 and 2 in Fig. 4.5(a) are determined at each floor level as discussed above, and these are resolved in X and Y directions. For large values of twist ϕ, the exact relationships are:

$$\delta_{1x} = y_1 . \sin\phi + x_1 . (1 - \cos\phi)$$
$$\delta_{1y} = x_1 . \sin\phi - y_1 . (1 - \cos\phi)$$
$$\delta_{2x} = y_2 . \sin\phi - x_2 . (1 - \cos\phi)$$
$$\delta_{2y} = x_2 . \sin\phi + y_2 . (1 - \cos\phi)$$

For small values of twist, it can be taken that:

$$\delta_{nx} \approx y_n . \phi$$
$$\delta_{ny} \approx x_n . \phi$$

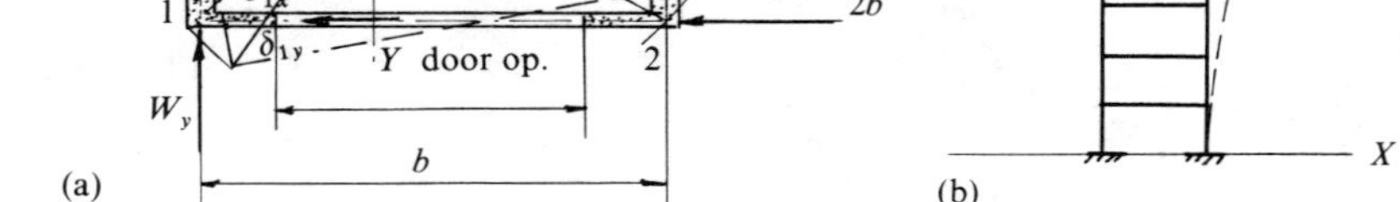

Fig. 4.5 Flexural stresses in perforated core resulting from torsion

Using the values of δ_x, resulting from torsional effect (see Fig. 4.5(a)), a matrix can be set out for evaluation of bending moments in the frame in the plane XZ containing openings. Alternatively, the values of the torsional forces W_x and W_y can be calculated from the condition of equilibrium:

$$W_x.a + W_y.b = W.e \quad \text{and} \quad W_x.a = W_y.b$$

from which

$$W_x = \frac{W.e}{2b} \quad \text{and} \quad W_y = \frac{W.e}{2a} \tag{4.7}$$

These forces form an equivalent torsional system of loads. Alternatively, the internal forces producing torsional shear flow are:

$$\frac{W.e}{2A}\cdot a \quad \text{and} \quad \frac{W.e}{2A}\cdot b \tag{4.8}$$

These are shown in Fig. 4.5(a). These forces can now be used for setting out the matrix for evaluation of bending moments in walls.

It should be noted that pure shear stresses cannot occur in cores stiffened by floor diaphragms which prevent the deformations of the sections taking place.

4.3 Bending of Perforated Cores

When the resultant lateral force W (Fig. 4.6) is passing through the shear centre of the lift-shaft, then by definition, the core is subjected to bending only and is free from torsional stresses and deformations. The deflections are then uni-directional, and because of the presence of floor diaphragms, both the solid and perforated walls must deflect equally at all floor levels. This feature is used here as the basis for the approximate analysis of bending of perforated cores.

The following assumptions are made:

(a) that the response of members to loads is linearly elastic,
(b) that the principle of superposition is valid, so that the effect of torsion, warping, and the shear lag can be treated separately, and
(c) that the action of loads is static.

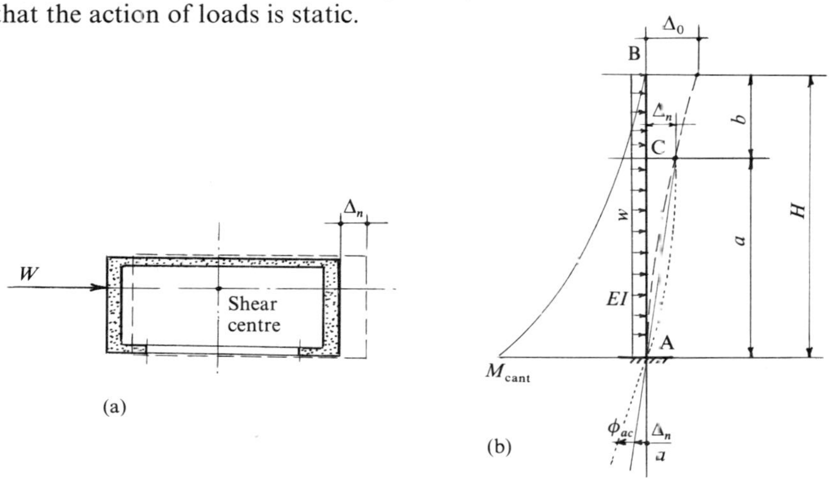

Fig. 4.6 Bending of perforated cores

Referring to Fig. 4.6(b) a cantilever AB of height H is subjected to a uniformly distributed load w and deflects by Δ_n at any distance a from the fixed end. Using the force–displacement method, this deflection can be calculated by assuming releases at A and C. The condition of compatibility of deformations at point A can then be written as:

$$\frac{\Delta_n}{a} + \theta_{ac} = \left(\frac{1}{2} wH^2\right) \cdot \frac{a}{3EI} + \left(\frac{1}{2} wb^2\right) \frac{a}{6EI}$$

from which:

$$\Delta_n = \frac{wa^4}{8EI}\left[2\left(\frac{H}{a}\right)^2 - \frac{4}{3}\left(\frac{H}{a}\right) + \frac{1}{3}\right] \tag{4.9}$$

and the maximum deflection at point B is of the well-known form:

$$\Delta_0 = \frac{wH^4}{8EI} \tag{4.10}$$

It will be convenient in the subsequent analysis to express deflection Δ_n in terms of Δ_0. From eqs. (4.9) and (4.10) we obtain:

$$\Delta_n = \Delta_0 \left(\frac{a}{H}\right)^2 \left[\frac{1}{3}\left(\frac{a}{H}\right)^2 - \frac{4}{3}\left(\frac{a}{H}\right) + 2\right] = A.\Delta_0 \tag{4.11}$$

in which

$$A = \frac{1}{3}\left(\frac{a}{3}\right)^2 - \frac{4}{3}\left(\frac{a}{H}\right) + 2 \tag{4.12}$$

The values of the non-dimensional coefficients A can be calculated for any number of points on the cantilever, and will usually be taken to coincide with the positions of floors. Figure 4.7 shows the diagram of these coefficients corresponding to ten divisions of the cantilever. For other number of floors eq. (4.12) can be used.

In the analysis by the force–displacement method, the values $\Delta_n - \Delta_{n+1}$ and $(\Delta_n - \Delta_{n+1}) - (\Delta_{n+1} - \Delta_{n+2})$ are required. These can be expressed in terms of Δ_0 as follows:

$$\Delta_n - \Delta_{n+1} = \Delta_0.(A_n - A_{n+1}) = B.\Delta_0 \tag{4.13}$$

and

$$(\Delta_n - \Delta_{n+1}) - (\Delta_{n+1} - \Delta_{n+2}) = \Delta_0(B_n - B_{n+1}) = C.\Delta_0 \tag{4.14}$$

The coefficients B and C are also non-dimensional and can be calculated for a given number of divisions on the cantilever. For ten divisions, these values are shown in Fig. 4.7.

The distribution of bending moments in the perforated part of the lift-shaft can now be found by setting out a matrix in its usual form, except that all deflections will now be expressed in terms of Δ_0, and that forces W are now statically indeterminate. The conditions of compatibility of deformations at the floor level 2 (see Fig. 4.8) are:

$$211' \ldots (m_{10} + m_{12}) \frac{L}{6EI_b} + m_{12} \frac{h}{3EI_c} - m_{21} \frac{h}{6EI_c} = \frac{\Delta 1 - \Delta 2}{h}$$

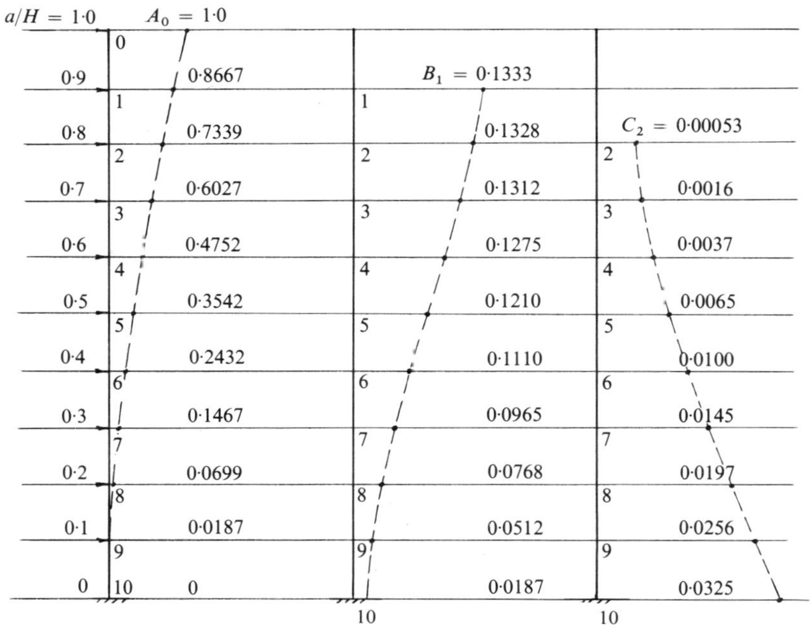

Fig. 4.7 Coefficients A, B, **and** C **for ten-storey cantilever**

or

$$(m_{10} + m_{12})\frac{L}{h} + 2m_{12} = B_2.\left(\Delta_0 \frac{6EI_c}{h^2}\right)$$

$$012\ldots \quad -m_{01}\frac{h}{6EI_c} + m_{10}\frac{h}{3EI_c} - m_{12}\frac{h}{3EI_c} + m_{21}\frac{h}{6EI_c} = \frac{\Delta 0 - \Delta 1}{h} - \frac{\Delta 1 - \Delta 2}{h}$$

or

$$-m_{01} + 2m_{10} - 2m_{12} + m_{21} = C_2\left(\Delta_0 \frac{6EI_c}{h^2}\right) \tag{4.15}$$

The equilibrium condition will now contain the statically indeterminate forces W to which the perforated part of the lift shaft will be subjected. This is in the form:

$$1\text{–}1'\ldots\frac{m_{12} + m_{21}}{h} - \frac{m_{10} + m_{01}}{h} = \tfrac{1}{2}W_1 \tag{4.16}$$

Based on eqs. (4.15) and (4.16) an element matrix can be set out as well as the overall matrix.

Example 4.2

The following example is used to illustrate the procedure and to discuss the results. Figure 4.8 shows a ten-storey lift shaft subjected to the overall pressure w per unit height, transmitted from a building to the lift shaft as point loads by the floor diaphragms. The unknown values are the bending moments in the perforated part of the lift-shaft and the forces W to which it is subjected.

Taking $E=28\times10^6$ kN/m² and for $I_t=6{\cdot}1074$ m⁴

$$\Delta_0 = \frac{w_t \,.\, H^4}{8EI_t} = 0{\cdot}000\ 592w$$

Taking pressure as 1 kN/m² and for the 12 m wide building $w=12$ kN/m run we obtain:

$$\Delta_0 = 7{\cdot}1 \text{ mm}$$

Based on eqs. (4.15) and (4.16) the overall matrix is as shown in Table 4.1. Figure 4.8 shows the distribution of bending moments and forces in the perforated part of the lift shaft. The solid part will therefore support the remaining part of the total wind pressure and will be in bending as a simple cantilever.

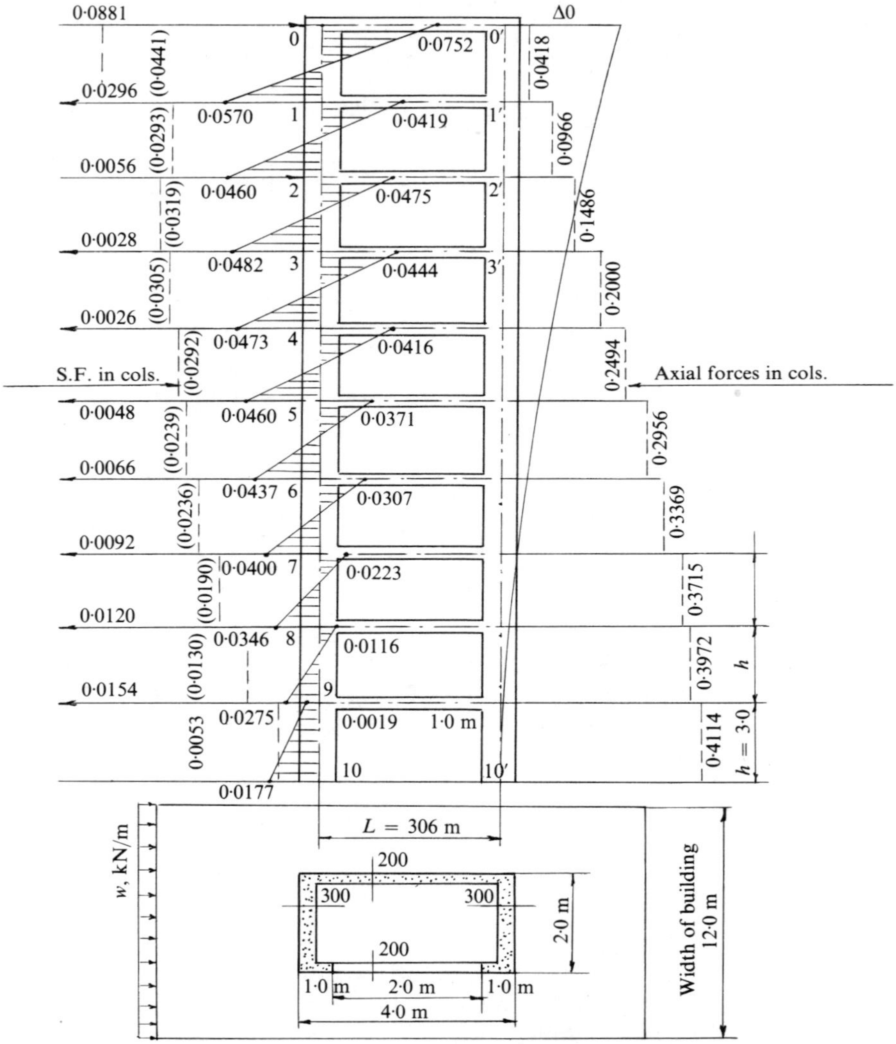

Fig. 4.8 Example of the analysis of perforated part of lift shaft (all moments and forces, W, are in terms of $\Delta_0(EI_c/h)$ and $\Delta_0=WH^4/8EI_t$)

Table 4.1 Matrix for the analysis of perforated part of lift shaft in Fig. 4.8

	0		1			2			3			4			5			6			7			8			9			10	
	0 1	W_0	1 0	1 2	W_1	2 1	2 3	W_2	3 2	3 4	W_3	4 3	4 5	W_4	5 4	5 6	W_5	6 5	6 7	W_6	7 6	7 8	W_7	8 7	8 9	W_8	9 8	9 10	W_9	10 9	
1 0 0′	2·152	0	−0·5																												$0{\cdot}13\Delta_0(EI/h)$
0–0′	1	−1·5	1																												0
2 1 1′	0	0	1·152	2·152	0	−0·5																									0·1328
0 1 2	−0·5	0	1	−1	0	0·5																									0·00053
1–1′	−1	0	−1	1	−1·5	1																									0
3 2 2′																															0·1312
1 2 3							F^e																								0·0016
2–2′																															0
4 3 3′																															0·1275
2 3 4										F^e																					0·0037
3–3′																															0
5 4 4′																															0·1210
3 4 5													F^e																		0·0065
4–4′																															0
6 5 5′																															0·1110
4 5 6																F^e															0·0100
5–5′																															0
7 6 6′																															0·0965
5 6 7																			F^e												0·0145
6–6′																															0
8 7 7′																															0·0768
6 7 8																						F^e									0·0197
7–7′																															0
9 8 8′																															0·0512
7 8 9																									F^e						0·0256
8–8′																															0
10 9 9′																															0·0187
8 9 10																												F^e			0·0325
9–9′																															0
10 9																												−0·5	0	+1	0·0187

Referring to Fig. 4.8 it is interesting to note that in this example, only the upper part of the perforated wall is subjected to the positive lateral forces, and the lower parts are subjected to negative forces, although of relatively small magnitude. Bending moments are high, but only in the upper part of the perforated wall, and diminish in the lower part. Comparatively high axial forces develop in columns. These are unexpected results, and they indicate that even for symmetrical bending, the response of a perforated lift-shaft is not easily predictable or in the form of a simple cantilever and that closer investigation is warranted.

4.4 Warping Stresses in Closed Tubes

It is not always possible to design a core or a lift-shaft so that the resultant lateral force will pass through the shear centre at all floor levels simultaneously, and most cores are therefore subjected to unsymmetrical bending. This results in torsion and warping of cores other than of circular cross-section, and is due to the plane sections not remaining plane after bending. The exact analysis of warping stresses in cores is not yet fully resolved, particularly with regard to the effect of perforations. The following analysis assumes that the walls are not perforated, and is therefore approximate.

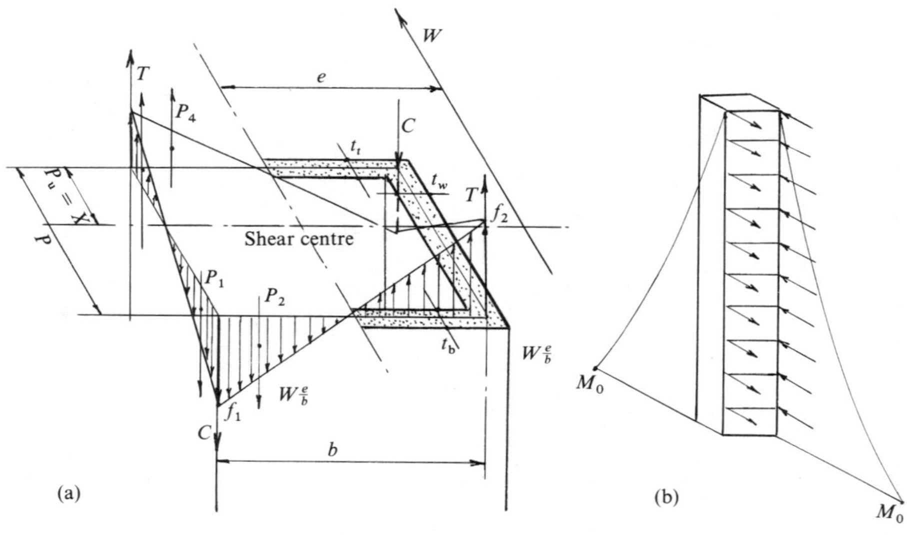

Fig. 4.9 Warping stresses in thin tubes

Figure 4.9 shows the distribution of warping stresses in a core without openings, based on the assumption that these stresses and deformations vary linearly between adjacent corners. The magnitude of these stresses is usually high and they should, therefore, be always considered.

If the total compression and tension forces were transferred to the corners of the box section, as shown in Fig. 4.9(a), their values could be expressed as follows:

$$C = P_1 + P_2 = \tfrac{1}{2}f_1 . n\, d . t_w + \tfrac{1}{2}f_1 . nb . t_t$$
$$T = P_3 + P_4 = \tfrac{1}{2}f_2(1 - n)\, d . t_w + \tfrac{1}{2}f_2 .(1 - n) . b . t_b$$

Using the conditions of equilibrium of longitudinal forces, the position of the neutral axis can be determined from:

$$n = \frac{1}{C}(\sqrt{C + 1} - 1) \tag{4.17}$$

in which
$$C = \frac{b(t_t + t_b)}{d.t_w + b.t_b}$$

and t with various suffixes refers to the thicknesses of walls at the top and bottom and of the web.

For equilibrium of moments, we can write:

$$\tfrac{1}{2}f_1.nd.t_w.d\left(1 - \frac{n}{3}\right) + \tfrac{1}{2}f_1.nb.t_t.d = M_0$$

from which the stresses at the corners of the box are:

$$f_1 = \frac{2M_0}{nd^2.t_w(1 - n/3) + nbd.t_t} \tag{4.18}$$

$$f_2 = f_1 \cdot \frac{1 - n}{n} \tag{4.19}$$

These stresses must be added to the stresses produced by the symmetrical bending, and those due to pure torsion.

4.5 Shear Lag in Cores

When members are subjected to shear stresses, the assumption in the Bernoulli–Euler bending theory, that plain sections remain plain after application of loads, is not strictly valid. Members subjected to shear stresses warp and the basic relationship of the bending theory $\sigma = M.y/I$ is not true, and in some cases may differ considerably from the actual stress distribution. Shear walls, and particularly box type structures, such as lift shafts and cores, are in this range. The deviation of the stress distribution based on the Bernoulli–Euler bending theory from the actual stresses is called the shear lag.

Shear lag in cores shows itself in non-uniform stress distribution across the width of the flanges. In the tension flange, maximum stress occurs at the flange–web junction with lower stresses at intermediate points, as shown in Fig. 4.10. Shear lag in the compression flange produces a reverse effect, i.e., reduces stress at the corners and increases it in the centre of the flange. The maximum effect of shear lag takes place at the extreme ends of members and particularly at the fixed end.

The exact solution of shear lag stresses is very complicated and approximate solutions are often used. One approach is based on the concept of the effective areas and effective second moments of area. Effectiveness coefficients α are used for various parts of the cross-section, which are based either on experience or on tests, e.g., it can be assumed that the section is transformed into lumped areas 1, 2, and 3, connected by webs. These lumped areas are assumed to take bending stresses and the webs shear stresses. Various lumped areas are then given an effectiveness coefficient α on which the values of the second moments of area are based. Stresses are then calculated from

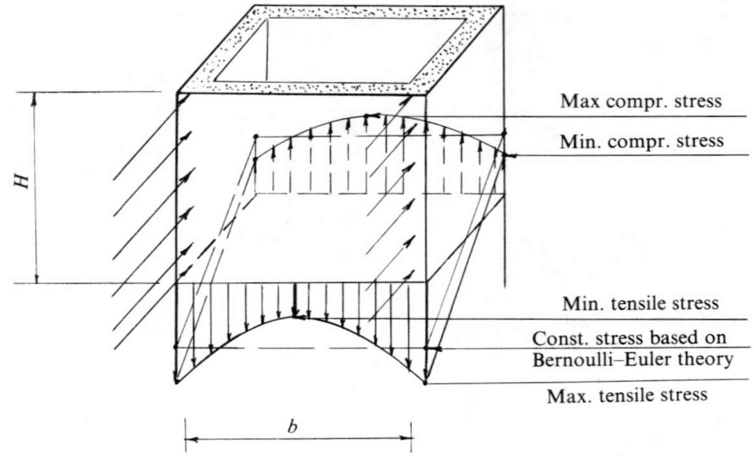

Fig. 4.10 Shear lag in box structures

the relationship:

$$\sigma = M.y/I_{\text{effective}}$$

Alternatively, Fig. 4.11(b) shows an approximate relationship between the ratio of the average stress/maximum stress in a box structure and the ratio of length to its flange width.

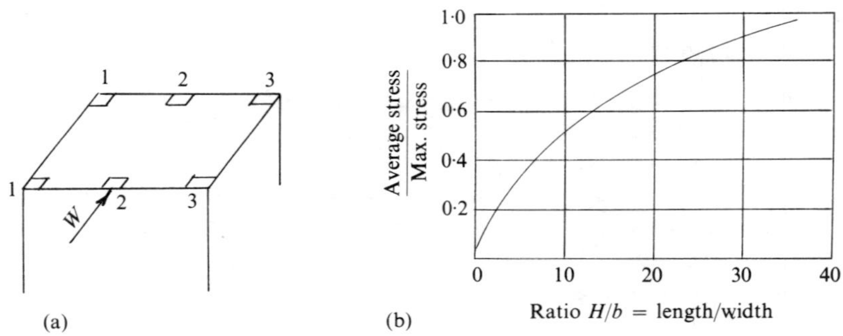

Fig. 4.11 Shear lag in structures

4.6 Bending of Frame-tube Structures

The expression 'frame-tube' is used to describe the form of construction of a tall building in which columns are used mainly on the periphery of the building and the floor span is usually the full width of the building (Fig. 4.12(a)). The floor beams are in most cases pin-connected to the walls but all external beams and columns form a rigid system of a rectangular perforated tube. This can be made in steel, concrete, or in composite materials. The lateral forces are transferred by the floor diaphragms to walls parallel to the direction of the wind force but, for symmetrical bending, a certain amount of cooperation is derived from the walls perpendicular to the direction of the wind force, i.e., while the main bending moments develop in the flanking walls,

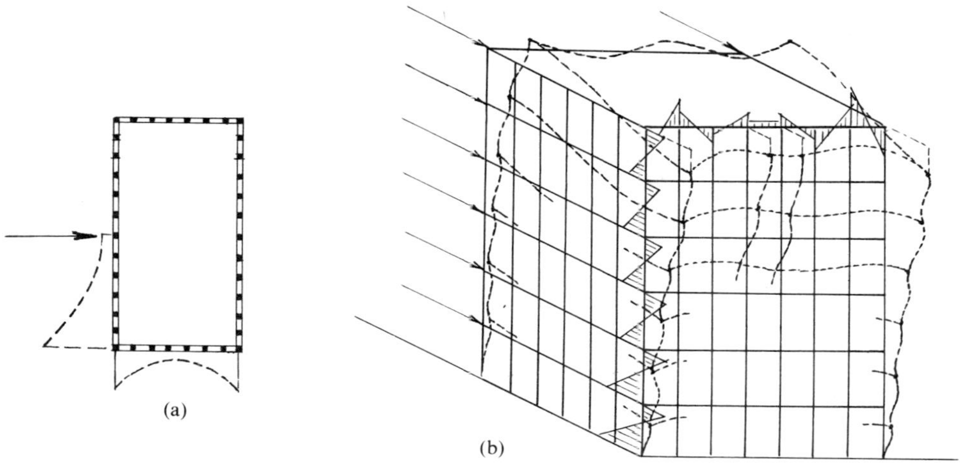

Fig. 4.12 **Frame-tube structures in bending**

some moments develop also in the walls perpendicular to the direction of the wind. This is because the corner columns are subjected to the highest axial forces and deformations which in turn produce bending moments in the walls perpendicular to the direction of the wind as shown diagrammatically in Fig. 4.12(b).

One other interesting aspect emerges in this form of construction. The 'loaded' frames develop bending moments in beams of such magnitude that the difference between shearing forces in beams on each side of the column diminishes towards the centre line of the frame. In consequence of this, the external columns develop maximum direct forces, and these diminish rapidly towards the centre of the building. A similar effect takes place in frames perpendicular to the wind force though for quite a different reason, i.e., it is due to the axial deformations and resulting bending moments which also diminish rapidly towards the centre line.

This pattern of the distribution of axial forces in the two sets of walls is often referred to as the 'shear lag' effect because it is produced by the shear forces in beams as discussed above.

The primary analysis of such structural systems for symmetrical bending does not add any new problems to those discussed previously and bending moments and deflections can be calculated in a normal way but including the effect of axial deformations of columns which affect moments in other walls.

References

1. R. Rosman, 'Torsion of Perforated Concrete Shafts', *J. Struct. Div. ASCE*, **95** (May 1969).
2. Nutan Kumar Subedi, *Hull-core Structures for High Rise Buildings*, Ph.D. Thesis, University of Strathclyde (1971).
3. V. Z. Vlasov, *Thin-walled Elastic Beams*, OTS 61-11400, Office of Technical Services, U.S. Dept of Commerce, Israel Program for Scientific Translations, Catalogue No. 428, Wash., D.C. (1959).
4. R. Dabrowski, 'Torsion Bending of Thin-walled Members of Non-Deformable Closed Cross Section', *Report of the School of Engineering*, Columbia University (September, 1963).
5. R. Dabrowski, 'Warping Torsion of Curved Box Girders with Non-Deformable Cross Section', *Der Stahlbau*, May 1965 (in German).

6. S. R. Abdel-Samad, *Analysis of Multicell Box Girders with Diaphragms*, thesis presented to the University of Illinois, at Urbana, Ill., in February, 1967, in partial fulfilment of the requirements for the degree of Doctor of Philosophy in Civil Engineering.
7. F. Resinger, *Die dünnwandigen Kastenträger*, Stahlbau-Verlag, Köln (1959).
8. B. Richmond, (Twisting of Thin-walled Box Girders', *Proc. I.C.E.*, paper 6868 (April, 1966).
9. D. C. Dalton and B. Richmond, 'Twisting of thin-walled Box Girders of Trapezoidal Cross-section', *Proc. I.C.E.*, paper 7031 (1969).

5 Analysis of Indeterminate Floor Systems in Tall Buildings

5.1 Introduction

Most floor systems in tall buildings are of simple beam and slab construction and present no difficulties in the analysis or design. However, occasionally, the system of the floor is of complex statically indeterminate form. Some of these forms are indicated in Fig. 5.1. These are:

(a) Interconnected floor beams, either in plan (Fig. 5.1(a)) or in section, when two or more floors may be tied in various ways.
(b) Two-way ribbed floors, in which ribs intersect at right angles (Fig. 5.1(b)), or
(c) Diagrid floors, in which beams intersect at any angle and are usually spaced long distances apart to reduce the weight of the floor. The advantage of this system lies in overall reduction in weight of the floor, in utilizing strength of short beams in providing elastic support to long beams, and utilizing torsional stiffnesses of beams and flexural stiffnesses of perimeter beams (Fig. 5.1(c)).

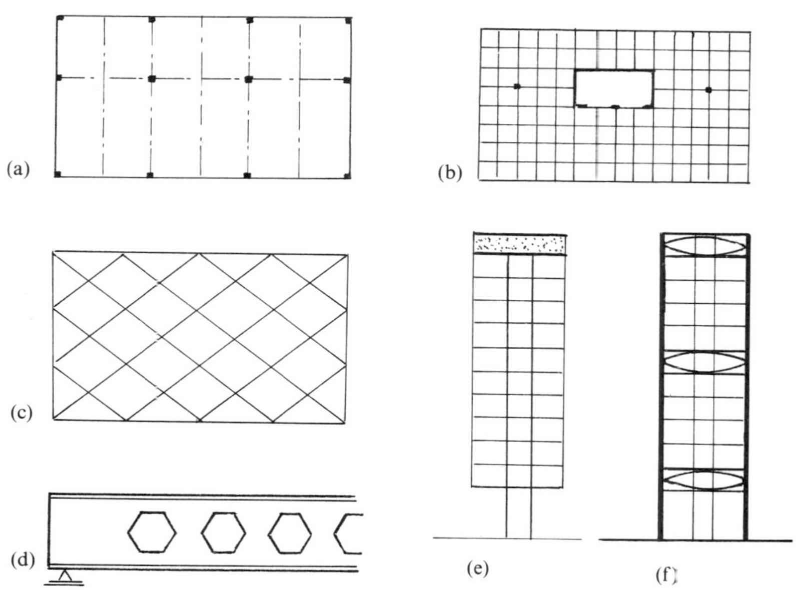

Fig. 5.1 Indeterminate floor systems

(d) Flat plates and lift slab constructions, which have been frequently used in buildings up to 10 or 16 floors. In these systems, column-to-slab connections are weak, and such buildings require stiffening by shear walls, lift shafts, infilled panels, or similar.
(e) Vierendeel or castellated girders as shown in Fig. 5.1(d).
(f) Floors suspended from roof girders (Fig. 5.1(e).
(g) Catenary arch floors, etc. (Fig. 5.1(f)).

In the following, some of these systems are analysed using the force–displacement method.

5.2 Interconnected Floor Beams

Various arrangements of interconnected beams frequently occur in the design, as shown in Fig. 5.2 in which beams 4–5–6 and 10–11–12 are elastically supported, while other beams such as 7–8–9 are rigidly supported by columns. In such floors beams can be subjected to bending moments and deflections of different magnitude. The analysis by the force–displacement method does not present difficulties and does not require any new formulation, e.g., assuming that the floor in Fig. 5.2 is subjected to uniformly distributed load w, the matrix can be set out as shown in Table 5.1. Figure 5.2 shows the distribution of bending moments and shear forces. It can be seen that the distribution of moments in beams is considerably affected by the support conditions.

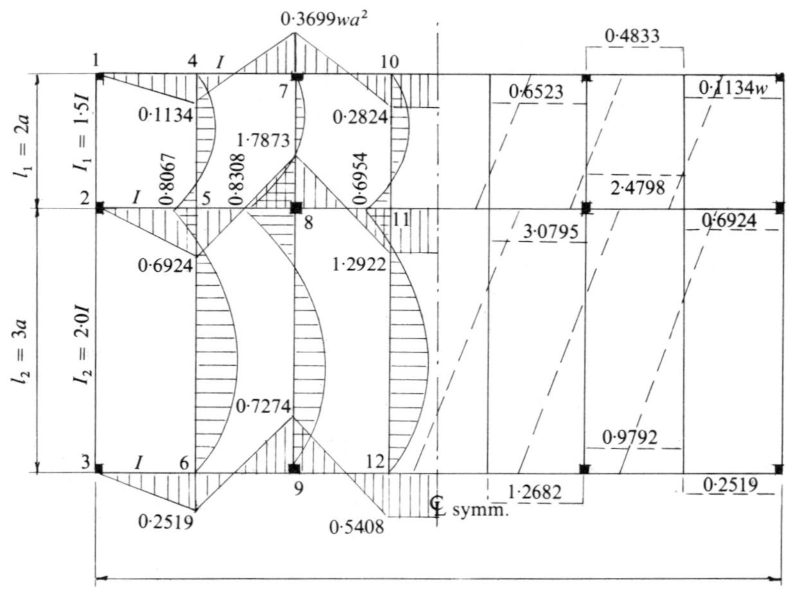

Fig. 5.2 Interconnected floor beams

5.3 Beam on Elastic Foundation

A beam supported by an elastic medium has received considerable attention in technical literature and has been analysed by Timoshenko, Heteny, and other authors.

Table 5.1

	4		7	10		5			8	11			6		9	12		$wa^2/6EI$
	4 1	Δ4	7 4	10 7	Δ10	5 4	5 2	Δ5	8 5	11 8	11 10	Δ11	6 3	Δ6	9 6	12 9	Δ12	
1 4 7	4	−2	−1															0
EC. 4	2	0	1			−0·5												$w.a$
4 7 10	−1	−1	4	−1	−1													0
7 10 10′	0	0	−1	5	−1													0
EC. 10	0	0	1	1	0						−0·5							$w.a$
2 5 8						0	4	−4	−1									0
4 5 6		0·375				4·25	0	−0·625	0					0·25				−0·3531wa
EC. 5						0·833	2	0	1									2·5wa
5 8 11							−1	−2	4	−1	0	−2						0
8 11 11′							0	0	−1	5	0	−2						0
10 11 12					0·375				0	0	4·25	−0·625					0·25	−3·5312wa
EC. 11									1	1	0·833	0						2·5wa
3 6 9													4	−2	−1	0	0	0
EC. 6						−0·333							2	0	1	0	0	1·5wa
6 9 12													−1	−1	4	−1	−1	0
9 12 12′															−1	5	−1	0
EC. 12′											−0·333				1	1	0	1·5wa

In the following, an analysis is made using the force–displacement method as applied to a particular case of a floor in which a spreader beam is elastically supported by a series of closely spaced beams as shown in Fig. 5.3. Such floors are frequently used in assembly halls, garages, warehouses, etc., and the purpose of the spreader beam is to distribute loads other than uniform, to prevent excessive deflection of one beam with respect to others, and to reduce the effect of twist in beams.

Spreader beams are seldom analysed in practice, particularly when the floor is designed for uniformly distributed load. However, when loads are other than uni-

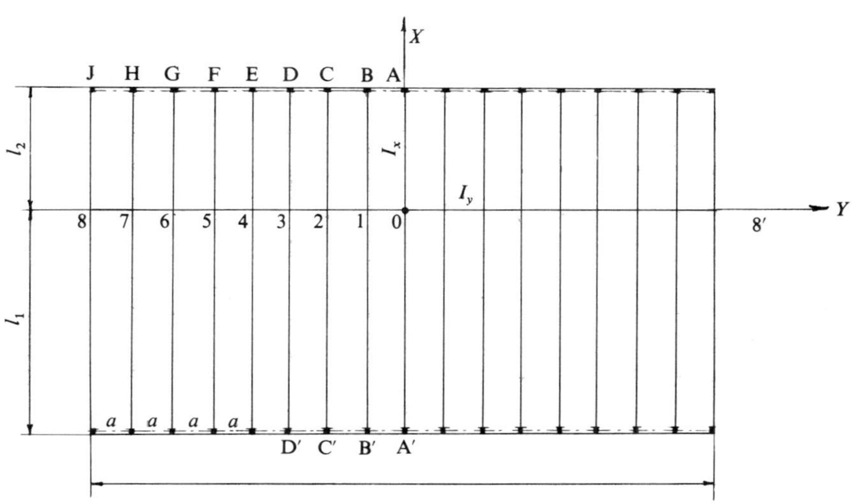

Fig. 5.3 Analysis of a beam on elastic foundation

Table 5.2 Matrix for the analysis of the floor shown in Fig. 5.4

$$r_1 = \frac{a}{l}\cdot\frac{I_1}{I_2}; \quad r_2 = \frac{l}{a}$$

	0			1			2			3			4			5			6			7			8			9			
	0X	0Y	Δ0	1X	1Y	Δ1	2X	2Y	Δ2	3X	3Y	Δ3	4X	4Y	Δ4	5X	5Y	Δ5	6X	6Y	Δ6	7X	7Y	Δ7	8X	8Y	Δ8	9X	9Y	Δ9	
0 A	2	0	−1	0	0	0																									0
0 1	0	$2r_1$	$-r_2$	0	r_1	r_2																									0
0	1	r_2	0	0	$-r_2$	0																									$0{\cdot}5Wl$
1 B		0	0	2	0	−1	0	0	0																						0
0 1 2		r_1	r_2	0	$4r_1$	$2r_2$	0	r_1	r_2																						0
1		$-r_2$	0	2	$2r_2$	0	0	$-r_2$	0																						0
2 C																															0
1 2 3								F^e																							0
2																															0
3 D																															0
2 3 4											F^e																				0
3																															0
4 E																															0
3 4 5														F^e																	0
4																															0
5 F																															0
4 5 6																	F^e														0
5																															0
6 G																															0
5 6 7																				F^e											0
6																															0
7 H																															0
6 7 8																							F^e								0
7																															0
8 J																															0
7 8 9																										F^e					0
8																															0
9 K																															0
8 9 10																															0
9																															0

form, the distribution of bending moments, shear forces, and deformations in the whole system may need to be evaluated accurately for economy of construction and to eliminate uncertainty of the performance of the system.

The setting out of the matrix does not require any new formulation to that already discussed previously. However, a simple numerical example is used to indicate the pattern of bending moments and deflections in the floor beams. Torsional moments are omitted because these are usually small in this type of floor. In this example a floor shown in Fig. 5.4 is analysed for the effect of one concentrated load placed at the point 0. The floor is assumed to be symmetrical with respect to X and Y axes. In the setting out of the matrix the following abbreviations are used:

$$r_1 = \frac{a.I_x}{l.I_y} \quad \text{and} \quad r_2 = \frac{l}{a}$$

The matrix, Table 5.2, is set out in a general form, but in the example it is assumed that $l/a = 5$ and $I_x/I_y = 1$, so that $r_1 = 0{\cdot}2$ and $r_2 = 5$. Figure 5.4 shows the distribution of bending moments in the main beams as well as in the spreader beam and deflections of the nodal points. It is interesting to note that in this case some beams (J–J′ and

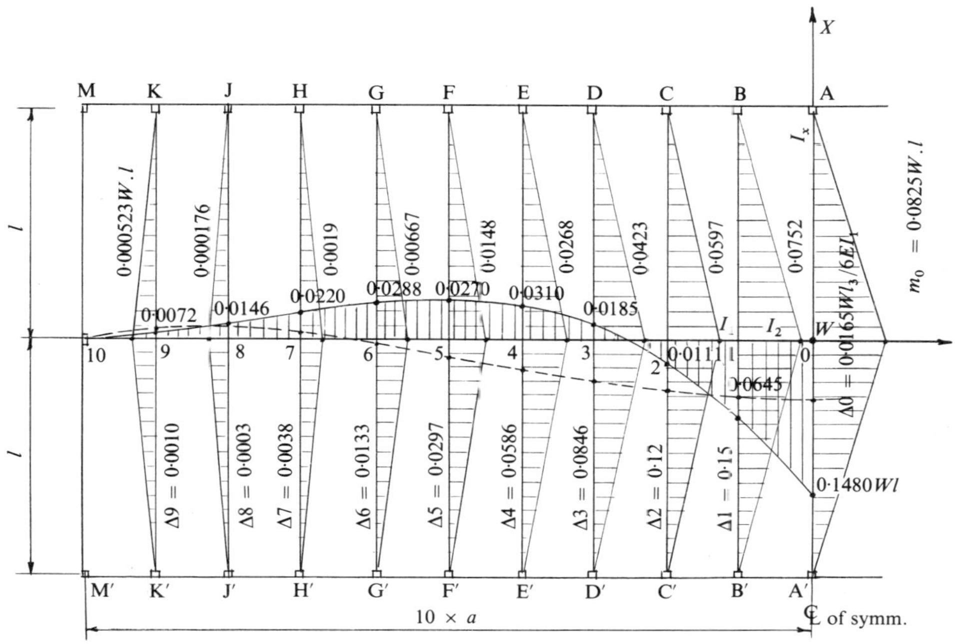

Fig. 5.4 Example of the analysis of a beam on elastic foundation

K–K′) are subjected to 'hogging' moments while the rest of the beams are subjected to 'sagging' moments. Bending moments in the spreader beam also change from sagging to hogging in the manner well known from other solutions of beams on elastic foundation. The maximum bending moment in the spreader beam is nearly twice as high as in the main beam A–A′. This is an indication of a considerable effect of the spreader beam in distributing point loads to other beams, i.e., the maximum bending moment in the central beam (A–A′) is only $\frac{1}{6}$ of the total moment. The deflections are mostly downward, with the exception of the part of the floor from the point 7 to 10, which deflects upwards. Other arrangements of loads or patterns of floors can of course be analysed in a similar way.

5.4 Analysis of Vierendeel Girders

Vierendeel girders are sometimes used as floor beams and these are highly statically indeterminate. Using the force–displacement method, the analysis of such beams can be made in the same way as for frames or shear walls. However, the beam shown in Fig. 5.5 is used to illustrate the procedure. This is a castellated beam and is analysed

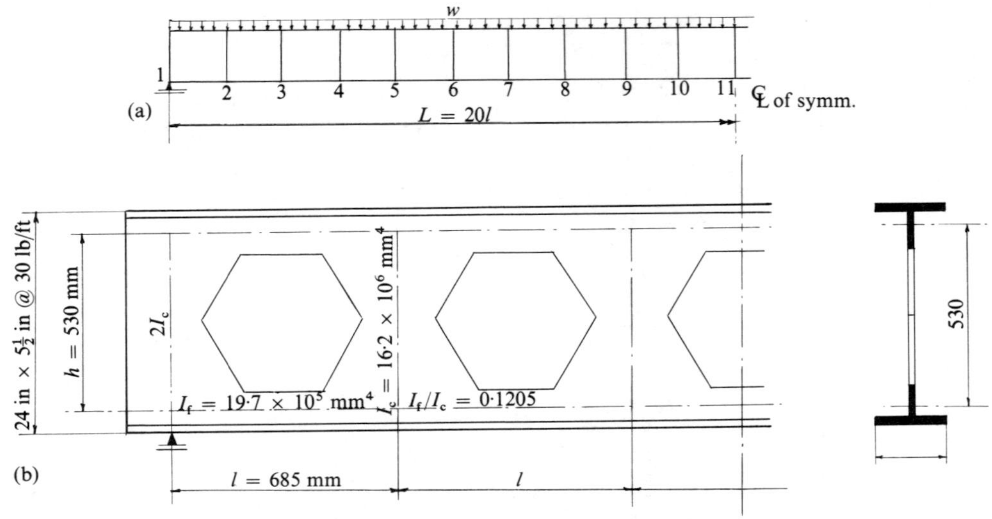

Fig. 5.5 Example of the analysis of a Vierendeel girder

here as a Vierendeel girder. All dimensions are shown in Fig. 5.5(a) and (b). The matrix shown in Table 5.3 is set out for a symmetrical girder and is in an explicit form, in terms of the following coefficients:

$$r_1 = \frac{f_c^{11}}{f_b^{11}}, \qquad r_2 = \frac{f_c^{12}}{f_b^{11}}, \qquad r_3 = \frac{f_b^{12}}{f_b^{11}} \quad \text{and} \quad a = 1 + r_1 - r_2$$

in which the suffixes b and c refer to the beam and column respectively. The flexibility coefficients f should, for greater accuracy of the analysis, include the effect of variable second moments of area of beams and columns. However, in this example, these are

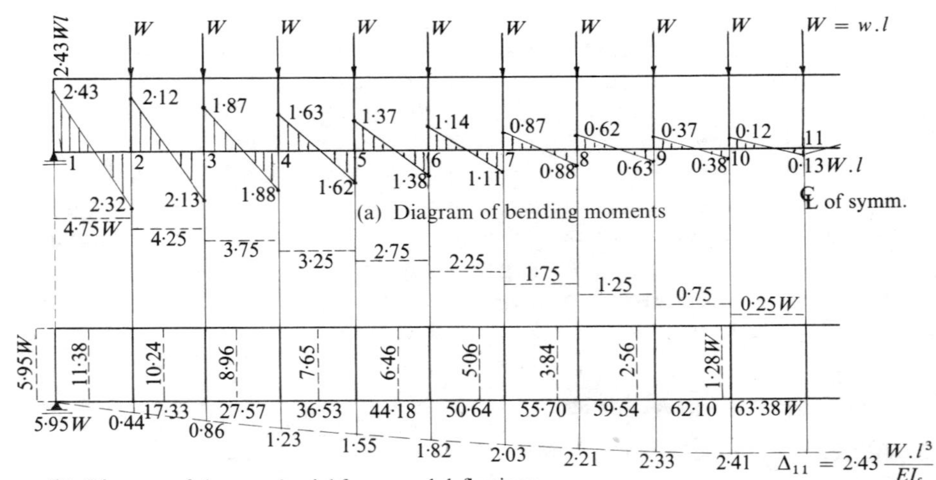

Fig. 5.6 Diagram of bending moments, deflections, shear and axial forces in the castellated beam in Fig. 5.5

Table 5.3 Matrix for the analysis of castellated beam in Fig. 5.5

	1	2			3			4			5			6			7			8			9			10			11		
	1 2	2 1	2 3	Δ2	3 2	3 4	Δ3	4 3	4 5	Δ4	5 4	5 6	Δ5	6 5	6 7	Δ6	7 6	7 8	Δ7	8 7	8 9	Δ8	9 8	9 10	Δ9	10 9	10 11	Δ10	11 10	Δ11	
1′ 1 2	a	$-r_3$	0	−1																											0
1 2 3	$-r_3$	1	$-r_3$	−2	1	0	1																								0
1 2 2′	$-r_3$	a	$a-1$	−1	0	0	0																								0
2–2′	1	1	−1	0	−1	0	0																								0·5Wl
2 3 4			$-r_3$	0	1	$-r_3$	−2	1	0	1																					0
2 3 3′			$-r_3$	0	a	$a-1$	−1	0	0	0																					0
3–3′			1	0	1	−1	0	−1	0	0																					0·5Wl
3 4 5																															0
3 4 4′									F^e																						0
4–4′																															0·5Wl
4 5 6																															0
4 5 5′												F^e																			0
5–5′																															0·5Wl
5 6 7																															0
5 6 6′															F^e																0
6–6′																															0·5Wl
6 7 8																															0
6 7 7′																		F^e													0
7–7′																															0·5Wl
7 8 9																															0
7 8 8′																					F^e										0
8–8′																															0·5Wl
8 9 10																															0
8 9 9′																								F^e							0
9–9′																															0·5Wl
9 10 11																															0
9 10 10′																											F^e				0
10–10′																															0·5Wl
10 11																											$-r_3$	1	1	−1	0
11–11′																											1	0	1	0	0

Note: moments are in terms of $W.l$ and deflections in terms of $Wl^3 f_b^v$.

taken of average values as indicated in Fig. 5.5. The values of r and a become:

$$r_1 = \frac{h.I_b}{l.I_c} = 0{\cdot}9408, \qquad r_2 = \tfrac{1}{2}r_1, \qquad r_3 = 0{\cdot}5 \quad \text{and} \quad a = 1{\cdot}4704$$

Figure 5.6 shows the distribution of bending moments, shear and axial forces, as well as deflections. The degree of accuracy of the analysis depends on the assumption with regard to the position of the centres of flanges and on the evaluation of the second moments of area of beams and columns.

It is interesting to note that, when analysing such girders as simple beams, the maximum axial forces in flanges do not differ appreciably when compared with the more exact analysis, e.g., in the example included, the maximum tension would be 64·02W, against 63·38W when analysed as a Vierendeel girder. However, the analysis of such girders as simple beams would not give any indication of the distribution of bending moments and shear forces in the remaining members of the girder, and would therefore not be safe. Similarly, an appreciable error could be made in the evaluation of deflections.

5.5 'Diagrid' Floors

In diagrid floors, beams are arranged to intersect diagonally at any angle, and external beams provide a major part of the load-carrying capacity of the floor. End supports can be of any form and this usually governs the symmetry of the system and the size of the matrix. Torsion is always present in these systems although the maximum values are near the corners only. Such floors can support heavy, and even concentrated, loads. The analysis can be made using the force–displacement method by first allowing the structure to deflect to its final equilibrium position and introducing releases at all junctions. A set of compatibility and equilibrium conditions can then be written in the usual way.

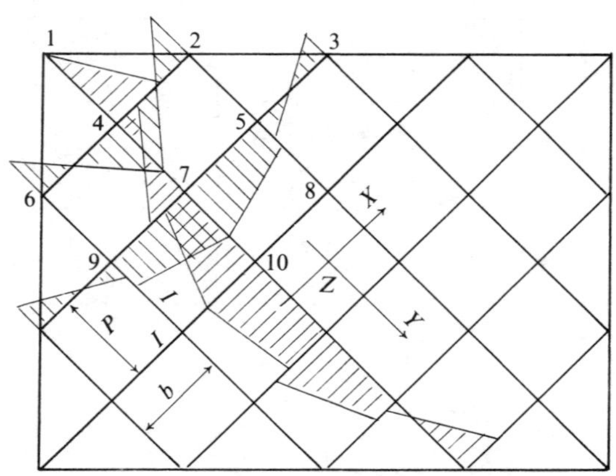

Fig. 5.7 'Diagrid' floor

In general, at any point, such as point 7 in Fig. 5.7, there are seven unknown values, i.e., four flexural bending moments (m_{74}, m_{75}, m_{79}, and m_{710}), two torsional moments (T_{75} and T_{710}), and one deflection (Δ_7), although at points of symmetry such as points 8, 9, and 10 there are only three values unknown. An element force–

displacement matrix at point 7 can be set out using two flexural conditions of compatibility of deformations (marked *F*.4–7–10 and *F*.5–7–9), two mixed compatibility conditions, involving flexural and torsional moments (marked *F*.47–*T*.75 and *F*.57–*T*.710), two conditions of the equilibrium of moments in the planes *XZ* and *YZm*, and one condition of equilibrium of forces marked *Ec*.7. These are as follows.

F. 4–7–10	$-m_{47}\cdot\frac{a}{6EI} + m_{74}\cdot\frac{a}{3EI} + m_{710}\cdot\frac{a}{3EI} + m_{107}\frac{a}{6EI}$	$\frac{\Delta_7 - \Delta_4}{a} - \frac{\Delta_{10} - \Delta_7}{a}$
F. 5–7–9	$m_{57}\cdot\frac{b}{6EI} + m_{75}\cdot\frac{b}{3EI} + m_{79}\cdot\frac{b}{3EI} + m_{97}\cdot\frac{b}{6EI}$	$\frac{\Delta_7 - \Delta_5}{b} + \frac{\Delta_7 - \Delta_9}{b}$
F. 47 – *T*. 75	$-m_{47}\cdot\frac{a}{6EI} + m_{74}\cdot\frac{a}{3EI} + T_{75}.\phi_{75}.b +$ $+\left[\frac{\Delta_8 - \Delta_5}{a} - m_{85}\cdot\frac{a}{6EI} - m_{57}\cdot\frac{a}{3EI}\right]$	$= \frac{\Delta_7 - \Delta_4}{a}$
F. 79 – *T*. 710	$m_{97}\cdot\frac{b}{6EI} + m_{79}\cdot\frac{b}{3EI} + T_{710}.\phi_{710}.a +$ $+\left[\frac{\Delta_{10} - \Delta_8}{b} - m_{108}\cdot\frac{b}{6EI} - m_{810}\cdot\frac{b}{3EI}\right]$	$= \frac{\Delta_7 - \Delta_9}{b}$
$\sum MXZ=0$	$m_{79} - m_{75} + T_{74} - T_{710}$	$= 0$
$\sum MYZ=0$	$m_{74} - m_{710} + T_{79} - T_{75}$	$= 0$
Ec. 7	$\frac{m_{79} - m_{97}}{b} + \frac{m_{75} - m_{57}}{b} + \frac{m_{107} - m_{710}}{a} - \frac{m_{74} - m_{47}}{a}$	$= W_7$

The torsional flexibility coefficients ϕ are given by eq. (3.3) (p.106). An overall matrix can now be set out based on the above element.

5.6 'Waffle' Floors

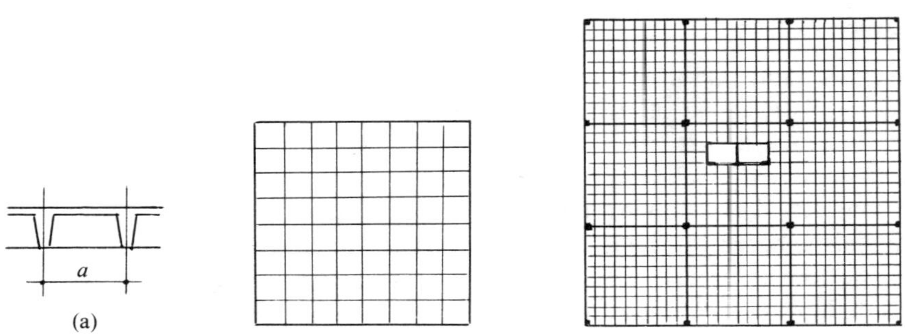

(a) (b) (c)

Fig. 5.8 Waffle floors

Waffle floors are formed by two sets of parallel, closely spaced ribs at right angles spanning in opposite directions, Fig. 5.8. Ribs are usually spaced at equal distances for repetitive use of formwork, and the floor panels are supported either on walls or columns and are simply-supported or continuous. Such floors are usually designed for uniformly distributed loads as two-way slabs. However, a more accurate analysis may be required when either the loads are not uniform or the plan configuration and support conditions are irregular.

In the 'exact' analysis the unknown values at each intersection are: bending moments in two directions, deflection, and torsional moments particularly near the corners. These values can be calculated by setting out a matrix as in Fig. 5.7 but, because of the great number of intersections of ribs, the matrix size can be very large. However, by omitting torsional moments, the matrix size can be reduced considerably though, of course, at the cost of slightly reducing the accuracy of the solution.

The setting out of the matrix does not require any new development and because of regularity of panels in most cases the matrix can be written directly by inspection of the floor diagram and without first writing the equations of compatibility of deformations and the equilibrium of forces. Although the element matrices F_1, F_2, F_3, and F_4 (see Table 5.4) are diagonally repetitive, the main force–displacement matrix is not entirely symmetrical because of different boundary conditions of various joints. A simple example is included to emphasize these points and to indicate a pattern of bending moments, shear forces, and deflections.

Figure 5.9 shows a quarter of a floor panel assumed to be symmetrical, supported continuously at all boundaries and loaded with one point load at the centre of the floor at point 12. The matrix (see Table 5.4), for evaluation of bending moments in the direction X and Y and deflections is written directly based on Fig. 5.9. To simplify the matrix, the compatibility equations are multiplied by the value of $6EI/a$ and the equilibrium equations by a. Figure 5.9(a) shows the distribution of bending moments and deflections. It should be noted that the moments are in terms of $W.a$, and deflections in terms of Wa^3/EI. Figure 5.9(b) shows the distribution of shearing forces in beams as well as the magnitude of interacting forces at joints. It is interesting to note that, in this case, shearing forces are at their maximum at the centre of the panel, and diminish towards the supports. This is not an unusual effect and is caused by the distribution of load to other beams. Similarly, interacting forces diminish from the centre towards the outer beams.

5.7 Cantilever Stairs

Cantilever stairs are relatively difficult to analyse exactly mainly because of three-dimensional distribution of stresses and deformations in all members. In addition, the form of construction and the plan may vary as well as sizes of members and boundary conditions. However, several analyses are available based on various assumptions. In the following an analysis is made using the force–displacement method. This is based on the usual linearly-elastic assumption and particularly that the deformations are small so that the equilibrium conditions can be satisfied in the original configuration. The loads are assumed to be static and at first it is also assumed that the effects of axial deformations and of beams being cranked are omitted. These, however, are included later.

Figure 5.10 shows a cantilever staircase of a spinal beam construction encastre at

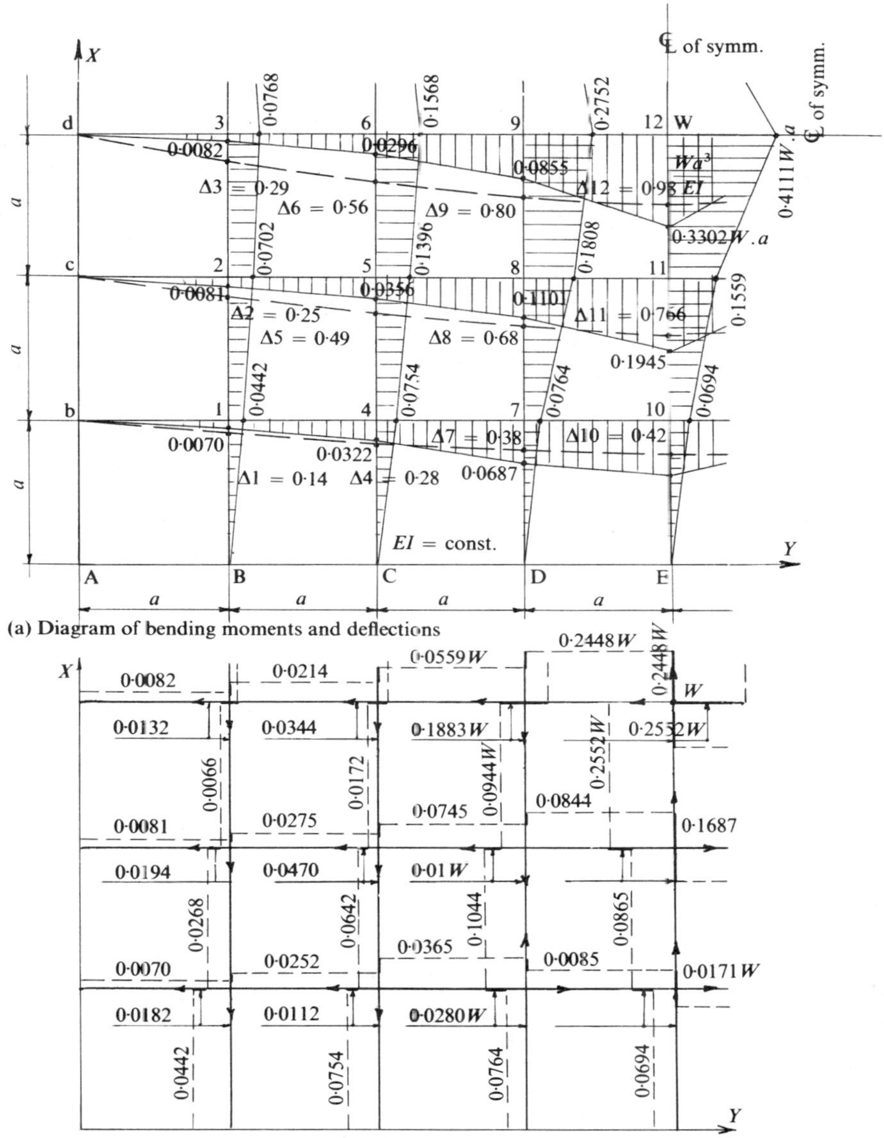

(a) Diagram of bending moments and deflections

(b) Diagram of shearing and interacting forces at joints

Fig. 5.9 Example of the analysis of a 'waffle' plate

A and D in all three directions. (Hinges at A and D are possible in Y direction and the analysis can then be adjusted by taking m_{ay} and $m_{\alpha y}$ equal to zero.) A perspective view is also shown in Fig. 5.11 to facilitate the derivation. All dimensions and sizes of members are shown on these diagrams together with the distribution of bending moments and deflections in the chosen system of coordinates. The system of loading (not indicated) can be of any form, although the uniformly distributed load is most frequently used in the design. Although in practice member sizes are often kept constant, to retain the general nature of the analysis, Figs. 5.10 and 5.11 show the staircase in which all sizes of members vary.

Table 5.4 Analysis of a 'waffle' plate shown in Fig. 5.9

	1			2			3			4			5			6			7			8			9			10			11			12			
	$1X$	$1Y$	$\Delta1$	$2X$	$2Y$	$\Delta2$	$3X$	$3Y$	$\Delta3$	$4X$	$4Y$	$\Delta4$	$5X$	$5Y$	$\Delta5$	$6X$	$6Y$	$\Delta6$	$7X$	$7Y$	$\Delta7$	$8X$	$8Y$	$\Delta8$	$9X$	$9Y$	$\Delta9$	$10X$	$10Y$	$\Delta10$	$11X$	$11Y$	$\Delta11$	$12X$	$12Y$	$\Delta12$	
B 1 2	4	0	−2	1	0	1					0	0																									0
b 1 4	0	4	−2	0	0	0					1	1																									0
EC. 1	2	2	0	−1	0	0					−1	0																									0
1 2 3	1	0	1	4	0	−2	1	0	1																												0
C 2 5	0	0	0	0	4	−2	0	0	0					F_4																							0
EC. 2	−1	0	0	2	2	0	−1	0	0																												0
2 3				1	0	1	2	0	−1																												0
d 3 6				0	0	0	0	4	−2								F_4																				0
EC. 3				−2	0	0	2	2	0																												0
C 4 5																																					0
1 4 7		F_4										F_1								F_4																	0
EC. 4																																					0
4 5 6																																					0
2 5 8					F_4									F_2									F_4														0
EC. 5																																					0
5 6																																					0
3 6 9								F_4								F_3										F_4											0
EC. 6																																					0
D 7 8																																					0
4 7 10											F_4										F_1								F_4								0
EC. 7																																					0
7 8 9																																					0
5 8 11														F_4									F_2									F_4					0
EC. 8																																					0
8 9																																					0
6 9 12																	F_4								F_3										F_4		0
EC. 9																																					0
E 10 11																																					0
7 10																				$2F_4$										F_1							0
EC. 10																																					0
10 11 12																																					0
8 11																							$2F_4$									F_2					0
EC. 11																																					0
11 12																																					0
9 12																										$2F_4$								F_3			0
EC. 12																																					$0{\cdot}5Wa$

Note: moments are in terms of $W.a$ and deflections in terms of $Wa^3/6EI$.

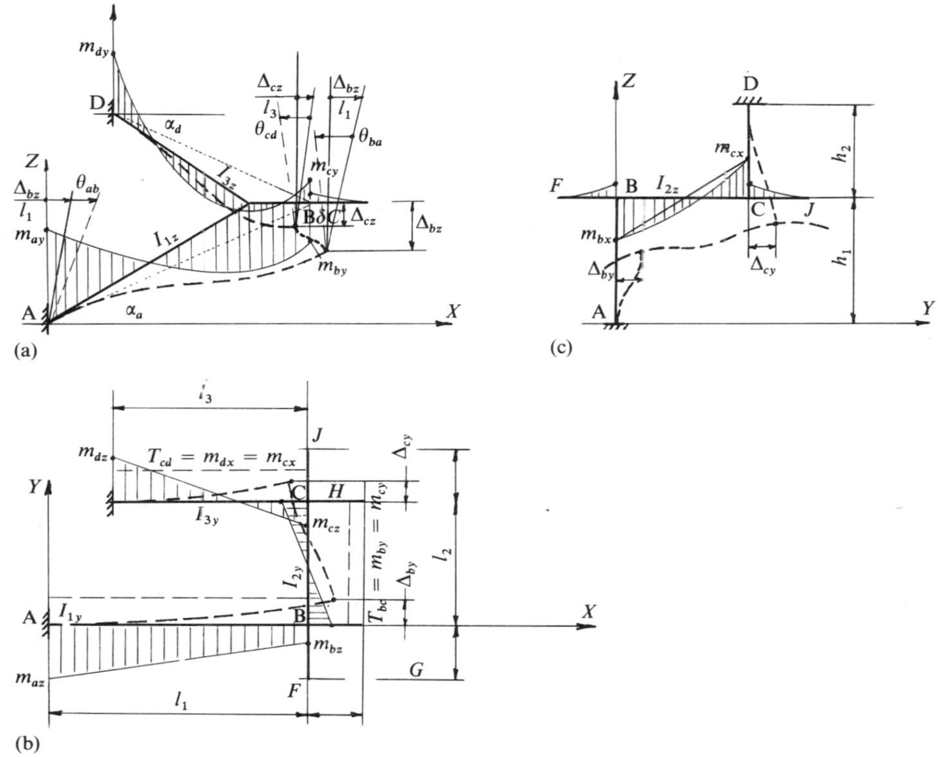

Fig. 5.10 Analysis of cantilever stairs

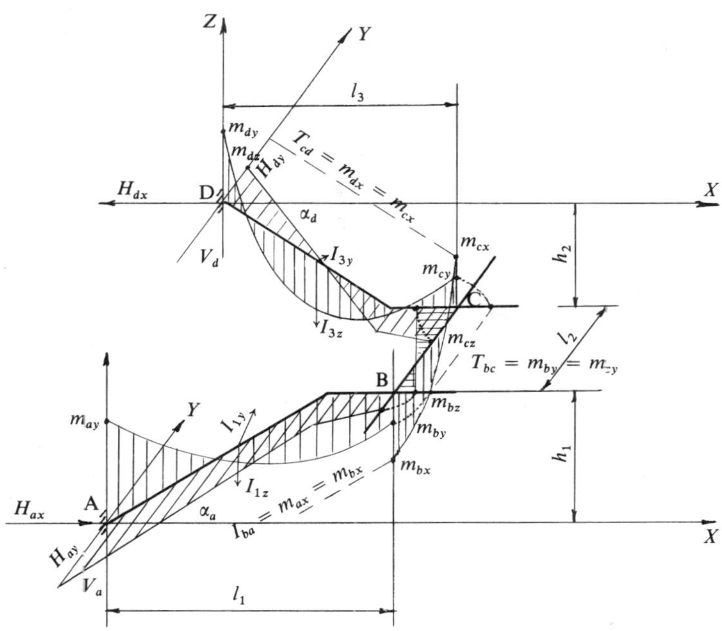

Fig. 5.11 Perspective view of a cantilever staircase

When loads are applied, the points B and C deflect with respect to all three co-ordinates but because of the assumed unaltered lengths of members and the deflections assumed to be small, only the deflections in the Z and Y direction are statically indeterminate. The components of the deflections Δb and Δc in the X direction can be related to the deflections Δbz and Δcz by geometry, thus:

$$\Delta bx = \Delta bz.\tan \alpha_a, \qquad \Delta cx = \Delta cz.\tan \alpha_d$$

At the points A, B, C, and D, bending moments are acting with respect to all three coordinates and are as follows:

A	B	C	D
m_{ax}	m_{bx}	m_{cx}	m_{dx}
m_{ay}	m_{by}	m_{cy}	m_{dy}
m_{az}	m_{bz}	m_{cz}	m_{dz}

However, not all these moments are statically indeterminate and for equilibrium of moments at the junctions B and C we can write that:

$$m_{ax} = T_{ab} = m_{bx}$$
$$m_{dx'} = T_{cd} = m_{cx}$$

in which the letter T is used to denote torsional moments.

The staircase shown in Fig. 5.10 contains therefore nine unknown bending moments and three deflections. For the purpose of the setting out of the required matrix, the structure is assumed to deflect to its final configuration and releases are then introduced at A, B, C, and D. The following compatibility and equilibrium conditions can be written which must be satisfied simultaneously (no detailed explanation is required as the setting out follows the usual pattern).

$$A_{XZ}; \quad m_{ay}\cdot\frac{l_1}{3EI_{1z}.\cos \alpha_a} + m_{by}\cdot\frac{l_1}{6EI_{1z}.\cos \alpha_a} = \theta_{ab} + \frac{\Delta_{bz}}{l_1}$$

$$A_{XY}; \quad m_{az}\frac{l_1}{3EI_{1y}.\cos \alpha_a} + m_{bz}\cdot\frac{l_1}{6EI_{1y}.\cos \alpha_a} = \frac{\Delta_{by}}{l_1}$$

$$B_{XZ}; \quad m_{ay}\frac{l_1}{6EI_{1z}.\cos \alpha_a} + m_{by}\cdot\frac{l_1}{3EI_{1z}.\cos \alpha_a} + T_{bc}.\phi_{bc}.l_2 + \left[m_{cy}\cdot\frac{l_3}{3EI_{3z}.\cos \alpha_d} - m_{dy}\cdot\frac{l_3}{6EI_{3z}.\cos \alpha_d}\right] = \theta_{ba} - \frac{\Delta_{bz}}{l_1}$$

$$B_{XY}; \quad m_{az}\frac{l_1}{6EI_{1y}} + m_{bz}.\left(\frac{l_1}{3EI_{1y}} + \frac{l_2}{3EI_{2y}}\right) - m_{cz}\cdot\frac{l_2}{6EI_{2y}} = \frac{\Delta_{bx} + \Delta_{cx}}{l_2} - \frac{\Delta_{by}}{l_1}$$

$$B_{YZ}; \quad m_{bx}\cdot\frac{l_2}{3EI_{2z}} - m_{cx}\frac{l_2}{6EI_{2z}} + T_{ab}.\phi_{ab}.l_1 = \frac{\Delta_{bz} - \Delta_{cz}}{l_2} - \theta_{bc}$$

$$\sum V_{XZ}; \quad \frac{m_{ay} - m_{by}}{l_1} + \frac{m_{cy} + m_{dy}}{l_3} = V_b^0 + V_c^0$$

$$\sum M_{XZ};\frac{m_{bz}+m_{cz}}{l_2}\cdot(h_1+h_2)+m_{ay}+m_{dy} = V_b^o.l_1+V_c^0.l_3'$$

$$C_{XY};\quad -m_{dz}\frac{l_3}{6EI_{3y}}+m_{cz}\left(\frac{l_3}{3EI_{3y}}+\frac{l_2}{3EI_{2y}}\right)-m_{bz}\cdot\frac{l_2}{6EI_{2y}} = \frac{\Delta_{bx}+\Delta_{cx}}{l_2}-\frac{\Delta_{cy}}{l_3}$$

$$C_{YZ};\quad m_{cx}\frac{l_3}{3EI_{3z}}-m_{bx}\frac{l_3}{6EI_{3z}}+T_{cd}.\phi_{cd}.l_3 = \frac{\Delta_{bz}-\Delta_{cz}}{l_2}+\theta_{cb}$$

$$\sum V_{XY};\frac{m_{az}-m_{bz}}{l_1}-\frac{m_{cz}+m_{dz}}{l_3} = 0$$

$$D_{XZ};\quad m_{ay}\frac{l_3}{3EI_{3z}.\cos\alpha_d}-m_{cy}\cdot\frac{l_3}{6EI_{3z}.\cos\alpha_d} = \theta_{dc}+\frac{\Delta_{cz}}{l_3}$$

$$D_{XY};\quad -m_{dz}\cdot\frac{l_3}{3EI_{3y}.\cos\alpha_d}+m_{cz}\cdot\frac{l_3}{6EI_{3y}.\cos\alpha_d} = \frac{\Delta_{cy}}{l_3}$$

These equations can now be rearranged and presented in a matrix form.

Multiplying compatibility equations by $6EI_{1\cdot z}/l_1$ and equilibrium equations by l_1, a matrix can be set out in terms of non-dimensional coefficients. The load vector contains of course all loads as well as the actual values of E, I_{1z}, and l_1. All bending moments obtained from this matrix must be multiplied by the applied loads and l_1 and the deflections by loads and the parameter $l_1^3/6EI_{1\cdot z}$. The solution therefore is in terms of the non-dimensional coefficients.

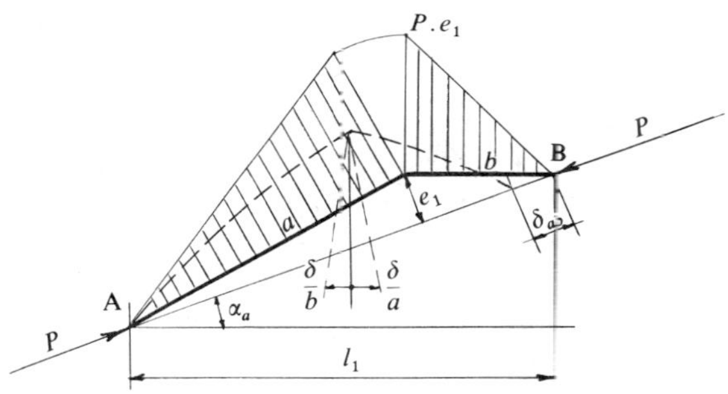

Fig. 5.12

A slight improvement in the analysis can be made by including the effect of axial deformations of beams and the cranks in the spans AB and CD. When a force P (see Fig. 5.12) is applied in the direction AB, then the deflection of the cranked beam can be calculated as follows. For the beam AB in bending only:

$$P.e_1\left(\frac{a}{3EI}+\frac{b}{3EI}\right)=\frac{\delta}{a}+\frac{\delta}{b}$$

or

$$\delta = P.e_1.ab/3EI$$

and including axial compression we can write that:

$$\delta_{ab} = P.e_1\frac{ab}{3EI_{1z}}+\frac{P.l_1}{EA_1} \tag{5.1}$$

Now, the value of the force P is expressed in terms of the statically indeterminate bending moments:

$$P = \frac{m_{bz} + m_{cz}}{l_2}$$

Similarly, for the beam CD (in bending and in tension) (Fig. 5.13):

$$\delta_{cd} = P.e_2 \cdot \frac{c.d}{3EI_{3z}} + \frac{P.l_3}{EA_3} \tag{5.2}$$

The values (5.1) and (5.2) can now be introduced into the main matrix and may increase the accuracy of the solution. The moments $P.e_1$ and $P.e_2$ must be added to the final diagram of bending moments.

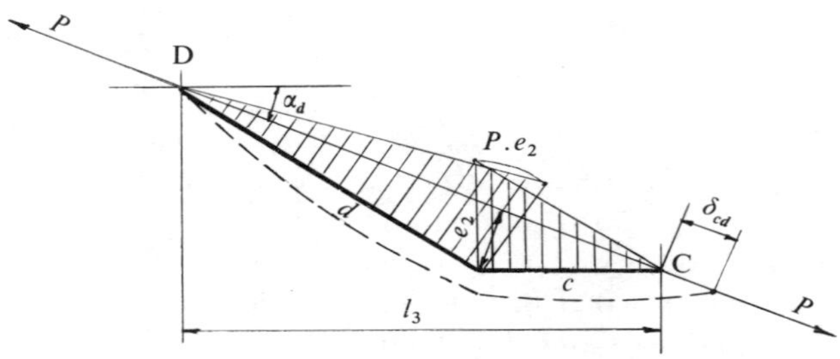

Fig. 5.13

5.8 Deflection of a Stepped Balcony Slab

Balcony slabs are often stepped down in relation to the main floor slabs as a precaution against the ingress of rainwater (Fig. 5.14). Such slabs can deflect considerably more than when the floor slab is at the same level as the balcony slab and particularly when the connecting beam is weak and flexible. However, the deflection of such slabs is rarely calculated accurately because of the large amount of work required relative to

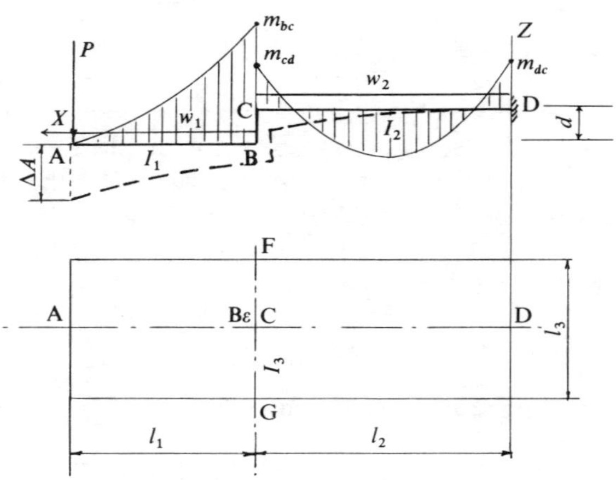

Fig. 5.14 Stepped balcony slab

the minor magnitude of the problem. In the following, an explicit expression is developed for this deflection which may reduce the amount of effort required.

Generally, the total deflection ΔA of the free end of the cantilever (Fig. 5.14) is a function of six separate effects, i.e.:

(a) the deflection of the slab AB acting as a cantilever encastre at B,
(b) the effect of the stepped connection between the slabs,
(c) the effect of rotation of the end C of the span CD,
(d) the effect of loads on the span CD and end conditions at D,
(e) the effect of flexural deflection of beam GF, and
(f) the effect of torsional stiffness of the beam GF (provided that this is restrained against torsional displacements at G and F).

These are calculated as follows:

(a) The effect of loads on the span AB considered as fixed at B:

for
$$m_{ba} = \tfrac{1}{2}w_1 l_1^2 + P.l_1$$

the deflection at A is calculated from:

$$\frac{\delta_a}{l_1} = m_{ba}\cdot\frac{l_1}{3EI_1}$$

or
$$\delta_a = \frac{w_1.l_1^4}{6EI_1} + \frac{P.l_1^3}{3EI_1} \qquad (5.3)$$

(b) The effect of step between the slabs. The flexural rotation of beam GF with respect to X axis is:

$$\theta_3 = m_{ba}.d/2EI_{3x}$$

and
$$\delta_a = \theta_3.l_1 \qquad (5.4)$$

(c) The effect of rotation of end C of the span CD (assuming that the beam GF is torsionally restrained):

$$m_{cd} = m_{ba}\cdot\frac{f_{cd}}{f_{cd} + \phi_{gf}.l_3/2} \quad \text{and} \quad m_{dc} = \tfrac{1}{2}m_{cd}$$

and
$$\theta_{cd} = m_{bc}\frac{l_2}{3EI_2} - \tfrac{1}{2}m_{bc}\cdot\frac{l_2}{6EI_2} = m_{bc}\cdot\frac{l_2}{4EI_2}$$

so that
$$\delta_a = \theta_{cd}.l_1 \qquad (5.5)$$

(d) The effect of loads on the span CD (assumed here as uniformly distributed):

$$\theta_{cd} = -\frac{w_2.l_2^3}{24EI_2} + m_{dc}\cdot\frac{l_2}{6EI_2}$$

in which
$$m_{dc} = \tfrac{1}{8}w_2.l_2^2$$

so that
$$\delta_a = \theta_{cd}.l_1 \qquad (5.6)$$

(e) The effect of flexural deflection of beam GF with respect to Z axis. The load carried by beam GF is:

$$w_3 = P + w_1.l_1 + \tfrac{1}{2}w_2.l_2 + (m_{cd} - m_{dc})/l_2$$

and, for simply supported end conditions at G and F in ZX plane:

$$m_{gf} = \tfrac{1}{8}w_3 . l_3^2$$

so that

$$\delta_a = \frac{5m_{gf} . l_3^2}{48EI_{3y}} \tag{5.7}$$

(f) Finally, the effect of torsional stiffness of beam GF (assuming that the ends G and F are restrained against torsional displacements in the ZY plane):

$$\delta_a = -\left(m_{ba} \cdot \frac{\phi_{gf}}{f_{bc} + \phi_{gf}}\right) . \phi_{af} . l_3 \tag{5.8}$$

in which ϕ_{gf} is as shown on page 106, eq. (3.3). The final deflection is the sum of the expressions (5.3) to (5.7).

References

1. W. Fuchsteiner, 'Die Freitragende Wendeltreppe', *Beton und Stahlbetonbau* (Berlin), **49**, 258–59 (1954).
2. W. Fuchsteiner, 'Treppen', *Betonkalender 1959*. Vol. II, pp. 345–59. Verlag Whm. Ernst & Son, Berlin.
3. A. Siev, 'Analysis of Free-standing Multiflight Staircase', *Proc. ASCE*, **88**, Part I, 207–232 (1962).
4. K. Hirschfeld, *Die Baustatik*, pp. 171–220. Springer Verlag, Berlin (1939).
5. 'Baustatik', *Betonkalender 1960*, Vol. I, pp. 311–325, Verlag Whm. Ernst & Son, Berlin.
6. L. H. Mitchell and F. S. Shaw, 'Columnless Stairs', *Architectural Science Review*, 80–88 (July, 1962).

6 Analysis of Thermal Stresses and Deformations

6.1 Introduction

In modern architecture, exposed exterior columns, walls, and beams are often used, and are subjected to the effect of seasonal and daily variation of temperature. While the temperature inside the building stays relatively constant, outside members are often subjected to the effect of considerable temperature differential and tend to change their dimensions relative to the interior members. If these dimensional changes could take place completely freely, no stresses would develop. If, however, restraints in various forms are present, high stresses and deformations may develop, and even cracks may appear. The presence of high thermal stresses may reduce the capacity of members to resist the primary effects of gravity and wind forces. In addition, thermal movements can adversely affect partitions, finishes, services, floor alignments, and even stability.

Theoretically, thermal stresses would not occur in two extreme cases: if solids did not conduct heat at all, and if the heat conduction were infinite; matter would then have constant temperature. However, since all solids have heat conduction of finite values, dimensional changes take place and, in the presence of restraints, stresses develop. The magnitude of these stresses depends on the thermal properties of materials, on the magnitude and form of the temperature gradient, and on the mass and size of members.

Creep of concrete and cracks tend to alleviate thermal stresses, and, for steady state, temperature stresses tend theoretically to zero over a long period of time. However, few structures are exposed to steady state of temperature over a sufficiently long period for this effect to be appreciable. The most effective measure in reducing thermal stresses is to avoid a high degree of continuity and restraints, to reduce the degree of exposure of members, and to apply insulating materials.

The exact mathematical formulation of thermal stresses and deformations is, in the majority of cases, either entirely intractable or too complicated for practical application. This is because of the complex thermal properties of materials, variability of the ambient air temperature, effect of wind and moisture, dissipation of heat through the sides of members, interrelation between temperature, creep, and shrinkage, etc.

The main sources of temperature changes are ambient changes between day and night, and annual changes. Other sources of temperature changes are the heat of hydration of cement and various heating installations, particularly underfloor heating. The latter can cause serious deflections of floors as well as cracks. Generally, many cracks attributed to shrinkage may in fact be due to thermal stresses.

Because of these difficulties, mathematical formulations of thermal effects are based on approximate assumptions. These, however, appear to reflect the actual response of most structures adequately for practical purposes. The main assumptions on which the analyses of thermal stresses and deformations are made, are:

(a) that the temperature variation in the material of the member can be determined independently of the magnitude of strains and stresses;
(b) that the deformations are small and the analysis can be based on the original configuration of a structure;
(c) that the principle of superposition is valid and the material obeys Hooke's Law; and
(d) that the thermal stresses can be evaluated independently of the effect of creep and shrinkage.

The first assumption implies the omission of thermoelastic coupling terms in the heat conduction equations. The latter define the dependence of heat flow and thermal stresses and strains (e.g., at high temperatures, heat conduction changes because of altered strains).

The second assumption implies that no distinction is needed between the coordinates of a particle before and after deformation, and that the displacement gradients are small. The most common problems in which this assumption does not hold are those of buckling.

The third assumption implies that neither the temperature variations nor stresses are too large. This assumption may not always be true, but it holds in the majority of cases in normal conditions.

The last assumption is not strictly true, as there is a relationship between creep and temperature which is not yet fully resolved, and separate consideration of these two effects represents, therefore, an approximation.

The expansion or contraction is assumed to be proportional to temperature changes. This proportionality is expressed by the coefficient of linear thermal expansion, α, which is defined as the change in length which a unit length of a member undergoes when its temperature is changed by one degree. For concrete:

$$\alpha = 0{\cdot}000\,012 \text{ per } 1^\circ \text{ C},$$

and for steel:

$$\alpha = 0{\cdot}000\,010 \text{ per } 1^\circ \text{ C}.$$

It is also assumed that the Bernoulli–Euler assumption holds, i.e., that plane sections before bending remain plane after loading. The thermal Poisson's ratio is usually taken as equal to zero.

Model studies appear to confirm that the elastic methods can be used to predict thermal response of concrete.

In the following analyses of thermal effects on structures, three main problems will be considered:

(a) the first concerns the thermal properties of materials, and particularly concrete,
(b) the second concerns the response of statically determinate members to temperature variation, and
(c) the third concerns stresses and deformations in indeterminate structures, particularly beams and frames.

6.2 Heat Transfer and its Effect on Concrete Structures

A knowledge of the laws of heat transfer is required in the analysis of thermal stresses and deformations, and also in problems of heat insulation. These laws will be briefly reviewed but, for more comprehensive treatment of heat transfer, reference should be made to standard textbooks.

There are three different types of heat transfer: by conduction, by convection, and by radiation. However, in structural engineering, heat conduction is the most significant and only this form of heat transfer will be considered here.

When heat is applied to a member, it is partly absorbed followed by a flow of heat and a rise of temperature. For heat conduction to occur, there must be a temperature difference, and then the heat always flows in the direction of decreasing temperature. The rate of heat conduction is a property of a particular material and depends on the level of temperature and moisture content.

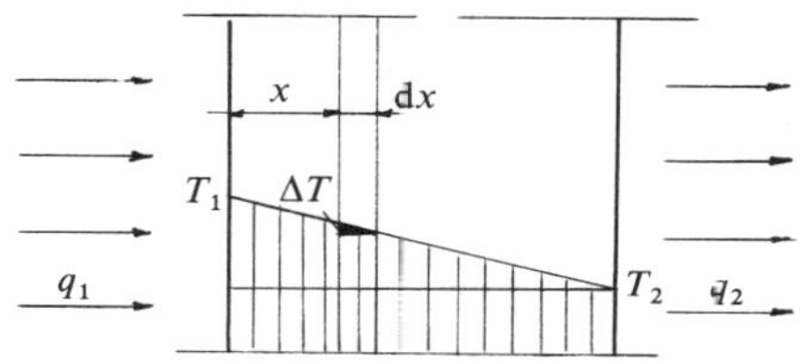

Fig. 6.1 Heat conduction in steady state

The term, 'a steady heat conduction', is used when the amount of heat absorbed is equal to the amount of heat dissipated, or, in other words, when the flow of heat is independent of time, and can occur with linear or curvilinear temperature gradient. If the amount of heat absorbed is not equal to the amount of heat dissipated, such flow is time-dependent and this state is called 'unsteady'. The transient state is a special form of unsteady state and describes a definite change between two limits.

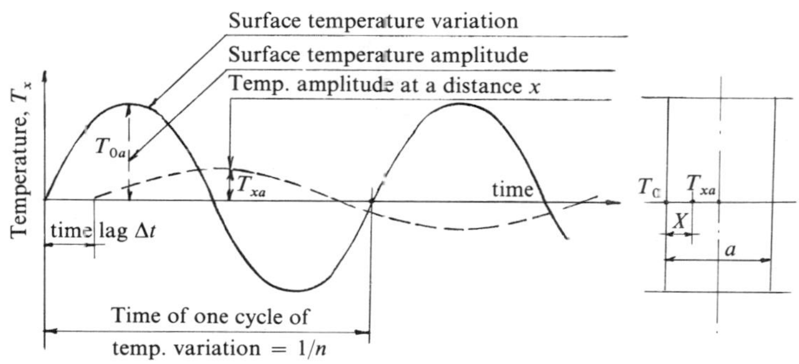

Fig. 6.2 Temperature variation inside a wall for sinusoidal variation of surface temperature

In buildings, the surface temperature does not stay constant for sufficiently long time to develop a steady state of heat transfer, but changes periodically with the daily and annual variation of temperature (Fig. 6.2). Observations of the response of concrete members to the outside temperature changes show two important phenomena:

(a) a time lag Δt, and
(b) attenuation (or damping) of intensity of temperature amplitude with the distance from the heated face of concrete (T_{xa}/T_{0a}), (Fig. 6.2).

The time lag is due to the thermal inertia of concrete in responding to the temperature change, while the attenuation indicates that the temperature in concrete does not rapidly reach the value of the temperature applied. Both phenomena have, therefore, a direct bearing on stresses and deformations in structures.

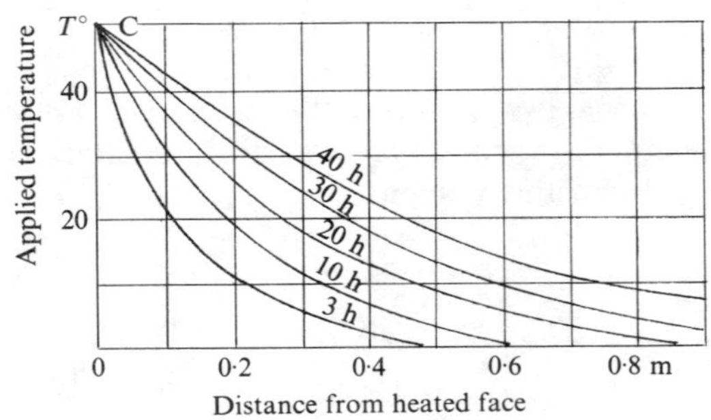

Fig. 6.3 Penetration of suddenly applied temperature at various time intervals for normal density concrete

Figure 6.3 shows average values of time lag in the arrival of surface temperature variation at various depths of concrete members for cycles of 1, 3, and 7 days. It can be seen from this diagram that rapid temperature changes penetrate very little into concrete, and only the amplitudes of slowly changing temperatures penetrate deep into the mass of concrete. The lower the thermal conductivity of concrete, the greater is the time lag and the attenuation of the temperature variation.

Figure 6.4 shows average values of attenuation of surface temperature with depth in a normal concrete in the range of 1 to 365 day cycles.

Thermal stresses and deformations in structures vary continuously with time and with oscillating ambient temperatures. Peak stresses may be of opposite sign, and it is difficult to assess when zero stress will occur at a particular point in a complex

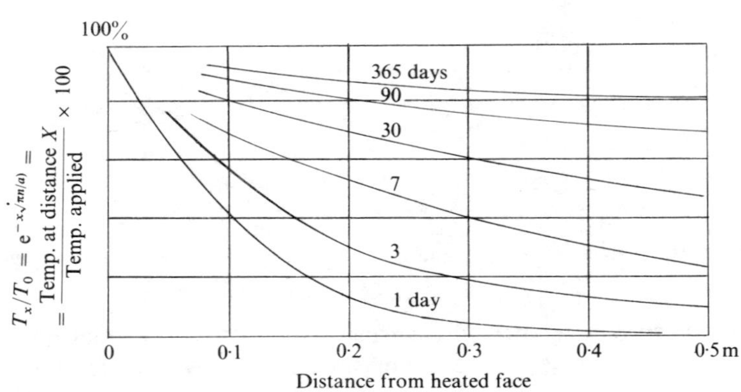

Fig. 6.4 The attenuation of surface temperature amplitudes at various depths of concrete mass

structure. Furthermore, zero stress at one point need not coincide with zero stress at other points.

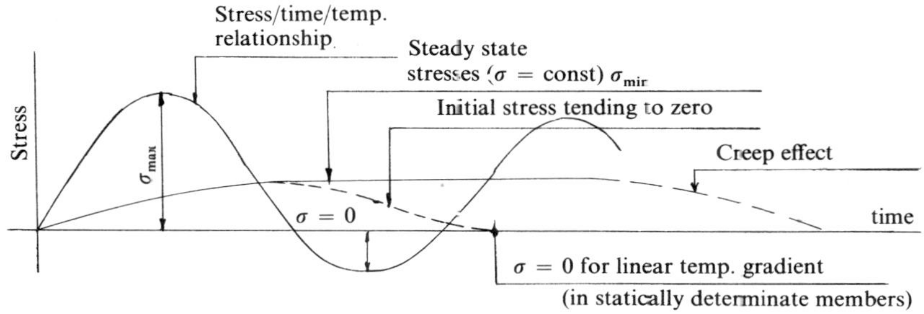

Fig. 6.5 Diagrammatic relationship between stresses and time of exposure of concrete members to temperature variation

Similarly, maximum deformations need not coincide with the maximum stresses and even at zero stress thermal deformations may be large or even at their maximum. It will be shown later that thermal stresses may not be even directly related to the lengths of spans, and deformations may not directly depend on stiffnesses of members. However, thermal stresses and deformations are additive to those from gravity and wind forces, and cracks may appear. Not infrequently such cracks open and close periodically.

Most analyses of temperature stresses are based on the steady state of heat conduction independent of time. However, both experiments and observations show that maximum stresses occur in the unsteady state and that there is a certain time when critical stress occurs (Fig. 6.5) although in complex structures, it is generally very difficult to determine this critical stress and its corresponding time. At one time, stresses may exceed the ultimate value, particularly in tension zone, and cracks may appear. If this maximum stress is below the value of the tensile strengths of concrete, the member may not crack and later, with reduced stresses, may have a better chance of resisting cracks. This is indicated diagrammatically in Fig. 6.5.

6.3 Thermal Stresses in Statically Determinate Members

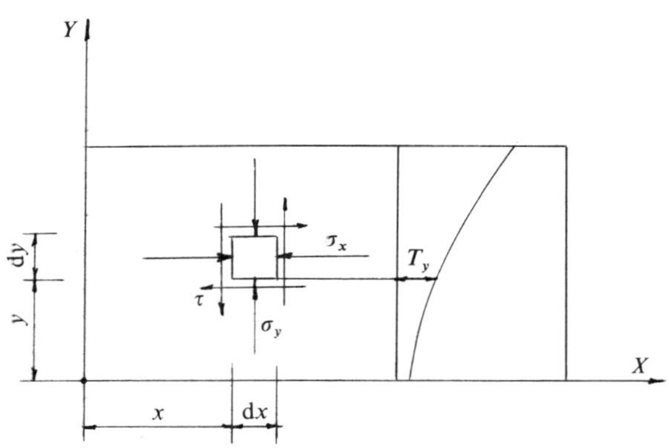

Fig. 6.6 Thermal stresses in plates

General, three-dimensional stress–strain relationships are discussed in a number of textbooks, such as by B. A. Boley and J. H. Weiner, H. S. Carslaw and J. C. Jaeger, and others. However, for two-dimensional and for linear structures, these relationships can be simplified to form a useful basis for the analysis of statically indeterminate structures, provided that the distribution of temperature within the member is known.

Based on the theory of elasticity, the stress–strain relations and the strain displacement relations in two dimensions are:

$$\varepsilon_x = \frac{1}{E}(\sigma_x - \nu.\sigma_y) + \alpha.T = \partial u/\partial x$$

$$\varepsilon_y = \frac{1}{E}(\sigma_y - \nu.\sigma_x) + \alpha.T = \partial v/\partial y$$

$$\gamma_{xy} = \frac{1}{G}\cdot\tau = \partial u/\partial y + \partial v/\partial x$$

in which $G = E/2(1 + \nu)$

The compatibility equation in terms of strains is:

$$\frac{\partial^2 \varepsilon_x}{\partial y^2} + \frac{\partial^2 \varepsilon_y}{\partial x^2} = \frac{\partial^2 \tau}{\partial x.\partial y}$$

and the equilibrium equations are:

$$\frac{\partial \sigma_x}{\partial x} + \frac{\partial \tau}{\partial y} + X = 0$$

$$\frac{\partial \sigma_y}{\partial y} + \frac{\partial \tau}{\partial x} + Y = 0$$

Using the equilibrium equations, the compatibility equation in terms of stresses becomes:

$$\left(\frac{\partial^2}{\partial x^2} + \frac{\partial^2}{\partial y^2}\right)(\sigma_x + \sigma_y + E\alpha T) = -(1 + \nu)\left(\frac{\partial X}{\partial x} + \frac{\partial Y}{\partial y}\right) \qquad (6.1a)$$

If the temperature variation is the function of one co-ordinate only (Fig. 6.7), the

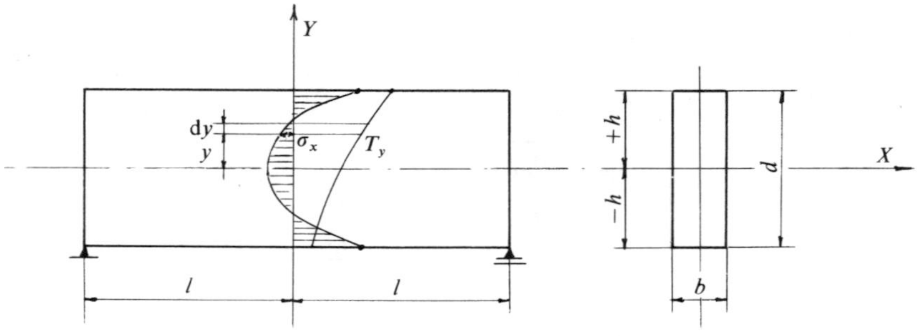

Fig. 6.7 Thermal stresses in beams

compatibility equation (6.1a) reduces to:

$$\frac{d^2}{dz^2}(\sigma_x + E\alpha T_y) = 0 \qquad (6.1b)$$

from which $\sigma_x = -E\alpha T + C_1 + C_2.y$

The stress is zero only if the function T_y is linear in y. However, if the temperature distribution is non-linear, different elements tend to expand by different amounts, and various elements exert a restraining action upon each other, resulting in continuous displacements. The system of strains is then accompanied by a corresponding system of self-equilibrating stresses.

The constants of integration C_1 and C_2 can be found from conditions of zero resultant forces and moments, i.e.,

$$\int_{-h}^{+h} \sigma_x.dy = 0 \quad \text{and} \quad \int_{-h}^{+h} \sigma_x.y.dy = 0$$

and eq. (6.1b) becomes:

$$\sigma_x = E.\alpha\left[-T_y + \frac{1}{2h}\int_{-h}^{+h} T.dy + \frac{3y}{2h^3}\int_{-h}^{+h} T.y.dy\right] \qquad (6.2)$$

or in the form:

$$\sigma_x = -E.\alpha.T + \frac{b}{A}\cdot N_T + b.y.M_T \qquad (6.3)$$

in which $N_T = \alpha E \int_{-h}^{+h} T.dy$ and $M_T = \alpha E \int_{-h}^{+h} T.y.dy$

The values of N_T and M_T have the dimensions of force and moment respectively.

The condition of zero stress at the ends of the beam cannot be satisfied because of insufficient number of constants of integration in eq. (6.1a). However, according to St Venant's principle, any local disturbance has little effect on sections away from this point.

The displacement components in the direction of X and Y respectively are:

$$U = \frac{x}{E}\cdot\left[\frac{b}{A}\cdot N_T + \frac{y.b}{I}\cdot M_T\right]$$

$$\delta_x = -\frac{bx^2}{2EI}\cdot M_T - \frac{\nu}{E}\left[\frac{b.y}{A}\cdot N_T - \frac{b.y^2}{2I}\cdot M_T\right] + \alpha\left(\frac{1-\nu}{E}\right)\int_0^y T.dy \qquad (6.4)$$

in which $A = 2bh$, $I = \frac{2}{3}bh^3$.

Equations (6.3) and (6.4) can be used to calculate stresses and deformations in statically determinate beams. However, before proceeding further, it will be found instructive to develop these relationships from first principles. This will indicate more clearly the actual meaning of terms and their limits of applicability.

6.3.1 *Symmetrical temperature distribution*

When a thin beam or a slab is subjected to a symmetrical temperature differential T_y (Fig. 6.8) which is an even function of y, the strain is $\varepsilon_y = -\alpha.T_y$. This strain can be entirely suppressed by applying at the ends of the beam a stress:

$$\sigma'_x = -\alpha . E . T_y \tag{6.5}$$

If the beam (or plate) is free to expand laterally, this stress will not produce any stress in the lateral direction and the resulting suppressing force at the ends of the beam will be:

$$F = b \int_{-h}^{+h} \alpha E . T . \mathrm{d}y$$

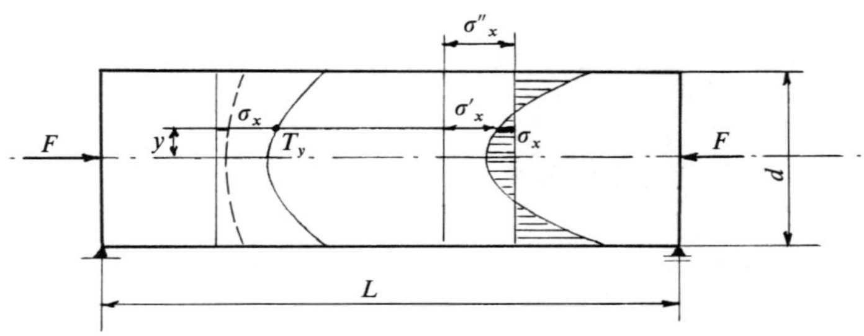

Fig. 6.8 Thermal stresses due to symmetrical temperature distribution

If this force is now 'released', at a sufficient distance from the ends, it will, according to St Venant, produce a uniform stress:

$$\sigma''_x = \frac{1}{2h} \int_{-h}^{+h} \alpha . E . T . \mathrm{d}y \tag{6.6}$$

so that the resultant thermal stress becomes:

$$\sigma_x = \sigma'_x + \sigma''_x = -\alpha E . T_y + \frac{1}{2h} \int_{-h}^{+h} \alpha E T_y . \mathrm{d}y \tag{6.7}$$

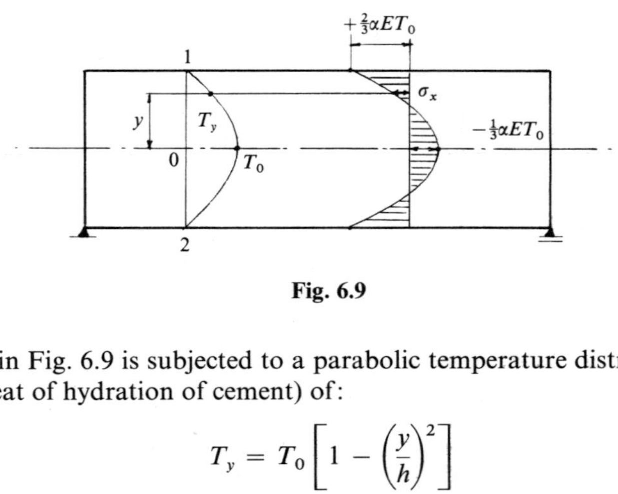

Fig. 6.9

If a beam in Fig. 6.9 is subjected to a parabolic temperature distribution (such as due to the heat of hydration of cement) of:

$$T_y = T_0 \left[1 - \left(\frac{y}{h} \right)^2 \right]$$

we obtain from eq. (6.7):

$$\sigma_x = \alpha E T_0 \left[\left(\frac{y}{h}\right)^2 - \frac{1}{3} \right]$$

from which: for $y = 0$, $\sigma_0 = -\tfrac{1}{3}\alpha . E . T_0$

and for $y = \pm h$, $\sigma_{1,2} = +\tfrac{2}{3}\alpha E T_0$

6.3.2 *Non-symmetrical temperature distribution*

If a simply-supported beam, such as in Fig. 6.10, is subjected to a non-symmetrical temperature distribution, we again suppress the free strain ε_x at the ends of the beam, and this will give rise to a resultant force F and moment M of the values:

$$F = -\int_{-h}^{+h} \alpha E T . \mathrm{d}y \quad \text{and} \quad M = -\int_{-h}^{+h} \alpha E T . y . \mathrm{d}y$$

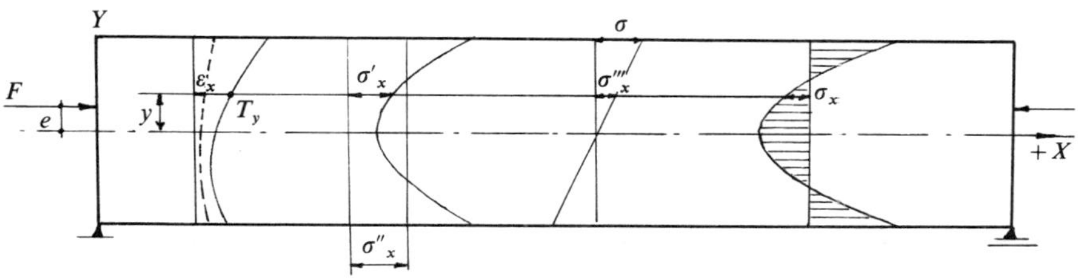

Fig. 6.10 Thermal stresses due to non-symmetrical temperature distribution

To satisfy the conditions of equilibrium, we must superimpose on compression σ'_x a uniform tension σ''_x as expressed by eqs. (6.5) and (6.6) in the previous case, and also stresses σ'''_x due to bending moment. The latter can be determined from the condition that the total moment of the forces must be equal to zero. The stress due to this bending moment is:

$$\sigma'''_x = \sigma_0 . y/h$$

and for equilibrium of moments we can write:

$$\int_{-h}^{+h} \left(\sigma_0 \frac{y}{h}\right) y . \mathrm{d}y - \int_{-h}^{+h} (\alpha E T) . y . \mathrm{d}y = 0$$

from which

$$\sigma_0 = \frac{3}{2h^2} \int_{-h}^{+h} \alpha E T . y . \mathrm{d}y$$

The total stress is therefore:

$$\sigma_x = \sigma'_x + \sigma''_x + \sigma'''_x$$

or

$$\sigma_x = -\alpha E T + \frac{1}{2h} \int_{-h}^{+h} \alpha E T . \mathrm{d}y + \frac{3y}{2h^3} \int_{-h}^{+h} \alpha E T . y . \mathrm{d}y \tag{6.8}$$

which is identical to eq. (6.2).

Equation (6.2) should therefore be used for unsymmetrical temperature while eq. (6.7) should be used for symmetrical temperature distribution within a section.

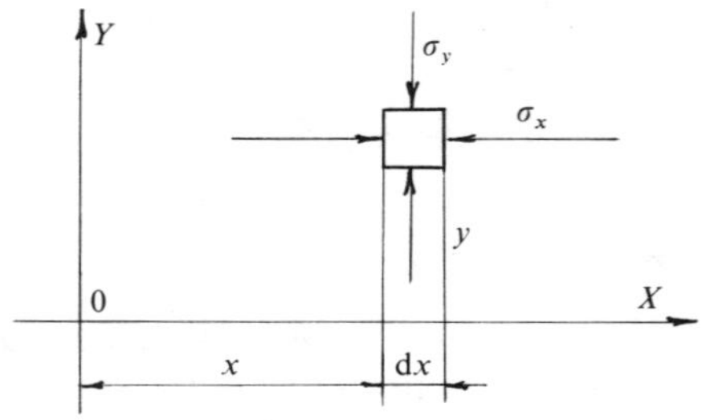

Fig. 6.11

In a similar way, we can obtain thermal stress equation for plates in the form (Fig. 6.11):

$$\sigma_x = \sigma_y = -\frac{\alpha ET}{1-\nu} + \frac{1}{2h(1-\nu)}\int_{-h}^{+h} \alpha ET + \frac{3y}{2h^3(1-\nu)}\int_{-h}^{+h} \alpha ET.y.\mathrm{d}y \tag{6.9}$$

Example 6.1

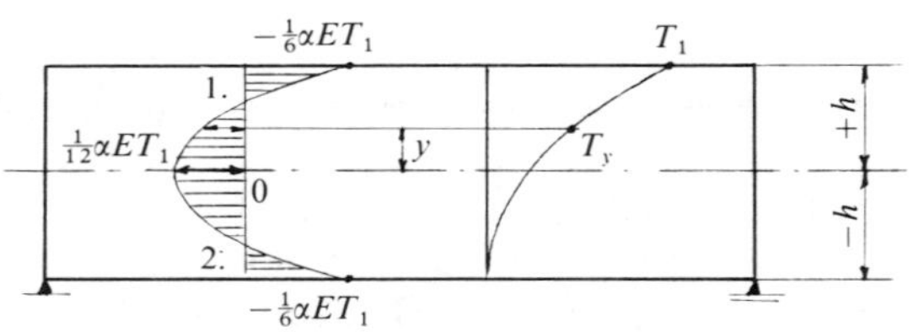

Fig. 6.12

For a beam in Fig. 6.12, subjected to a parabolic temperature distribution in accordance with the equation:

$$T_y = \frac{T_1}{4}\left[\left(\frac{y}{h}\right)^2 + 2\left(\frac{y}{h}\right) + 1\right]$$

we obtain from eq. (6.2):

$$\sigma_x = \tfrac{1}{4}\alpha ET_1\left[\frac{1}{3} - \left(\frac{y}{h}\right)^2\right]$$

from which: for $y = \pm h$, $\quad \sigma_{1\cdot 2} = -\frac{1}{6}\alpha.E.T_1$

and for $y = 0$, $\quad \sigma_0 = +\frac{1}{12}\alpha.E.T_1$

Example 6.2

If a beam is subjected to a linear temperature distribution, as shown in Fig. 6.13, then:

$$T_y = T_1 \cdot \frac{h + y}{2h}$$

and, introducing this value in eq. (6.2), we obtain:

$$\sigma_x = 0$$

for all values of y, i.e., a simply-supported, homogeneous beam when subjected to a linear temperature distribution is free from thermal stresses, although, as will be shown later, it is subjected to bending deformations.

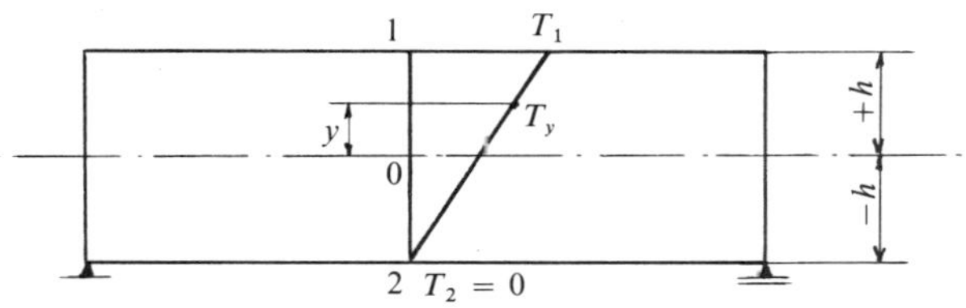

Fig. 6.13

6.3.3 *Effect of temperature change over a part of cross-section of a member*

It can be expected that in practice a temperature change may occur gradually over a part of the depth of member and the problem may then arise: what is the effect of such temperature distribution on stresses and deformations in the member or in the whole structure? Although the temperature distribution could be curvilinear, in the following analysis it is assumed that this is linear over a certain part of a cross-section as shown in Fig. 6.14. Equation (6.2) cannot be used because the temperature is not an even function of y. However, the analysis can be made as follows:

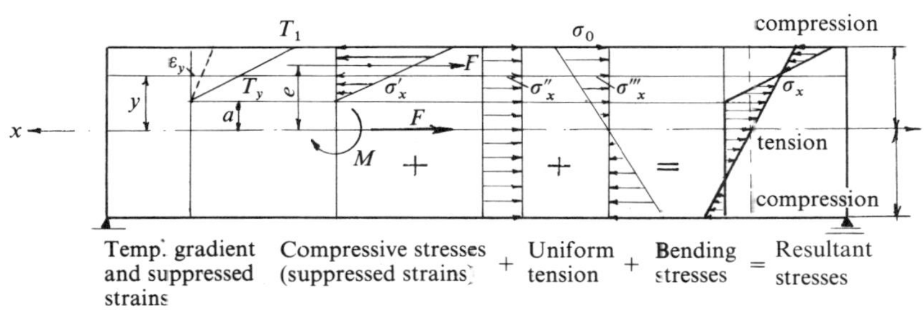

Fig. 6.14

If a simply-supported beam is subjected to a linear temperature distribution over part of the cross-section (Fig. 6.14), then, by completely suppressing strains, compressive stresses are induced of the value:

$$\sigma'_x = -\alpha . E . T_y$$

and the resultant force is:

$$F = \alpha E T_1 . \tfrac{1}{2}(h - a)b = \tfrac{1}{2}\alpha . E . T_1(h - a) . b$$

This force is now applied in the opposite direction, and will, therefore, produce an average tension of the value:

$$\sigma''_x = \frac{F}{2hb} = \frac{1}{4h}\cdot\alpha ET_1(h - a)$$

as well as bending stress σ_0 which can be calculated from the condition of equilibrium of moments as follows. The eccentricity of the force is:

$$e = a + \tfrac{2}{3}(h - a) = \tfrac{1}{3}(2h + a)$$

and for equilibrium:

$$M = F.e = \tfrac{1}{2}\sigma.bh \times \tfrac{2}{3} \times 2h$$

from which, by introducing the values of F and e, we obtain:

$$\sigma_0 = \alpha ET_1\cdot\frac{1}{4h^2}(h - a)(2h + a)$$

and

$$\sigma''' = \sigma_0.y/h$$

The final stress is therefore:

$$\sigma_x = \sigma'_x + \sigma''_x + \sigma'''_x$$

or

$$\sigma_x = -\alpha ET_1 + \alpha ET_1\cdot\frac{h - a}{h}\left[1 + \left(2 + \frac{a}{h}\right)\cdot\frac{y}{h}\right] \qquad (6.10)$$

The first term in this equation is valid only for the values of $+h>y>a$.
From eq. (6.10), we obtain:

(a) For $a = +h$, $\sigma_x = 0$ for all values of y
(b) for $a = 0$, $\sigma_x = -\alpha ET_y + \tfrac{1}{2}\alpha ET_1(\tfrac{1}{2} + y/h)$
(c) for $a = -h/2$, $\sigma_x = \tfrac{3}{2}\alpha ET_1(1 + \tfrac{3}{2}.y/h)$ and $\max \sigma_x = -\tfrac{15}{4}.\alpha ET_1$
(d) for $a = -h$, $\sigma_x = -\alpha ET_y + \alpha T_1(h + y)/2h = 0$ for all values of y.

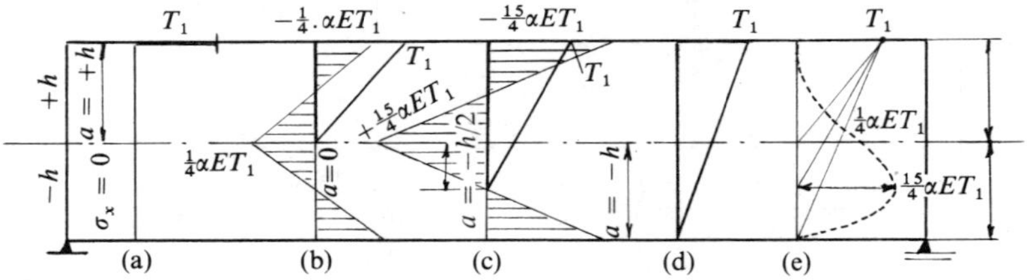

Fig. 6.15 Variation of temperature stresses vs. depth of temperature penetration *a*

Figure 6.15 shows the overall variation of stresses as a function of depth of temperature penetration a.

6.4 Analysis of Thermal Deformations in Beams

6.4.1 *Effect of linear temperature distribution*

When a simply-supported homogeneous beam is subjected to a linear temperature distribution, the axial deformation of an element dx as well as bending deformations of the external fibres (see Fig. 6.16) are equal to:

$$de = \alpha . T_1/2 . dx$$

and the angular change in an element is:

$$d\theta = de/h = \alpha\left(\frac{T_1}{2}\right) dx/h$$

from which

$$\theta = \int_0^{L/2} d\theta = \frac{\alpha . T_1 . L}{2d} \qquad (6.11)$$

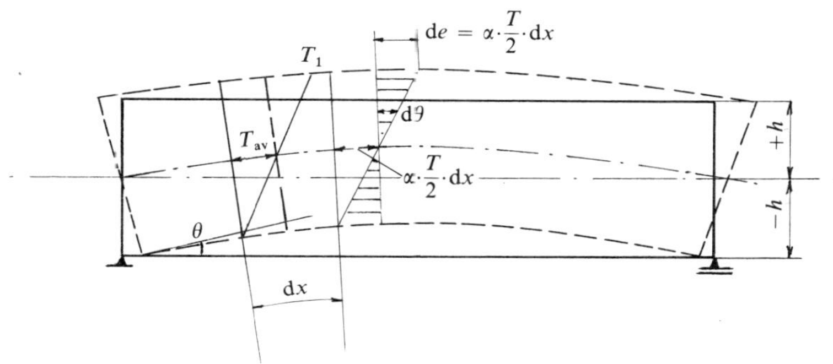

Fig. 6.16 Deformations in a beam due to a linear temperature distribution

Since the curvature of the beam is:

$$\frac{1}{R} = \frac{d\theta}{dx} = \frac{M}{EI} = \frac{\alpha . T_1}{d}$$

we obtain:

$$R = d/\alpha . T_1$$

and

$$M = \alpha . T_1 . EI/d \qquad (6.12)$$

The value of M represents a bending moment required to prevent the beam from bending, or an equivalent bending moment required to produce curvature similar to that due to change of temperature.

Equation (6.11) can be obtained alternatively by using the Unit Load Method from:

$$\theta = \int M . m . dx/EI$$

in which

$$M = \alpha T_1 EI/d \quad \text{and} \quad m_x = 1.(L - x)/L$$

Equation (6.12) can be alternatively obtained by the force–displacement method from:

$$M\left(\frac{L}{3EI} + \frac{L}{6EI}\right) = \theta = \frac{\alpha . T_1 . L}{2d}$$

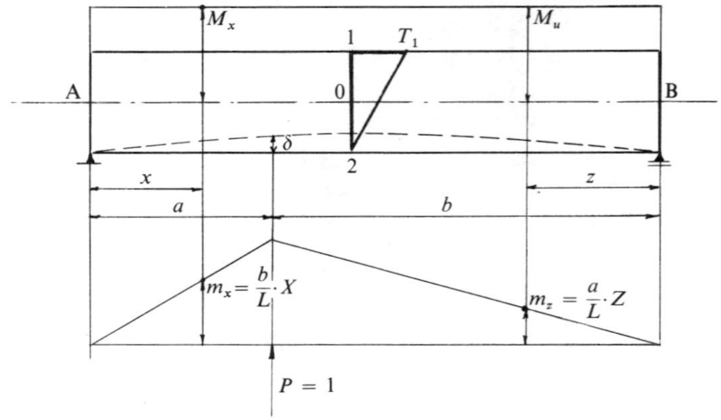

Fig. 6.17 Analysis of thermal deflection at any point in a beam

The deflection δ at any point on the beam can be calculated by the Unit Load Method using the values of:

$$M_x = \alpha T_1 . EI/d, \qquad m_x = \frac{b}{L}\, x, \quad \text{and} \quad m_z = \frac{a}{L} \cdot z$$

then

$$\delta = \int_A^B \frac{M.m.\mathrm{d}x}{EI} = \int_0^a \left(\frac{\alpha TEI}{d}\right)\left(\frac{b}{L}\, x\right) \mathrm{d}x + \int_0^b \left(\frac{\alpha TEI}{d}\right)\left(\frac{a}{L} \cdot z\right) . \mathrm{d}z$$

from which
$$\delta = \frac{\alpha . T_1 . a . b}{2d} \tag{6.13}$$

If $a = b = L/2$,

$$\max \delta = \frac{\alpha . T_1 . L^2}{8d} \tag{6.14}$$

or, by the force–displacement method:

$$M\left(\frac{a}{3EI} + \frac{a}{6EI} + \frac{b}{3EI} + \frac{b}{6EI}\right) = \frac{\delta}{a} + \frac{\delta}{b}$$

from which:
$$\delta = \alpha T_1 . a . b/2d$$

The linear deflection of a beam can also be calculated from:

$$\frac{1}{R} = \frac{\partial^2 y}{\partial x^2} = -\frac{b . M_t}{EI}$$

in which
$$M_t = \alpha E \int_{-h}^{+h} T . y . \mathrm{d}y$$

and
$$\max y = \alpha . T_1 . L^2/8d$$

which is identical with eq. (6.14).

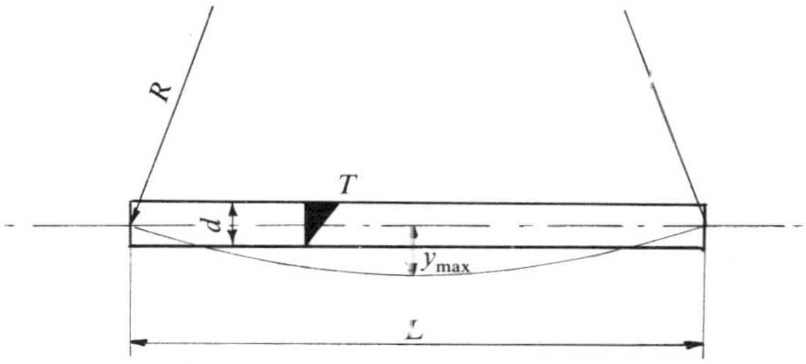

Fig. 6.18

Maximum deflection can also be calculated by the geometry of the deflected beam. For a constant bending moment, the curvature is constant (Fig. 6.18), and we can write:

$$(2R - y).y = (L/2)^2$$

Introducing $R = d/\alpha T$, we obtain (neglecting the value of y^2), the same expression as in eq. (6.14).

6.4.2 *Effect of curvilinear temperature distribution*

If the temperature distribution is curvilinear then, for the purpose of the analysis, this can be 'resolved' into a linear part (Fig. 6.19), and a curvilinear part with zero temperatures at the extreme fibres. The first part, being a linear temperature, does not

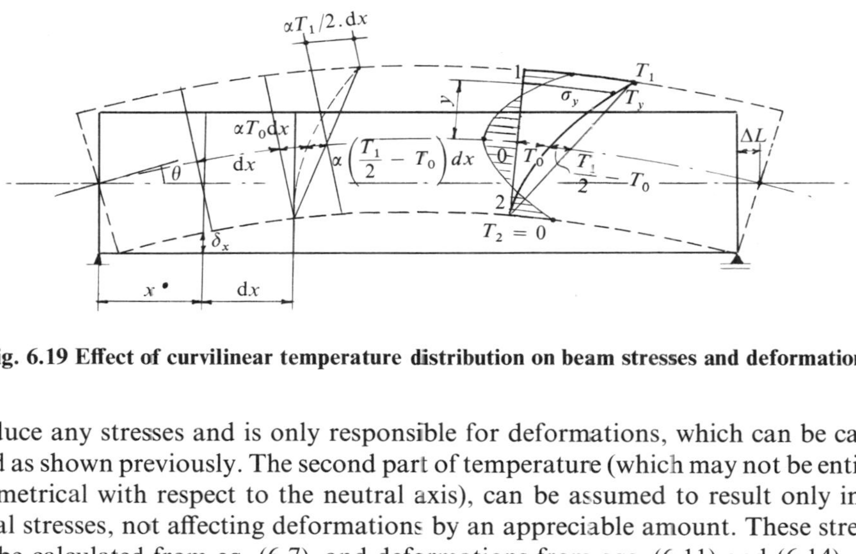

Fig. 6.19 Effect of curvilinear temperature distribution on beam stresses and deformations

produce any stresses and is only responsible for deformations, which can be calculated as shown previously. The second part of temperature (which may not be entirely symmetrical with respect to the neutral axis), can be assumed to result only in internal stresses, not affecting deformations by an appreciable amount. These stresses can be calculated from eq. (6.7), and deformations from eqs. (6.11) and (6.14), provided that the temperature distribution is an even function of y.

However, for a beam in which the temperature varies over a part of its cross-section as shown in Fig. 6.20, the angular deformation θ can be calculated from:

$$\theta = M.L/2EI \tag{6.15}$$

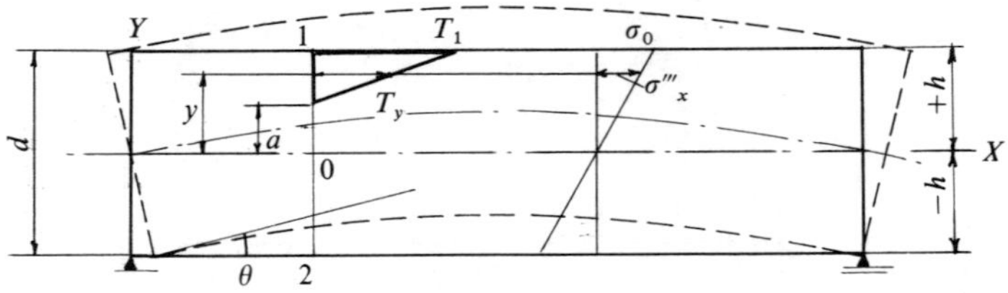

Fig. 6.20

in which
$$M = \int_{-h}^{+h} (\sigma . y/h) . y . \mathrm{d}y$$

or
$$M = \int_{-h}^{+h} \alpha . ET_1 \cdot \frac{1}{4h^2} \cdot (h - a)(2h + a) \frac{y^2}{h} \cdot \mathrm{d}y \tag{6.16}$$

Introducing eq. (6.15) into eq. (6.16), we obtain:

$$\theta = \frac{\alpha . T_1 . L}{d^3} [2h^2 - h.a - a^2] \tag{6.17}$$

Equation (6.17) relates the angular deformation, θ, to the depth of temperature penetration, a.

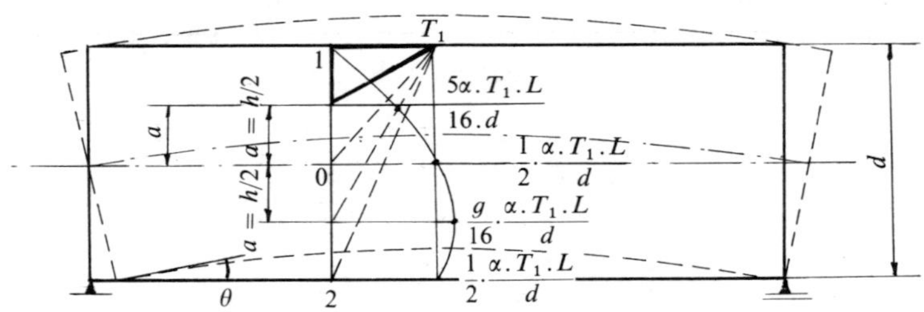

Fig. 6.21 The relationship between θ and a for a beam with linear temperature distribution

From eq. (6.17), taking $\mathrm{d}\theta/\mathrm{d}a = 0$, we obtain:

$$a = -\frac{h}{2} \quad \text{for which} \quad \theta \max = \frac{9}{16} \cdot \frac{\alpha T_1 L}{d}$$

For $\quad a = +h, \quad \theta = 0$

for $\quad a = +\frac{h}{2}, \quad \theta = \frac{5}{16} \cdot \frac{\alpha T_1 L}{d}$

for $\quad a = 0, \quad \theta = \frac{1}{2} \cdot \frac{\alpha T_1 L}{d}$

and for $\quad a = -h, \quad \theta = \frac{1}{2} \cdot \frac{\alpha T_1 L}{d}$

These values are indicated in Fig. 6.21.

Example 6.3

Calculate the temperature differential T_0 at which a beam carrying uniformly distributed load w will have no deflection at the centre of the beam, or a fixed-end beam will have zero moment at the centre.

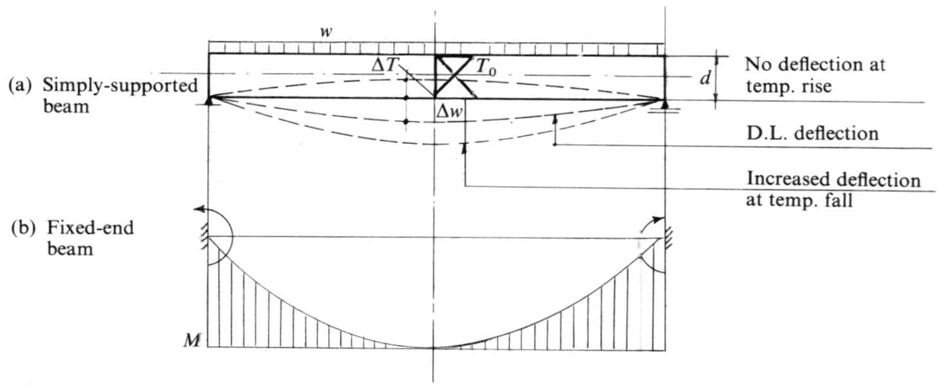

Fig. 6.22

Deflection due to temperature differential (see eq. (6.14)) is:

$$\Delta T = \frac{\alpha T_1 L^2}{8d}$$

Equating this value to the deflection due to a uniform load, we have:

$$\frac{\alpha . T_1 . L^2}{8d} = \frac{5w . L^4}{384EI}$$

from which:

$$T_0 = \frac{5}{48} \cdot \frac{wL^2}{\alpha EI . d}$$

The distribution of bending moments and stresses is not affected.

For a fixed-end beam, a temperature change may cause redistribution of bending moments, and even zero bending moment may take place and the beam will then work as a double cantilever when:

$$M_{\text{temp}} = M_{\text{udl}}$$

from which

$$T_1 = \frac{1}{8} \cdot \frac{w . L^2}{\alpha EId}$$

This is indicated diagrammatically in Fig. 6.22.

6.5 Cantilever Shear Wall subjected to a Linear Temperature Distribution

In the analysis of thermal stresses in tall building frames and shear walls, it can be useful to find first the deformations of a free standing solid cantilever, as shown in Fig. 6.23. When two opposite faces of such a cantilever are subjected to a linear tem-

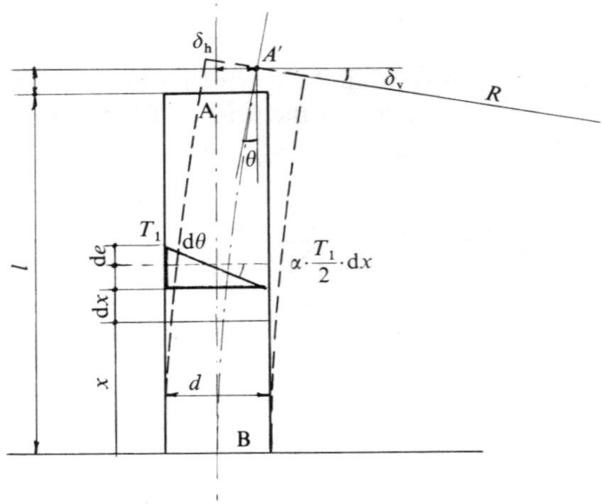

Fig. 6.23 Deformations of a cantilever due to a linear temperature distribution

perature distribution, throughout its width d, the dimensional changes θ, δ_v, and δ_h can be calculated as follows. The angular change dθ of an element dx is, from Fig. 6.23,

$$\mathrm{d}\theta = \frac{\mathrm{d}e}{d/2} = \frac{\alpha(T/2).\mathrm{d}x}{d/2} = \frac{\alpha.T_1.\mathrm{d}x}{d} \tag{6.18}$$

or the total angular change of a cantilever is:

$$\theta = \int_0^l \mathrm{d}\theta = \alpha.T_1.H/d \tag{6.19}$$

The curvature of the beam is expressed as:

$$\frac{\mathrm{d}\theta}{\mathrm{d}s} = \frac{1}{R} = \frac{M}{EI}$$

from which:

$$M = \alpha.T_1.EI/d \tag{6.20}$$

The same expression can be obtained using the force–displacement method from:

$$M.H/EI = \theta = \alpha.T_1.H/d$$

The value of M represents the bending moment required to restrain the cantilever from bending.

The radius of curvature is, from eq. (6.20):

$$R = d/\alpha.T_1 \tag{6.21}$$

The vertical translation of a free end can be calculated from:

$$\delta_v = \pm\alpha.T_{\mathrm{av}}.H \tag{6.22}$$

in which T_{av} is the temperature drop at the centre of the cantilever, which for a linear temperature distribution is $T_{\mathrm{av}} = \frac{1}{2}T_1$.

The horizontal translation δ_h can be calculated in a number of ways, e.g., by geometry:

$$\delta_h\,(2R - \delta_h) = (H + \delta_v)^2$$

from which, omitting a small value of $\delta_h{}^2$, and introducing eq. (6.21), we have

$$\delta_h = \alpha . T_1(H + \delta_v)^2/2d \tag{6.23}$$

Alternatively, using the expression:

$$\delta_h = \int_0^H x.\mathrm{d}\theta = \int_0^H x.\alpha T_1/d.\mathrm{d}x = \alpha T_1 . H^2/2d$$

or, in a convenient form:

$$(R - \delta_h)^2 = R^2 - s^2$$

from which

$$\delta_h = \tfrac{1}{2}s.\theta \tag{6.24}$$

in which s is the total length of the cantilever, $s = H + \delta_v$.

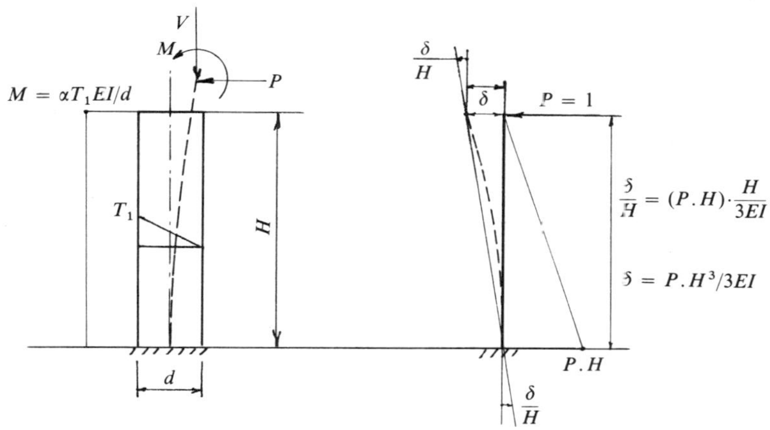

Fig. 6.24

To suppress all deformations of a cantilever subjected to a linear temperature variation, the following forces would have to be applied: V, P, and M as shown in Fig. 6.24.

The axial force, V, required to prevent deformation δ_v can be calculated from:

$$V.H/EA = \delta_v = \alpha\left(\frac{T_1}{2}\right).H$$

from which $V = \frac{1}{2}\alpha . T_1 . EA$.

The force H required to prevent horizontal deformation δ_h can be calculated from:

$$P.H^3/3EI = \delta_h = \alpha T_1 . H^2/2d$$

from which $$H = \tfrac{3}{2}.\alpha T_1 . EI/H.d$$

and bending moment to prevent flexure is as shown in Fig. 6.24.

6.6 Cantilever Shear Wall subjected to a Parabolic Temperature Distribution

If a cantilever is subjected to a parabolic temperature change expressed by the equation:

$$T_y = \tfrac{1}{4}T_1\left[\left(\frac{y}{h}\right)^2 + 2\left(\frac{y}{h}\right) + 1\right] \tag{6.25}$$

then, from eq. (6.10), we obtain:

$$\sigma_y = \tfrac{1}{4}\alpha E T_1\left[\frac{1}{3} - \left(\frac{y}{h}\right)^2\right] \tag{6.26}$$

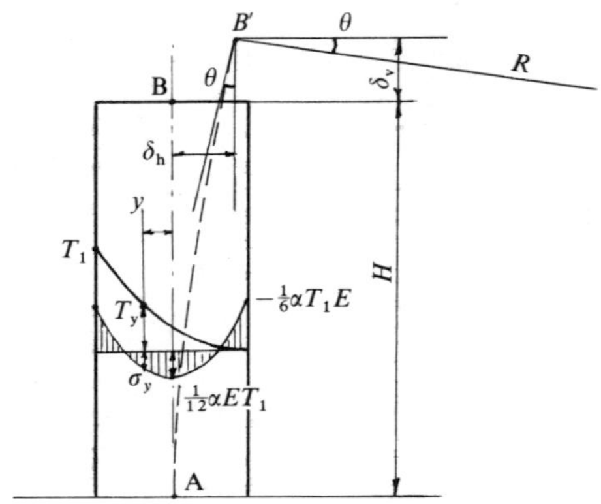

Fig. 6.25 Stresses and deformations in a cantilever due to a parabolic temperature distribution

The angular deformation θ can be calculated using the unit load method from:

$$\theta = \int_0^L M.m.\mathrm{d}x/EI \tag{6.27}$$

in which

$$M = \int_{-h}^{+h} T.y.\mathrm{d}x = \tfrac{1}{12}\alpha E T_1 d^2 \tag{6.28}$$

and $m=1$.

Introducing these values in eq. (6.27), we obtain:

$$\theta = \int_0^H \frac{[\frac{1}{12}\alpha E T_1 d^2].(1).\mathrm{d}x}{EI} = \tfrac{1}{12}\alpha T_1 d^2 H$$

or $\theta = \alpha T_1 H/d$ as for the uniform temperature distribution.

Similarly, the horizontal translation of the point B can be calculated from equation of unit load method by introducing:

$$m = 1.(H - x)$$

from which:

$$\delta_h = \int_0^H (\tfrac{1}{12}\alpha E T_1 d^2)(H - x)\,\mathrm{d}x/EI$$

or

$$\delta_h = \alpha T_1 H^2/2d$$

as for uniform temperature distribution. However, internal stresses will be set in as calculated in Fig. 6.12, i.e., of the value:

$$\sigma_y = \tfrac{1}{4}\alpha E T_1 . \left[\frac{1}{3} - \left(\frac{y}{h}\right)^2\right] \tag{6.29}$$

6.7 Parapet Beams subjected to Temperature Changes

Exposed parapet beams are particularly vulnerable to temperature changes because of the high degree of restraint provided by floor diaphragms and sometimes by walls or columns. The form of the temperature change varies and can be as shown in Fig. 6.26, i.e., a gradient can develop in the vertical or horizontal direction, or there may be a uniform temperature change throughout.

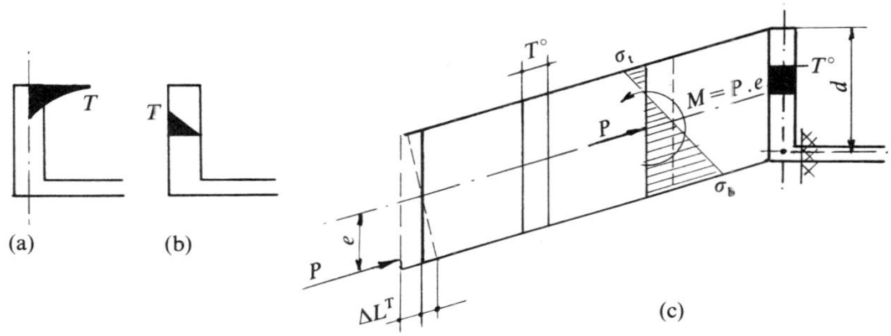

Fig. 6.26 Temperature stresses in parapet beams

If the temperature change is uniform, as shown in Fig. 6.26(c), then a unit length of the parapet, when free from the floor, would undergo a change in length by $\Delta L = \alpha . T$. Assuming that the floor slab suppresses this movement completely, the force P will develop at the floor level, such that:

$$\Delta L^P = P/EA$$

from which the restraining force is:

$$P = \alpha T . EA$$

The stresses in the parapet wall can then be calculated from the column formula:

$$\sigma_{t,b} = -\frac{P}{A}\left(1 \pm \frac{6e}{d}\right)$$

If the tensile stress so produced is larger than the tensile strength of the material of the parapet, then cracks may appear. Creep and yielding of the floor at the joint will, however, reduce the degree of restraint and, therefore, the stresses.

If the temperature gradient develops across the wall, as shown in Fig. 6.27, and the parapet is treated as a linear member, then the deformations at release with the floor are:

$$\theta = \alpha T_1 l/2b$$
$$\Delta l = \alpha T l$$

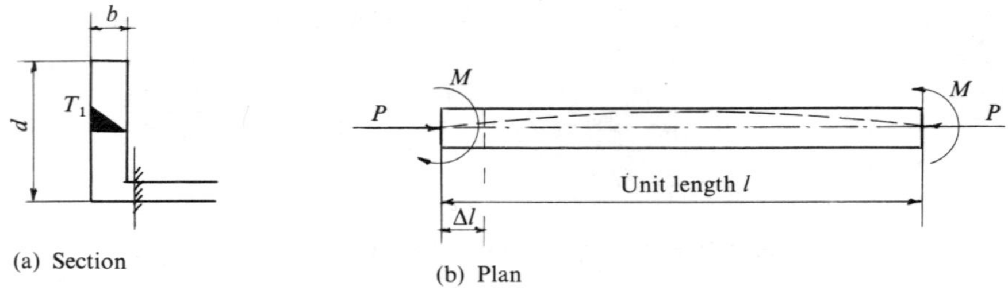

Fig. 6.27 Parapet wall subjected to a linear temperature distribution

The statically indeterminate values of the restraining bending moment M and force P at the floor level can be calculated from the conditions of compatibility of deformations at the joint with the floor, i.e.,

$$M.\alpha^m - P.\alpha^P = \theta$$
$$-M.\delta^m + P(\delta^P + \delta^s) = \Delta \tag{6.30}$$

in which α^m and δ^m are the angular and linear deflections due to unit bending moments $m=1$; α^P and δ^P are similar values due to the applied unit force $P=1$, and δ^s is the deformation of the floor due to applied force $P=1$. It can be shown that:

$$\alpha^m = l/2EI$$
$$\delta^m \text{ is negligible}$$
$$\alpha^P = bl/4EI$$
$$\delta^P = l/EA$$
$$\delta^s = \sim 0$$

Introducing the above values in eqs. (6.30) we obtain:

$$P = \tfrac{1}{2}\alpha T_1 EA$$

and

$$M = \alpha EI\left[\frac{T_1}{b} + \frac{bA}{4I}T_1\right] \tag{6.31}$$

6.8 Effect of Temperature Changes on Statically Indeterminate Structures

When statically indeterminate structures are subjected to a temperature change, stresses and deformations develop irrespective of whether the heat flow is steady or unsteady, and whether the temperature distribution is linear or curvilinear. In the following analysis, these stresses and deformations are determined based on the concept of the force–displacement method, in which, as shown earlier, the unknown values are generally bending moments and deformations considered simultaneously.

For a two-span beam in Fig. 6.28, subjected to a linear temperature distribution, a release is introduced at the support B and the angular discontinuity $(\theta_{ba}+\theta_{bc})$ is calculated using the eq. (6.11). The equation of the compatibility of deformations at the support B is then written in the usual form:

$$B_1 M_b\left(\frac{L_1}{3EI_1} + \frac{L_2}{3EI_2}\right) = \theta_{ba} + \theta_{bc} = \frac{\alpha T_1}{2}\left(\frac{L_1}{d_1} + \frac{L_2}{d_2}\right)$$

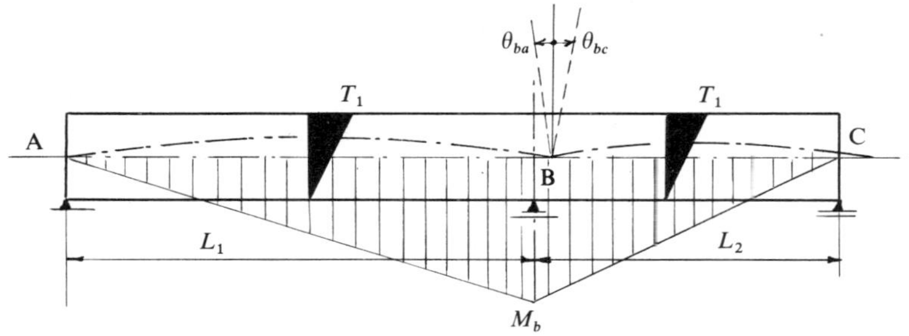

Fig. 6.28 Continuous beam subjected to a linear temperature distribution

If $d_1 = d_2$, then:

$$M_b = \frac{3EI\alpha T_1}{2d} \tag{6.32}$$

A similar answer can be obtained by introducing release at the support B so that the beam becomes simply-supported between A and C (Fig. 6.29). The statically indeterminate value is then the reaction R_b at the support B. This value can be calculated by the usual flexibility method from:

$$R_b . \delta_b^{P=1} = \Delta B \tag{6.33}$$

in which ΔB is the deflection at the point B in beam AC due to the temperature variation, as determined in eq. (6.13), and the value of $\delta^{P=1}$ represents the deflection at B due to a unit load applied at this point. This value can be first calculated from:

$$M_0 . \left(\frac{L_1}{3EI} + \frac{L_2}{3EI}\right) = \frac{\delta}{L_1} + \frac{\delta}{L_2}$$

from which:

$$\delta = \tfrac{1}{3} M_0 . (L_1 + L_2)/EI$$

or

$$\delta_b^{P=1} = \frac{(L_1 . L_2)^2}{3EI(L_1 + L_2)}$$

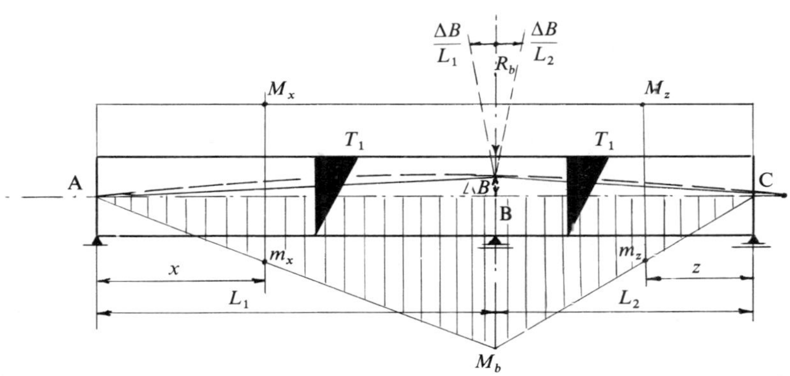

Fig. 6.29 Two-span beam subjected to a linear temperature distribution

Introducing the values of $\delta_b^{P=1}$, and ΔB from eq. (6.33) we obtain:

$$R_b \cdot \frac{(L_1 . L_2)^2}{3EI(L_1 + L_2)} = \frac{\alpha . T_1 . L_1 . L_2}{2d}$$

from which

$$R_b = 3EI . \alpha T_1 \cdot \frac{(1/L_1 + 1/L_2)}{2d} \tag{6.34}$$

and

$$R_a = R_b \cdot \frac{L_2}{L_1} = \frac{3EI . \alpha T_1}{2d . L_1} \tag{6.35}$$

and finally,

$$M_b = R_a . L_1 = \frac{3EI . \alpha T_1}{2d} \tag{6.36}$$

This is identical to eq. (6.32).

It should be noted that in a two-span beam, subjected to a linear temperature variation, bending moment M_b, (see eq. (6.36)) does not depend on the span, while the reactions are a function of spans (eqs. (6.34) and 6.35)). For a beam with two unequal spans, the bending moment M_b depends on the ratio of L_1/L_2 and not on the actual spans of beams.

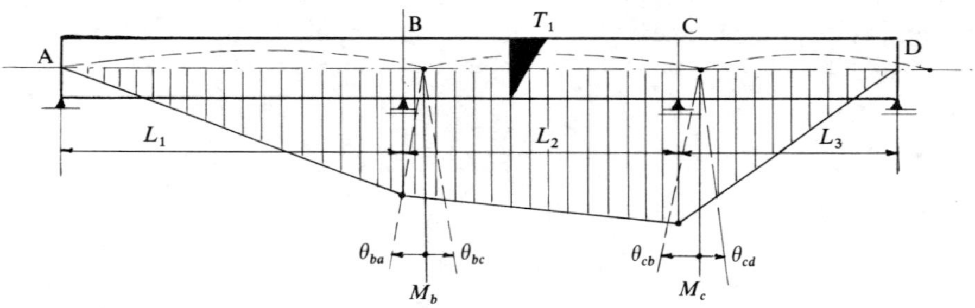

Fig. 6.30 Three-span beam subjected to a linear temperature distribution

For the three-span beam in Fig. 6.30, the statically indeterminate bending moments can be calculated from:

$$\text{B.} \quad M_b \left(\frac{L_1}{3EI} + \frac{L_2}{3EI} \right) + M_c \frac{L_2}{6EI} = \theta_{ba} + \theta_{bc}$$

$$\text{C.} \quad M_b \cdot \frac{L_2}{6EI} + M_c \left(\frac{L_2}{3EI} + \frac{L_3}{3EI} \right) = \theta_{cb} + \theta_{cd} \tag{6.37}$$

It can be seen again from eqs. (6.37) that, for continuous beams of more than two spans, statically indeterminate bending moments depend only on the ratios between respective spans, e.g., L_2/L_1, L_3/L_1, etc., but not on the actual values of respective spans. These moments, however, depend on stiffnesses of beams.

6.8.1 *Portal frames*

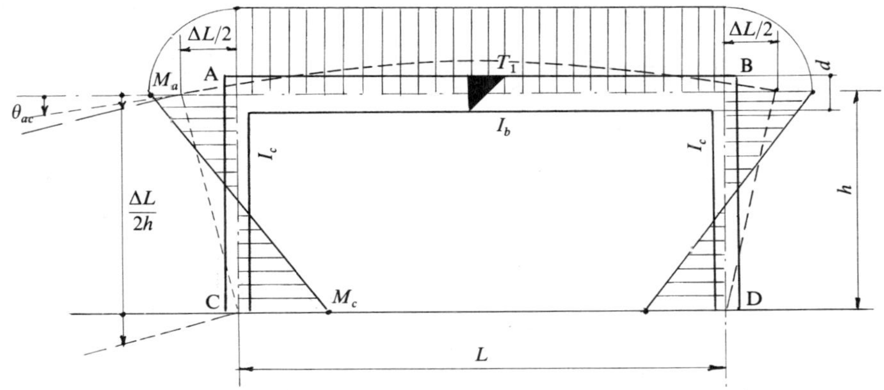

Fig. 6.31 The effect of linear temperature distribution on a portal frame

The effect of linear temperature distribution on a portal frame (Fig. 6.31) can be determined from:

$$\text{A.}\quad M_a\left(\frac{L}{2EI_b} + \frac{h}{3EI_c}\right) - M_c\cdot\frac{h}{6EI_c} = \frac{\Delta L}{2h} - \theta_{ab}$$

$$\text{C.}\quad -M_a\cdot\frac{h}{6EI_c} + M_c\cdot\frac{h}{3EI_c} = \frac{\Delta L}{2h} \tag{6.38}$$

in which: $\Delta L = \frac{1}{2}\alpha T_1 . L, \qquad \theta_{ab} = \alpha T_1 L/2d$

From eq. (6.38) we obtain:

$$M_a = \left(\frac{3}{2}\cdot\frac{\Delta L}{h} - 2\theta_{ab}\right)\cdot\frac{EI_b}{2L + nh}$$

$$M_c = \frac{3}{2}\left(1 + \frac{L}{nh}\right)\cdot\frac{\Delta L}{h} - \theta_{ab}\cdot\frac{EI_b}{2L + nh}$$

where $n = I_b/I_c$

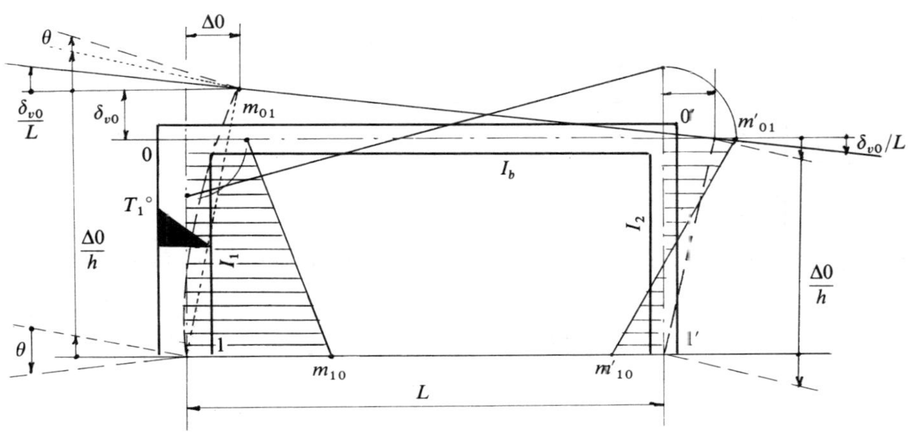

Fig. 6.32 The effect of linear temperature distribution in a column

If the linear temperature gradient affects one column of a portal frame only, as in Fig. 6.32, then the statically indeterminate bending moments and deflection can be calculated by the force–displacement method by setting-out a matrix expressing the requirements of compatibility and equilibrium, i.e., from the condition that the stresses must be 'self-equilibrating'.

$$\begin{bmatrix} \frac{h}{3EI_1}+\frac{L}{3EI_b} & \frac{h}{6EI_1} & -\frac{L}{6EI_b} & 0 & -\frac{1}{h} \\ -\frac{L}{6EI_b} & 0 & \frac{h}{3EI_2}+\frac{L}{3EI_b} & -\frac{h}{6EI_2} & -\frac{1}{h} \\ \frac{h}{6EI_1} & \frac{h}{3EI_1} & 0 & 0 & \frac{1}{h} \\ 0 & 0 & -\frac{h}{6EI_2} & \frac{h}{3EI_2} & -\frac{1}{h} \\ -1 & 1 & -1 & -1 & 0 \end{bmatrix} \begin{Bmatrix} m_{01} \\ m_{10} \\ m'_{01} \\ m'_{10} \\ \Delta \end{Bmatrix} = \begin{Bmatrix} \theta-\frac{\delta_{v0}}{L} \\ -\frac{\delta_{v0}}{L} \\ \theta \\ 0 \\ 0 \end{Bmatrix} \tag{6.39}$$

The following numerical example is used to illustrate the magnitude of stresses and deformations.

Example 6.4

In Fig. 6.33, column 0–1 is assumed to be subjected to a linear temperature change of $T=20°$ C and the bending moments and deflection are required. The following data are assumed: the coefficient of thermal expansion $\alpha=0{\cdot}000\,01$ per 1° C, and $E=21\times10^6$ kN/m².

From eq. (6.14): $\alpha T_1 h/2d=0{\cdot}000\,18$ and $\delta_{v0}=\frac{1}{2}\alpha Th=0{\cdot}000\,36$ and the matrix

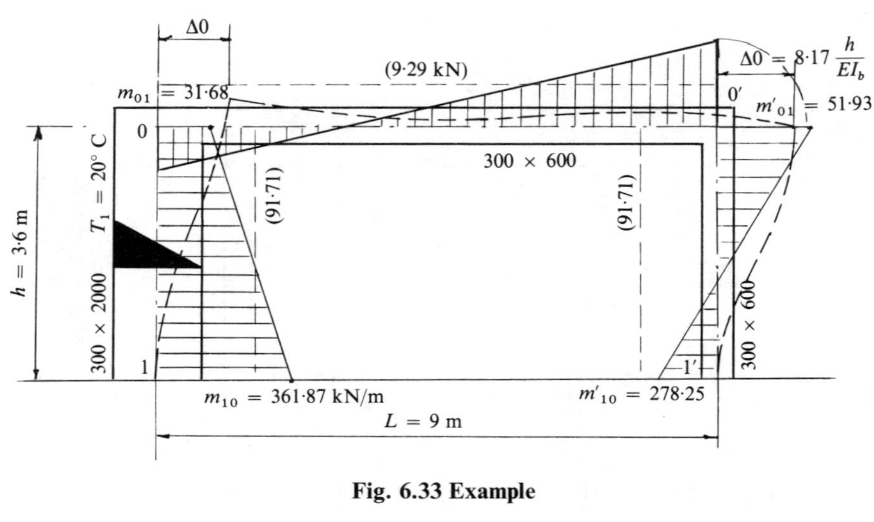

Fig. 6.33 Example

becomes:

$$\begin{bmatrix} 3{\cdot}0324 & -1{\cdot}5 & 0{\cdot}0162 & 0 & -1 \\ -1{\cdot}5 & 4{\cdot}2 & 0 & -0{\cdot}6 & -1 \\ 0{\cdot}0162 & 0 & 0{\cdot}0324 & 0 & +1 \\ 0 & -0{\cdot}0162 & 0 & 0{\cdot}0324 & -1 \\ 1 & 1 & -1 & 1 & 0 \end{bmatrix} \begin{Bmatrix} m_{01} \\ m'_{01} \\ m_{10} \\ m'_{10} \\ \Delta \end{Bmatrix} = \begin{Bmatrix} 15{\cdot}876 \\ -4{\cdot}536 \\ 20{\cdot}412 \\ 0 \\ 0 \end{Bmatrix} \tag{6.40}$$

from which bending moments and deflection can be calculated. These are shown in Fig. 6.33.

6.9 The Effect of Axial Forces arising from Temperature Changes

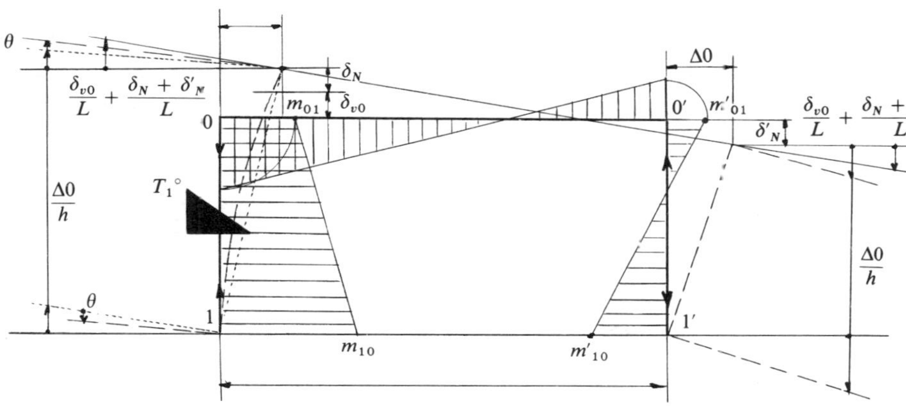

Fig. 6.34 Temperature stresses in a portal frame including the effect of axial deformations

The effect of axial forces in the analysis of thermal stresses is not usually very high but, if required, can be taken into account as follows: The axial deformations in columns 01 and 0′1′, in Fig. 6.34, can be expressed as:

$$\delta_N = \frac{N_1}{EA_1}\cdot h = \frac{h}{EA_1}\cdot\frac{m_{01} + m'_{01}}{L}$$

$$\delta'_N = \frac{N_1}{EA_2}\cdot h = \frac{h}{EA_2}\cdot\frac{m_{01} + m'_{01}}{L}$$

so that:

$$\frac{\delta_N + \delta'_N}{L} = \frac{h}{EL^2}\left(\frac{1}{A_1} + \frac{1}{A_2}\right)(m_{01} + m'_{01}) \tag{6.41}$$

or in a form suitable for introduction into the matrix:

$$\frac{\delta_N + \delta'_N}{L}\cdot\frac{EI_b}{h} = \frac{h^2}{E^2 I_b L^2}\left(\frac{1}{A_1} + \frac{1}{A_2}\right)(m_{01} + m'_{01})\frac{EI_b}{h}$$

Then, the compatibility equation at point 0 in Example 6.4 becomes:

$$m_{10} \cdot \frac{h}{6EI_1} + m_{01}\left(\frac{L}{3EI_b} + \frac{h}{3EI_1}\right) - m'_{01} \cdot \frac{L}{6EI_b} =$$

$$= \left[\frac{\Delta_0}{n} + \theta - \frac{\delta_{v0}}{L}\right] - \frac{h}{EL^2}\left(\frac{1}{A_1} + \frac{1}{A_2}\right)(m_{01} + m'_{01}) \quad (6.42)$$

Similar compatibility conditions can be set out at the points 0′, 1′, and 1.

6.10 Tall Building Frames and Shear Walls Subjected to Temperature Changes

The effect of temperature changes on tall building frames and shear walls can be analysed in a similar way to that shown in the previous example. However, this is further illustrated by the following examples.

Example 6.5

Figure 6.35 shows a six-storey frame in which it is assumed that the column 0–6 is subjected to a linear temperature change of $T=20°$ C. It is also assumed that $E=21$ kN/mm^2 and $\alpha=0{\cdot}000\ 01$ per 1° C.

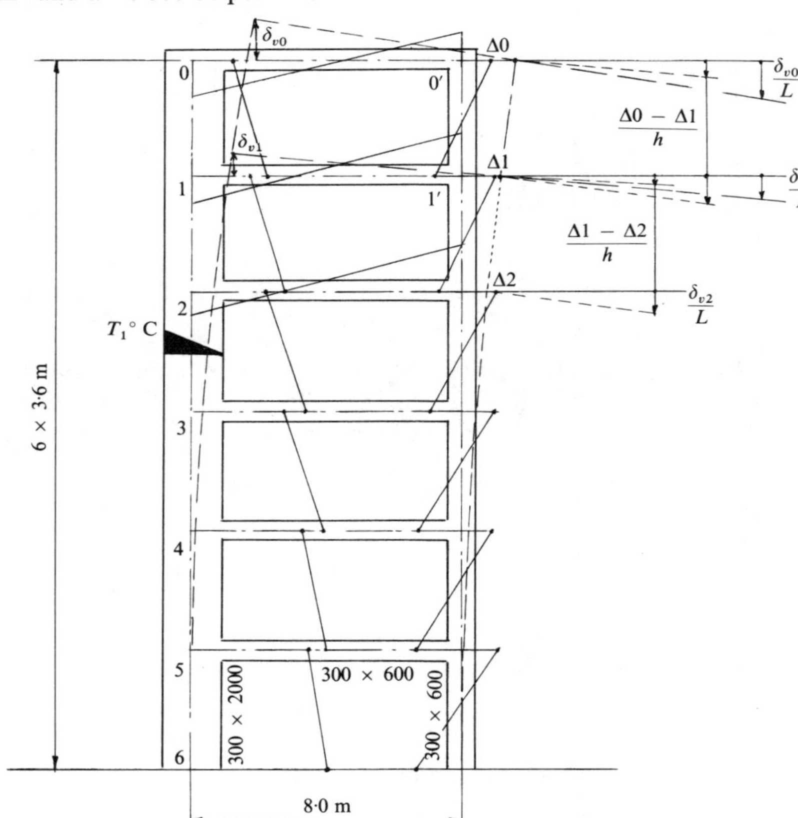

Fig. 6.35 The effect of linear temperature variation in a column on tall building frame

We proceed in a similar way to that in the case of lateral loads, by allowing the structure to deflect to its final position, defined by the deflections $\Delta 0$, $\Delta 1$, etc., and in this position we introduce releases at all joints. Then, the statically indeterminate bending moments are applied of such values as are necessary to satisfy simultaneously all conditions of compatibility of deformations and equilibrium of forces. However, before the matrix can be set out, the values of θ and δ_v must be calculated. These are as follows:

$$\theta = \alpha T_1 h/2d = 0{\cdot}0001 \times 20° \times 3{\cdot}6/4 = 0{\cdot}000\,18$$

$$EI_1\theta = 1{\cdot}134 \times 10^5 \times 0{\cdot}000\,18 = 20{\cdot}412$$

$\delta_{v0} = \frac{1}{2}\alpha T_1 H = 0{\cdot}002\,16;$ similarly:

$\delta_{v2} = 0{\cdot}001\,44,$ $\delta_{v3} = 0{\cdot}001\,08,$ $\delta_{v4} = 0{\cdot}000\,72,$ $\delta_{v5} = 0{\cdot}000\,36$

$\dfrac{EI}{L}\cdot\delta_{v0} = 27{\cdot}216;$ $\dfrac{EI}{L}\delta_{v1} = 22{\cdot}68;$ $\dfrac{EI}{L}\delta_{v2} = 18{\cdot}144;$

$\dfrac{EI}{L}\cdot\delta_{v3} = 13{\cdot}608;$ $\dfrac{EI}{L}\delta_{v4} = 9{\cdot}072;$ $\dfrac{EI}{L}\delta_{v5} = 4{\cdot}536$

Other values needed in the matrix are:

$$n = I_b/I_c = 0{\cdot}0054/0{\cdot}2 = 0{\cdot}027$$

$$EI_b = 21 \times 10^6 \times 0{\cdot}0054 = 1{\cdot}134 \times 10^5$$

$$EI_b/h = 1{\cdot}134 \times 10^5/3{\cdot}6 = 0{\cdot}315 \times 10^5$$

The element force–displacement matrix is set out for the joints 1 and 1′ based on the conditions of compatibility of deformations and equilibrium of forces:

2 1 1′ $$m_{21}.h/6EI_1 + m_{12}.h/3EI_1 - (m_{12} - m_{10}).L/3EI_b - (m'_{10} + m'_{12}).L/6EI_b = \frac{\Delta 1 - \Delta 2}{h} + \theta - \frac{\delta_{v1}}{L}$$

2′ 1′ 1 $$(m_{12} - m_{10})L/6EI_b + (m'_{10} - m'_{12})L/3EI_b + m'_{12}h/3EI_1 + m'_{21}.h/6EI_1 = \frac{\Delta 1 - \Delta 2}{h} - \frac{\delta_{v1}}{L}$$

0 1 2 $$-m_{01}.h/6EI_1 - m_{10}.h/3EI_1 - m_{12}.h/3EI_1 - m_{21}.h/6EI_1 = \left(\frac{\Delta 0 - \Delta 1}{h} - \theta\right) - \left(\frac{\Delta 1 - \Delta 2}{h} + \theta\right)$$

0′ 1′ 2 $$-m'_{01}.h/6EI_2 + m'_{10}.h/3EI_2 - m'_{12}.h/3EI_2 + m'_{21}.h/6EI_2 = \frac{\Delta 0 - \Delta 1}{h} - \frac{\Delta 1 - \Delta 2}{h}$$

1–1′ $$\left[\frac{m_{21} - m_{12}}{h} - \frac{m_{10} - m_{01}}{h}\right] - \left[\frac{m'_{21} + m'_{12}}{h} - \frac{m'_{10} + m'_{01}}{h}\right] = 0$$

Table 6.1 shows the main matrix and Fig. 6.36 shows the distribution of bending moments, shear and axial forces, as well as deflections.

Table 6.1 Matrix for the analysis of a shear wall for the effect of linear temperature distributi

	0		0′	1			1′		2			2′		3		
	0 1	Δ0	0′ 1′	1 0	1 2	Δ1	1′ 0′	1′ 2′	2 1	2 3	Δ2	2′ 1′	2′ 3′	3 2	3 4	Δ
0	3·0324	−1	−1·5	0·0162	0	+1	0									
0′	−1·5	−1	4·2	0	0	+1	−0·6									
0–0′	1	0	1	−1	0	0	1									
2 1 1′	0	0	0	−3·0	3·0324	−1	−1·5	−1·5	0·0162	0	+1	0				
2′ 1′ 1	0	0	0	1·5	−1·5	−1	3·0	4·2	0	0	+1	−0·6				
0 1 2	0·0162	+1	0	0·0324	0·0324	−2	0	0	0·0162	0	+1	0				
0′ 1′ 2′	0	−1	−0·6	0	0	+2	1·2	−1·2	0	0	−1	0				
1–1′	−1	0	−1	1	1	0	−1	1	−1	0	0	1				
3 2 2′																
3′ 2′ 2																
1 2 3										F^e						
1′ 2′ 3′																
2–2′																
4 3 3′																
4′ 3′ 3																
2 3 4															F^e	
2′ 3′ 4																
3–3′																
5 4 4′																
5′ 4′ 4																
3 4 5																
3′ 4′ 5																
4–4′																
6 5 5′																
6′ 5′ 5																
4 5 6																
4′ 5′ 6																
5–5′																
6 5																
6′ 5′																

3′		4			4′		5			5′		6	6′	$\times \frac{EI_b}{h}$
2′	3′ 4′	4 3	4 5	Δ4	4′ 3′	4′ 5′	5 4	5 6	Δ5	5′ 4′	5′ 6′	6 5	6′ 5′	
														−6·804
														−27·215
														0
														−2·268
														−22·68
														40·824
														0
														0
														2·268
														−18·144
														40·824
														0
														0
														6·804
														−13·608
														40·824
														0
														0
														11·340
														−9·072
				F^e										40·824
														0
														0
														15·876
														−4·536
								F^e						40·824
														0
														0
								0·0162	1	0	0	0·0324	0	4·536
								0	0	0	−0·6	0	1·2	0

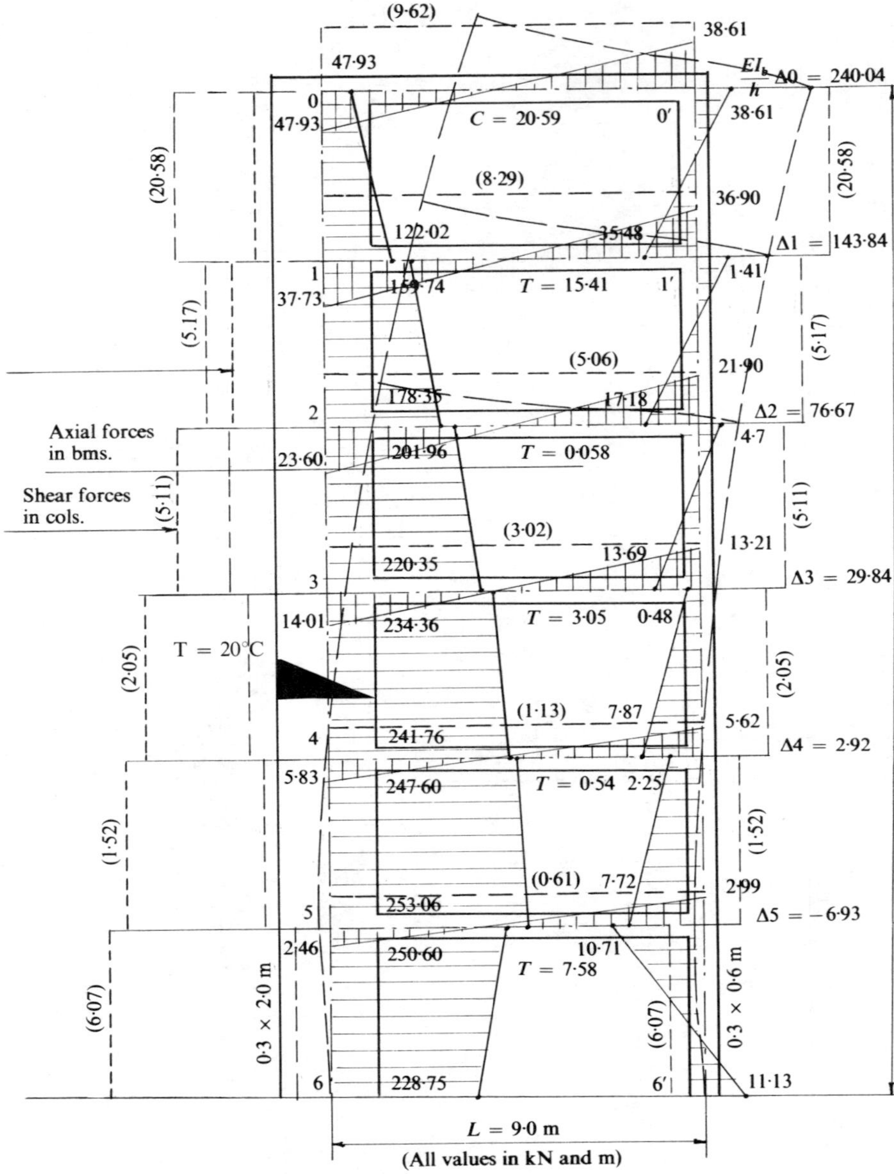

Fig. 6.36 Distribution of bending moments, shear and axial forces due to linear temperature variation in column

Example 6.6

In this example, a frame shown in Fig. 6.37 of constant cross-section throughout is analysed for the effect of linear temperature variation in the column. It is assumed, as in the previous example, that $T = 20°$ C, $E_c = 21$ kN/mm^2, and $\alpha = 0{\cdot}000\,01$ per 1° C.

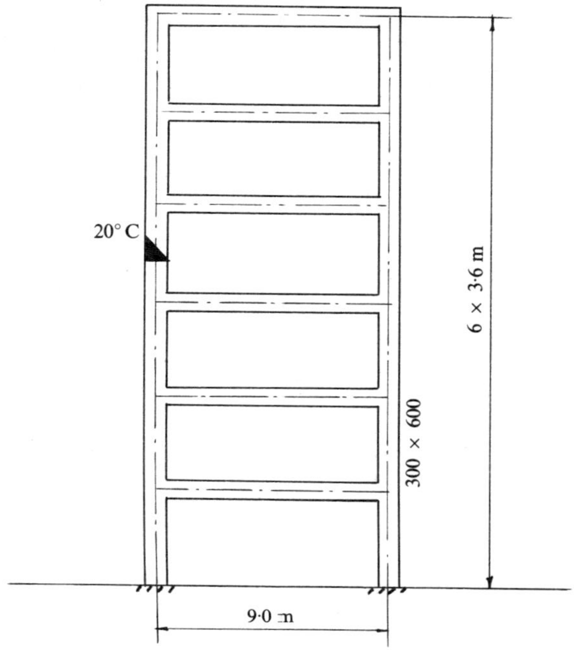

Fig. 6.37 Example of the analysis of temperature stresses and deformations

Table 6.2 shows the setting out of the force–displacement matrix and Fig. 6.38 shows the distribution of moments, forces, and deformations.

It is interesting to note that, in this case, the pattern of moments and shear forces differs considerably from that in the previous example. It appears that it is possible for some beams to be subjected to tension (as well as bending moments), while alternate beams may be subjected to axial compression as shown in Fig. 6.38. This may explain why, in practice, some floors are observed to crack while others appear to be immune to cracks. However, no general conclusions can be drawn based on one example only, and the problem warrants further study.

One other interesting conclusion can be drawn from this example, i.e., that the effect of temperature changes in a column is felt most strongly in the upper floors and in the column directly exposed to the temperature change. A far side column shows bending moments greatly reduced except at the uppermost level. The magnitude of bending moments in both columns appears to fluctuate very little. Similarly, the axial forces in columns increase towards the base, but extremely slowly. Axial forces in the beams reach very high values but only at the two upper levels. Only the top beam shows large shearing forces.

Deflections appear to increase rapidly in the upper floors and, in the example, reach a maximum horizontal component of 8·48 mm, and vertical component of 2·16 mm. It appears, therefore, that the horizontal movement can be expected to be much larger than the vertical, although observations show that the vertical differential movement alone can be sufficient to cause cracks in rigid partitions and in plasters.

Table 6.2 Matrix for the analysis of a frame in Fig. 6.37 for the effect of linear temperature variation in a column

	0		0′	1			1′		2			2′		3			3′		4			4′		5			5′		6	6′	$\frac{EI}{h}$
	0 1	$\Delta 0$	0′ 1′	1 0	1 2	$\Delta 1$	1′ 0′	1′ 2′	2 1	2 3	$\Delta 2$	2′ 1′	2′ 3′	3 2	3 4	$\Delta 3$	3′ 2′	3′ 4′	4 3	4 5	$\Delta 4$	4′ 3′	4′ 5′	5 4	5 6	$\Delta 5$	5′ 4′	5′ 6′	6 5	6′ 5′	
1 0 0′	4·2	−1	−1·5	0·6	0	1	0																								40·824
1′ 0′ 0	−1·5	−1	4·2	0	0	1	−0·6																								−21·216
0–0′	1	0	1	−1	0	0	1																								0
2 1 1′	0	0	0	−3·0	4·2	−1	−1·5	−1·5	0·6	0	1	0																			45·36
2′ 1′ 1	0	0	0	1·5	−1·5	−1	3·0	4·2	0	0	1	−0·6																			−22·68
0 1 2	0·6	1	0	1·2	1·2	−2	0	0	0·6	0	1	0																			136·08
0′ 1′ 2′	0	−1	−0·6	0	0	2	1·2	−1·2	0	0	−1	0·6																			0
1–1′	−1	0	−1	1	1	0	−1	1	−1	0	0	1																			0
3 2 2′																															49·896
3′ 2′ 2																															−18·144
1 2 3											F^e																				136·08
1′ 2′ 3′																															0
2–2′																															0
4 3 3′																															54·432
4′ 3′ 3																															−13·608
2 3 4																F^e															136·08
2′ 3′ 4′																															0
3–3′																															0
5 4 4′																															58·968
5′ 4′ 4																															−9·072
3 4 5																					F^e										136·08
3′ 4′ 5′																															0
4–4′																															0
6 5 5′																															63·504
6′ 5′ 5																															−4·536
4 5 6																									F^e						136·08
4′ 5′ 6′																															0
5–5′																															0
6 5																									0·6	1	0	0	1·2	0	68·04
6′ 5′																									0	−1	0	−0·6	0	1·2	0

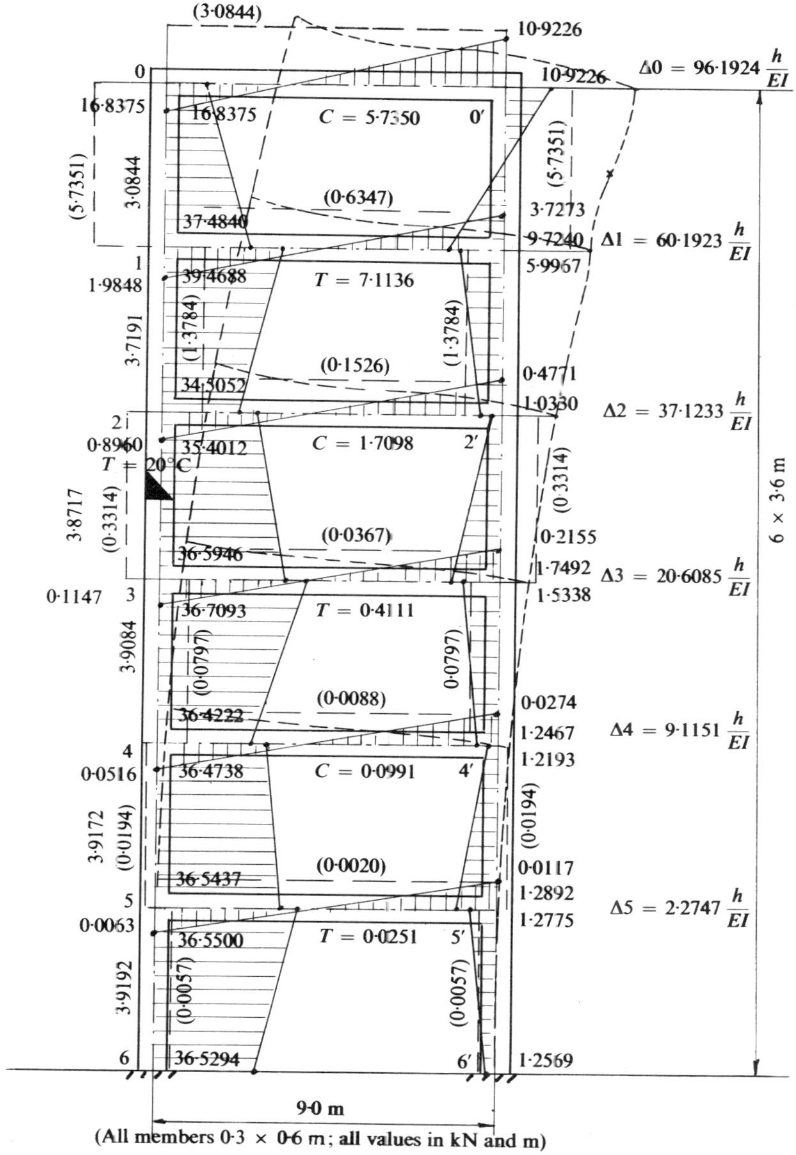

Fig. 6.38 Example of analysis of temperature stresses and deformations

Example 6.7

This example is introduced to illustrate the effect of change of temperature of the roof beam on the tall frame which was analysed in the previous example (Fig. 6.38). Assuming that $E=21 \times 10^6$ kN/m^2, $\alpha=0{\cdot}000\ 01$ per 1° C, and retaining the temperature change T° C in the explicit form, the force–displacement matrix can be set out in a similar way to that in Example 6.6. The following value is required in the matrix:

$$\frac{\Delta L}{2h} EI = \frac{0{\cdot}000\ 09}{2 \times 3{\cdot}6} \times T^\circ \times 0{\cdot}1134 \times 10^6 = 1{\cdot}415T^\circ$$

Because of symmetry of the frame and temperature effect, the frame will theoretically not deflect horizontally except for the deformation $\Delta L/2 = \alpha TL/2$ on each side of the roof beam. This deformation will be only slightly suppressed by the column shear forces. Omitting this effect and also the effect of axial forces in the remaining part of the frame, the matrix can be set out as shown in Table 6.3. Figure 6.39 shows the distribution of moments and forces in the frame in terms of T.

Table 6.3 Matrix for the analysis of frame shown in Fig. 6.39

	0	1		2		3		4		5		6	
	0 1	1 0	1 2	2 1	2 3	3 2	3 4	4 3	4 5	5 4	5 6	6 5	
1 0 0′	2·7	−0·6											$1{\cdot}415T^\circ$
2 1 1′	0	4·5	−3·3	−0·6									0
0 1 2	−0·6	1·2	1·2	−0·6									$1{\cdot}415T^\circ$
3 2 2′					F^e								0
1 2 3													0
4 3 3′							F^e						0
2 3 4													0
5 4 4′									F^e				0
3 4 5													0
6 5 5′											F^e		0
4 5 6													0
6 5											−0·6	1·2	0

Assuming, e.g., that $T = 25^\circ$ C, we obtain $\Delta L = 22{\cdot}5$ mm and the maximum stress $2{\cdot}55$ N/mm^2. The former represents a distortion of 1/300 which could cause cracks in partitions if these were tightly built with columns.

This analysis confirms many observations that high stresses and deformations can develop in the upper floors when the roof is subjected to high temperature variations. This is greatly promoted by lack of roof insulation. Figure 6.39 shows that stresses rapidly attenuate in the lower floors. It is also interesting to note that bending moments in beams oscillate from negative to positive.

6.11 Summary of Unusual Effects of Temperature Changes on Structures

The unusual effects of temperature changes on structures, discussed above, can be summarized as follows:

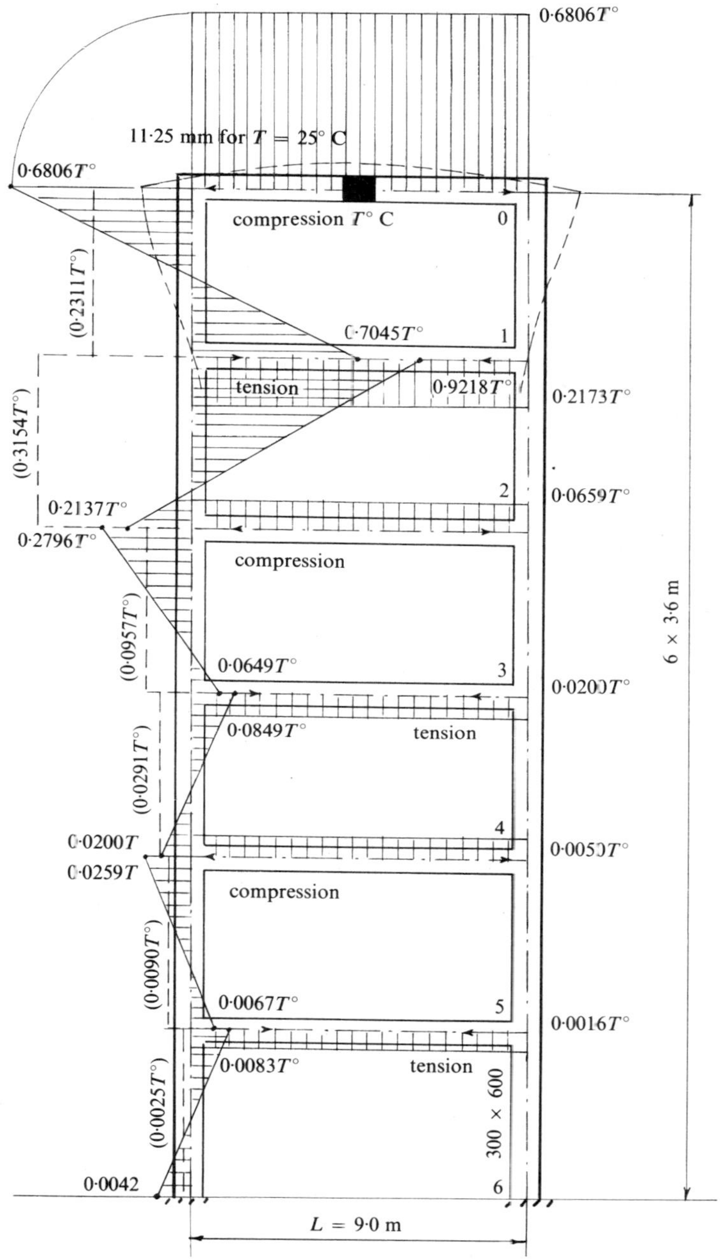

Fig. 6.39 Example of the analysis of thermal stresses due to the uniform temperature change in a roof beam

(a) A beam may be subjected to high stresses but may be free from strains, as shown in Fig. 6.40(a).
(b) A beam may be subjected to bending deformations but may be free from stresses (Fig. 6.40(b)).

(c) In statically determinate beams subjected to a linear temperature variation, the equivalent bending moment producing a curvature similar to the effect of temperature is equal to:

$$M = \alpha T_1 . EI/d$$

i.e., this moment is independent of the span but depends on the beam stiffness *EI*. Similarly, the deflection of a beam subjected to the linear temperature variation is:

$$\max \delta = \alpha T_1 . L^2/8d$$

i.e., deflection does not depend on the stiffness of the beam *EI* although it is a function of the span and the depth of beam in the first power only, i.e., continuous beams, whether short or long are equally vulnerable to temperature variations. Similar effect can be shown to exist in frames, contrary to the views sometimes expressed that only very tall buildings are vulnerable to thermal changes.

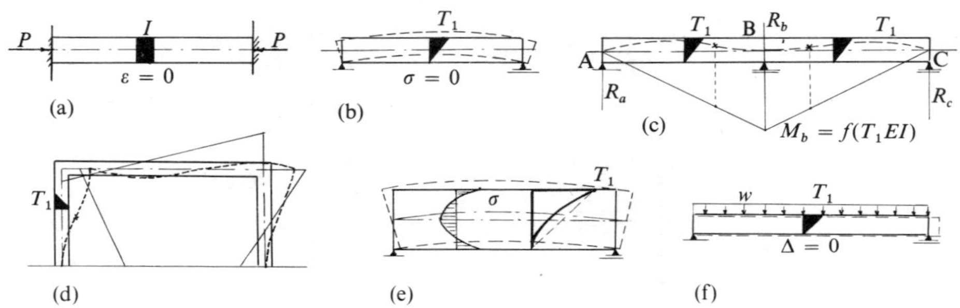

Fig. 6.40

(d) In continuous beams, thermal moments do not depend on the actual lengths of beams but on their ratio, and depend on actual stiffnesses of beams (while in the case of gravity loads, bending moments depend on the actual lengths of beams, but only on the ratio of the stiffnesses of beams). However, the reactions depend on the actual lengths of beams (Fig. 6.40(c)).

(e) Points of contraflexure due to thermal changes need not coincide with the zero bending moments, as in the case of gravity loads (Fig. 6.40(c) and (d)).

(f) In the case of curvilinear temperature variation, in statically determinate structures, stresses are caused only by the difference between the linear temperature change and a curve with zero temperatures at the extreme fibres. The deformations of such elements depend on the linear component of the curvilinear temperature distribution (Fig. 6.40(e)).

(g) A beam subjected to external load may have zero deflection when combined with a certain temperature effect (or a double deflection when the temperature sign is reversed), while internal stresses may not be affected.

These effects of temperature on concrete structures, together with the inelastic response of concrete, make interpretation of site measurements of strains very difficult, except for those due to transient effects.

References

1. B. A. Boley and J. H. Weiner. *Theory of Thermal Stresses*, John Wiley & Sons Limited, New York (1964).
2. S. Timoshenko and J. N. Goodier. *Theory of Elasticity*, McGraw-Hill Book Co. Limited, New York (1951).
3. W. H. McAdams. *Heat Transmission*, McGraw-Hill, London (1954).
4. H. S. Carslaw and J. C. Jaegar. *Conduction of Heat in Solids*, Oxford University Press, London (1959).
5. P. J. Schneider. *Conduction Heat Transfer*, Addison-Wesley, Cambridge, Mass. (1955).
6. B. E. Batewood. *Thermal Stresses*, McGraw-Hill, London (1957).
7. C. R. Lee, A. J. Chabowski and J. R. Miles. 'Behaviour of Model Concrete Structures under Temperature Loading', *Pressure Vessels Conference*, London (1967).

7 Analysis of Vibrations of Elastic Structures

7.1 Introduction

This chapter contains a short review of the basic principles of free and forced, steady-state vibrations of single-degree-of-freedom systems, such as: undamped systems and systems having viscous and structural damping. These systems are often used in assessing the response of tall buildings to vibrations.

The basic requirement for vibration to take place is the existence of mass, stiffness, and exciting force. Vibratory systems, such as tall buildings, possess all these characteristics and the means for storing potential energy by their stiffness, means for storing kinetic energy by their mass, and also a mechanism by which the energy is dissipated. In this review, mainly the lumped systems are considered, represented by idealized springs, masses, and dampers, each having a single function to perform. In tall buildings, both the translational and rotational motions must be considered and are significant.

The analyses of frequencies are made either by the method of equilibrium or by the principle of conservation of energy, using Newton's second law of motion. In the first approach, the dynamic forces are equalized to the internal resisting actions expressed in terms of frequency, while in the second the kinetic energy in the position of zero displacement is equated to the internal strain energy at maximum amplitude. However, explicit solutions are possible only in the most simple problems when mass and stiffness distribution can be expressed as continuous mathematical functions.

Figure 7.1(a) shows a linear spring subjected to a force F, in which the change in

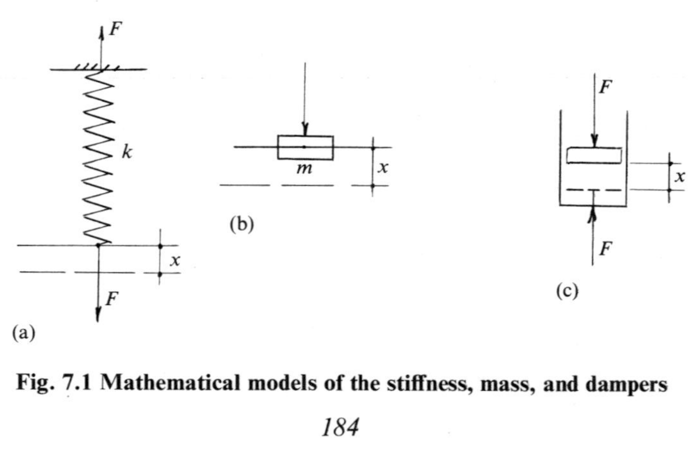

Fig. 7.1 Mathematical models of the stiffness, mass, and dampers

length x is proportional to the force, i.e.,

$$F = kx$$

Such a spring is assumed to have no mass, so that the force acting on one end is equal and opposite to the force acting on the other end. The value of k is a constant of proportionality and is called the spring constant or the stiffness of a member or of a structure.

A mass is a rigid body (Fig. 7.1(b)), whose acceleration $\ddot{x}$ according to Newton's second law of motion is proportional to the resultant F of all forces acting on the mass; i.e.,

$$F = m\ddot{x}$$

In viscous dampers (Fig. 7.1(c)), the applied force is assumed to be proportional to the relative velocity of its position, i.e.,

$$F = c.\dot{x}$$

in which c is the damping coefficient. The damper is assumed to have no mass.

In rotational systems, the kinetic energy is stored by inertia. The stiffness and damping coefficients are related to angular displacement and angular velocity, so that:

The spring law becomes	$M = k_r\alpha$
damping law	$M = c_r\dot{\alpha}$
inertia law	$M = I\ddot{\alpha}$

Equations of rotational motion can be written in an analogous way to the translational equations.

7.2 Single-Degree-of-Freedom Systems

Actual vibrating systems are for the purpose of analysis represented by a mass moving in a linear or circular path as shown in Fig. 7.2. The motion of the mass is confined to a single quantity x. Free vibration of the mass takes place when the mass is displaced from its equilibrium position and then is allowed to vibrate free from any external forces. Vibration is forced when an external force is continuously acting on the mass.

Free vibrations without damping are studied by using Newton's equation, in which the force $F=m\ddot{x}$ exerted by the mass m on the spring is equal and opposite to

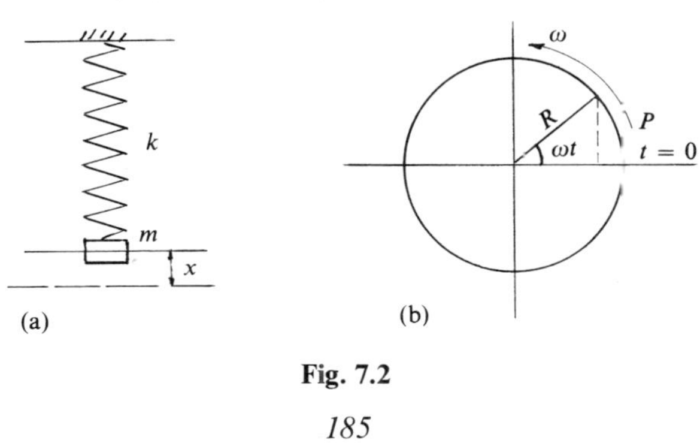

Fig. 7.2

the force kx applied by the spring on the mass, i.e.,

$$m\ddot{x} + kx = 0 \tag{7.1}$$

The solution of eq. (7.1) is in the form:

$$x = A.\sin\sqrt{\frac{k}{m}}\cdot t + B.\cos\sqrt{\frac{k}{m}}\cdot t \tag{7.2}$$

in which the term $\sqrt{(k/m)}$ is the angular natural frequency defined by:

$$\omega_{\text{n}} = \sqrt{\frac{k}{m}} \quad \text{in rad/s} \tag{7.3}$$

For sinusoidal oscillations of the mass, the time interval to complete one cycle is called the period:

$$T = \frac{2\pi}{\omega_{\text{n}}} \tag{7.4}$$

The reciprocal of the period is the natural frequency:

$$f_{\text{n}} = \frac{1}{T} = \frac{\omega_{\text{n}}}{2\pi} = \frac{1}{2\pi}\sqrt{\frac{k}{m}} = \frac{1}{2\pi}\sqrt{\frac{k.g}{W}} \tag{7.5}$$

in which $W = mg$ is the weight of the rigid body forming the mass of the system.

Static deflection can be used to calculate the frequency of a simple mass–spring system. The spring constant k results from the deflection of the system from gravity force of its mass, so that:

$$\Delta_{\text{st}} = \frac{mg}{k}$$

and eq. (7.5) becomes:

$$f_{\text{n}} = \frac{1}{2\pi}\cdot\sqrt{\frac{g}{\Delta_{\text{st}}}} \tag{7.6}$$

This relation applies only when the system is linear and elastic.

Equation (7.2) can be re-written using eq. (7.3) as:

$$x = A.\sin\omega_{\text{n}}t + B.\cos\omega_{\text{n}}t = C.\sin(\omega_{\text{n}}t + \theta)$$

where

$$C = (A^2 + B^2)^{\frac{1}{2}} \quad \text{and} \quad \theta = \tan^{-1}(B/A)$$

The angle θ is called the phase angle.

Equation (7.5) shows that the period of vibrations of one-degree-of-freedom systems without damping depends only on the magnitude of the weight W and the spring constant k, and is independent of the magnitude of the amplitude of oscillations.

Similarly, the frequency of torsional vibrations of shafts, when there is no external exciting torque and frictional resistance is neglected, is:

$$f = \frac{1}{2\pi}\sqrt{\frac{c}{J}}$$

where $c = I_{\text{p}}G/L$ is the torsional rigidity of the shaft and G is the shear modulus of elasticity.

7.3 Rayleigh's Method

An approximate solution of the fundamental frequency of vibration developed by Rayleigh is based on the energy method, that at resonance a system will exchange energy between potential (stored due to deformation or position), and kinetic due to motion of the system. To determine critical frequency, it is assumed that the vibrating system is conservative, i.e., with no damping.

When a deflection curve is assumed, the maximum kinetic energy and maximum potential energy of a structure vibrating at a constant circular frequency ω, are expressed as a function of ω and elastic properties of the structure. The maximum kinetiç energy of the system is:

$$\mathrm{KE} = \frac{1}{2}\cdot\frac{\omega}{g}\cdot\int_0^L W.y^2.\mathrm{d}x \tag{7.7}$$

in which w is weight of an element per unit length,

y is deflection of this element at a distance x.

Similarly, the potential energy of strain in bending is:

$$\mathrm{PE} = \int_0^L \frac{M^2.\mathrm{d}x}{2EI} \tag{7.8}$$

Assuming that deflections are small, $M = EI(\mathrm{d}^2y/\mathrm{d}x^2)$ and substituting this value in eq. (7.8), we obtain:

$$\mathrm{PE} = \frac{1}{2}\int_0^L EI\left(\frac{\mathrm{d}^2y}{\mathrm{d}x^2}\right)^2 \mathrm{d}x \tag{7.9}$$

If there is no energy loss, eqs. (7.7) and (7.9) can be equated and solved for ω:

$$\omega^2 = g\cdot\frac{\int_0^L EI(\mathrm{d}^2y/\mathrm{d}x^2)^2\,\mathrm{d}x}{\int_0^L W.y^2.\mathrm{d}x} \tag{7.10}$$

To solve this equation, a modal shape y must be known as a function of the coordinate x and the solution in a closed form is therefore possible in some simple cases only. Generally, if the assumed mode is exact, then the frequency is exact. For practical purposes, it is often assumed that the modal shape is of the same form as the static deflection curve. This always over-estimates the frequency. In the static case, the inertia acceleration (due to gravity) is constant, while in dynamic cases, it is a product of the radius and square of the angular velocity, and is therefore variable.

Equation (7.10) is often used in the form:

$$\omega^2 = \frac{g\sum_1^n W_i.y_i}{\sum W_i.y_i^2}$$

so that

$$f_\mathrm{n} = \omega/2\pi$$

Alternatively, the period of vibration is expressed as:

$$T = 2\pi\sqrt{\left(\frac{\sum W_i.y_i^2}{g.\sum W_i.y_i}\right)} \tag{7.11}$$

The deflected mode of a structure y_i can be assumed to be that resulting from the static analysis of a structure as discussed previously and indicated in Fig. 7.3.

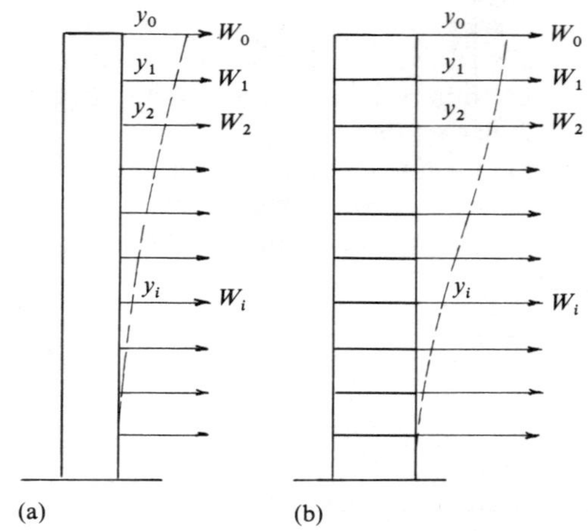

Fig. 7.3

Rayleigh's method is one of the simplest forms based on energy and can provide a quick answer to the fundamental frequency of a structure. An assumed mode in terms of a single arbitrary displacement leads to the energy equation from which the natural frequency is calculated. Even large errors in assumed mode result only in small errors in the frequency. However, for higher modes, the Rayleigh method may give large errors. The Ritz method, based on Rayleigh's energy method, may then be used and gives a higher degree of accuracy. This method, however, can be very laborious. Stotola's method converges rapidly but at higher frequencies may involve large errors. Other methods used are those of Myklestad and Holzer, but these are more suitable for computers than for manual work.

Example of calculations of frequencies of frames and shear walls

Figures 7.4(a) to (h) show frames and shear walls which were analysed earlier for the distribution of bending moments and deformations due to the uniformly distributed lateral pressure. Assuming that all floors in each case are of constant weight, then the deflection profiles previously calculated in terms of any load W can be used in conjunction with Rayleigh's formula, eq. (7.11). Taking: $W=200$ kN, $E=28$ kN/mm^2, $g=9{\cdot}81$ m/s^2, and introducing the coordinates of deflections into Rayleigh's energy expression, we obtain the following frequencies:

case (a)	(see Fig. 1.7)	$f = 0{\cdot}68$ Hz
case (b)	(see Fig. 8.24)	$f = 0{\cdot}51$ Hz
case (c)	(see Fig. 2.25)	$f = 5{\cdot}20$ Hz
case (d)	(see Fig. 1.9)	$f = 0{\cdot}90$ Hz
case (e)	(see Fig. 1.26)	$f = 0{\cdot}62$ Hz
case (f)	(see Fig. 2.4)	$f = 2{\cdot}90$ Hz
case (g)	(see Fig. 2.7)	$f = 7{\cdot}70$ Hz
case (h)	(based on eq. (7.5))	$f = 4{\cdot}20$ Hz for $d = g$ m

The above examples show that the natural frequencies depend on the overall stiffness of a shear wall as well as on its boundary conditions. The presence of flexible connections or hinges at the base increases deflections and reduces the frequency. The introduction of infill increases the frequency considerably. A solid cantilever wall on its own (Fig. 7.4(h)) can have very low frequency, but when it is connected to even a very slender column, as in Fig. 7.4(d), has considerably increased frequency.

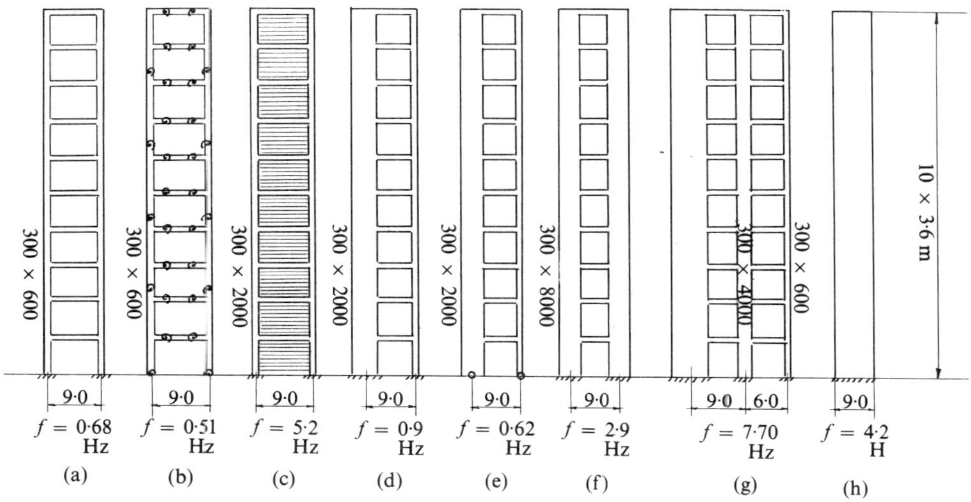

Fig. 7.4 Examples of calculations of frequencies of shear walls and frames

7.4 Ritz Method

In Rayleigh's method, the fundamental frequency of vibrations of elastic systems is determined based on the chosen shape of the deflection curve. This is equivalent to introducing constraints which reduce the system to one having a single degree of freedom, and the resulting frequency is a little higher than its exact value.

A better approximation of the fundamental frequency as well as the frequencies of higher modes can be obtained using the Ritz method, which is an extension of Rayleigh's energy method. In the Ritz method, the deflection curve, representing the mode

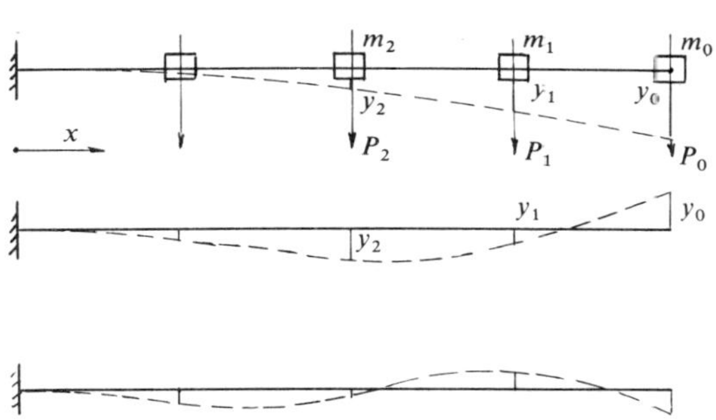

Fig. 7.5

of vibration, has to be taken with several parameters or arbitrary functions describing the shape of the vibrating element or a structure. These functions must satisfy only the boundary conditions for slope, deflection, and bending moments and need not be adjusted for variable mass or stiffness distribution. For the required N number of modes of vibration the same number of arbitrary coefficients must be introduced in the assumed function, e.g., for a solid cantilever beam, as shown in Fig. 7.5, it is convenient to use the expression:

$$y = A\left(1 - \cos\frac{\pi x}{2L}\right) + B\left(1 - \cos\frac{3\pi x}{2L}\right) + C\left(1 - \frac{5\pi x}{2L}\right) + \text{etc.}$$

For distributed mass and stiffness, it is convenient to base further procedure on the Hamilton principle which states that the total energy is unaffected by a small variation in the displacement–time function. Time, therefore, must be introduced into equations of the strain energy, U, as well as kinetic energy, V, so that:

$$\int_{t_1}^{t_2} (\delta U - \delta V)\, dt = 0$$

It can be shown that, for a cantilever beam, the first two angular frequencies are:

$$\omega_1 = 3{\cdot}52\sqrt{(EI/mL^4)}$$
$$\omega_2 = 24{\cdot}4\sqrt{(EI/mL^4)}$$

e.g., for a cantilever in Fig. 7.4(h) of $d=9$ m, and $W=200$ kN per floor, we obtain:

$$f_1 = 4{\cdot}25 \text{ Hz}, \qquad f_2 = 28{\cdot}0 \text{ Hz}$$

For complex arrangements, such as tall frames or shear walls, higher order frequencies and modes of vibrations can be evaluated using Ritz method as follows:

By Newton's second law of motion, we can write that:

$$\begin{aligned} P_1 &= -m_1 . \omega^2 . y_1 \\ P_2 &= -m_2 . \omega^2 . y_2, \text{ etc.} \end{aligned} \tag{7.12}$$

where P_1, P_2, etc., are the elastic restoring forces arising from the distortion of a structure; y_1, y_2, etc., are dynamic deflections under the action of these loads, m_1, m_2, etc., are masses at points 1, 2, etc., and ω is the circular frequency of vibration ($\omega_n = 2\pi . f_n$).

The deflections of a structure under the action of loads P_1, P_2, etc., can be written as:

$$\begin{aligned} y_1 &= a_{11}(-P_1) + a_{12}(-P_2) + \cdots \\ y_2 &= a_{21}(-P_1) + a_{22}(-P_2) + \cdots \end{aligned} \tag{7.13}$$

in which all a_{12}, etc., are the flexibility coefficients, i.e., deflections of a structure due to the unit loads placed at the positions of masses m_1, m_2, etc.

Substituting values for P from eqs. (7.12) and (7.13), we obtain a matrix for evaluation of displacements y in the general form:

$$\begin{bmatrix} a_{11}m_1\omega^2 - 1 & a_{12}m_2\omega^2 & a_{13}m_3\omega^2 & \cdots \\ a_{21}m_1\omega^2 & a_{22}m_2\omega^2 - 1 & a_{23}m_3\omega^2 & \cdots \\ a_{21}m_1\omega^2 & a_{32}m_2\omega^2 & a_{33}m_3\omega^2 - 1 & \cdots \\ \cdots & \cdots & \cdots & \cdots \end{bmatrix} \begin{Bmatrix} y_1 \\ y_2 \\ y_3 \\ \vdots \end{Bmatrix} = 0 \qquad (7.14)$$

However, since the right-hand side of this matrix is zero, the displacements are also zero except for the case when the determinant of the matrix is zero. In the latter case, the ratios of $y_1 : y_2 : y_3$, etc., can also be obtained. This is sufficient to define the shape of the vibrating system and its frequencies.

The determinant is a function of ω^2 and n values of ω can be obtained which will make the determinant equal to zero, i.e., the analysis will yield n modes of free vibrations at n frequencies, each associated with a particular amplitude ratio $y_1 : y_2$, etc. The mode of lowest frequency is the fundamental mode of vibration.

Denoting:

$$\alpha = \frac{m.h}{EI}\omega^2$$

the determinant c can be written as:

$$\begin{bmatrix} a_{11}\alpha - 1 & a_{12}\alpha & a_{13}\alpha & \cdots \\ a_{21}.\alpha & a_{22}\alpha - 1 & a_{23}\alpha & \cdots \\ a_{31}.\alpha & a_{32}.\alpha & a_{33}.\alpha - 1 & \cdots \\ \cdots & \cdots & \cdots & \cdots \end{bmatrix} = 0 \qquad (7.15)$$

or in the form:

$$A = (\alpha B - I) = 0$$

or

$$|B - \lambda I) = 0 \qquad (7.16)$$

in which I is the identity matrix and λ represents the roots of the determinant. The eigenvalues λ and therefore α can be obtained either by Jacobi's or by Householder's method.[10,11] The latter, although requiring more storage than Jacobi's method, is faster and saving in time increases with the order of the matrix. An additional saving can be obtained if not all the eigenvalues are required, while Jacobi's method computes all the eigenvalues simultaneously.

Generally, the procedure can be very laborious and time-consuming, particularly in evaluating the flexibility coefficients for a whole structure, but furnishes higher frequencies as well as modes of vibrations. The procedure is illustrated by the following two examples

Example 7.1

In this example, the frame shown in Fig. 1.7 is again used to calculate the frequencies and the modes of vibration by the Ritz method. At first it is assumed that the masses are concentrated at four levels as shown in Fig. 7.6.

Taking $m_0 = 400$ kN/g, $m_3 = 1{\cdot}5m_0$, $m_6 = 1{\cdot}5m_0$, $m_9 = m_0$ and $E = 28 \times 10^6$ kN/m^2, we can denote:

$$\alpha = \frac{400h}{EI.g}$$

The deflections (equal to the flexibility coefficients) can be denoted as:

$$\delta = c.h/EI$$

and these must be calculated for the unit lateral forces placed separately at points 0, 3, 6, and 9.

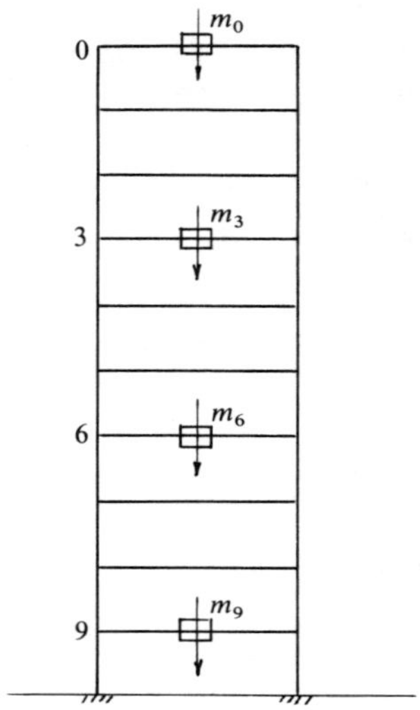

Fig. 7.6 Example

The general form of the matrix (7.14) becomes now

$$\begin{bmatrix} c_{00}.\alpha - 1 & 1{\cdot}5c_{30}.\alpha & 1{\cdot}5c_{60}.\alpha & c_{90}.\alpha \\ c_{03}.\alpha & 1{\cdot}5c_{33}.\alpha - 1 & 1{\cdot}5c_{63}.\alpha & c_{93}.\alpha \\ c_{06}.\alpha & 1{\cdot}5c_{36}.\alpha & 1{\cdot}5c_{66}.\alpha - 1 & c_{96}.\alpha \\ c_{09}.\alpha & 1{\cdot}5c_{39}.\alpha & 1{\cdot}5c_{69}.\alpha & c_{99}.\alpha - 1 \end{bmatrix} \begin{Bmatrix} y_0 \\ y_3 \\ y_6 \\ y_9 \end{Bmatrix} = 0$$

The evaluation of the coefficients c involves a standard procedure and is not shown here. Introducing these values into the above matrix, we obtain:

$$\begin{bmatrix} 25{\cdot}3\alpha - 1 & & & \\ 31{\cdot}3\alpha & 22{\cdot}5\alpha - 1 & & \\ 12{\cdot}75\alpha & 12{\cdot}3\alpha & 12\alpha - 1 & \\ 1{\cdot}5\alpha & 1{\cdot}95\alpha & 1{\cdot}65\alpha & 0{\cdot}9\alpha - 1 \end{bmatrix} = 0$$

The eigenvalues of this are: 61·6968, 5·7767, 0·6748, and −7·4482, and the latent roots α are:

$$\alpha_1 = 0{\cdot}0162$$
$$\alpha_2 = 0{\cdot}1231$$
$$\alpha_3 = 1{\cdot}4819$$
$$\alpha_4 = 3{\cdot}1343$$

For the assumed value of:

$$\alpha = \frac{400h}{gEI}\cdot\omega^2$$

we have:

$$\omega = 29{\cdot}7\sqrt{\alpha} \qquad f = \omega/2\pi$$

The corresponding frequencies are therefore:

$$f_1 = 0{\cdot}6 \text{ Hz} \qquad f_2 = 1{\cdot}97 \text{ Hz} \qquad f_3 = 15{\cdot}8 \text{ Hz}$$

Example 7.2

If the same frame is analysed for frequencies and modes of vibrations assuming that the masses are distributed at all ten floors as shown in Fig. 7.7, then the matrix becomes:

$$\begin{bmatrix}
25{\cdot}3\alpha-1 & & & & & & & & & \\
24\alpha & 22{\cdot}9\alpha-1 & & & & & & & & \\
21{\cdot}1\alpha & 22\alpha & 21\alpha-1 & & & & & & & \\
20{\cdot}9\alpha & 20{\cdot}4\alpha & 18{\cdot}6\alpha & 15\alpha-1 & & & & & & \\
16{\cdot}2\alpha & 15\alpha & 14{\cdot}8\alpha & 13\alpha & 13\alpha-1 & & & & & \\
11{\cdot}5\alpha & 10{\cdot}5\alpha & 10{\cdot}6\alpha & 10{\cdot}1\alpha & 11\alpha & 10\alpha-1 & & & & \\
8{\cdot}5\alpha & 8\alpha & 8\alpha & 8{\cdot}2\alpha & 9\alpha & 9\alpha & 8\alpha-1 & & & \\
5\alpha & 4{\cdot}8\alpha & 4{\cdot}8\alpha & 4{\cdot}8\alpha & 5\alpha & 6\alpha & 7\alpha & 6\alpha-1 & & \\
3\alpha & 3\alpha & 2{\cdot}8\alpha & 2{\cdot}7\alpha & 2{\cdot}5\alpha & 4\alpha & 4\alpha & 4\alpha & 3\alpha-1 & \\
1{\cdot}5\alpha & 1{\cdot}4\alpha & 1{\cdot}3\alpha & 1{\cdot}3\alpha & 1{\cdot}2\alpha & 1{\cdot}1\alpha & 1{\cdot}1\alpha & 1{\cdot}0\alpha & 1{\cdot}0\alpha & 0{\cdot}9\alpha-1
\end{bmatrix}$$

The solution of this determinant furnishes the following values for α:

$$\alpha_0 = 0{\cdot}0092, \quad \alpha_1 = 0{\cdot}0676, \quad \alpha_2 = 0{\cdot}3114, \quad \alpha_3 = 0{\cdot}5069$$

The corresponding values of frequencies are:

$$f_1 = 0{\cdot}655 \text{ Hz} \qquad f_2 = 2{\cdot}23 \text{ Hz} \qquad f_3 = 16{\cdot}05 \text{ Hz}$$

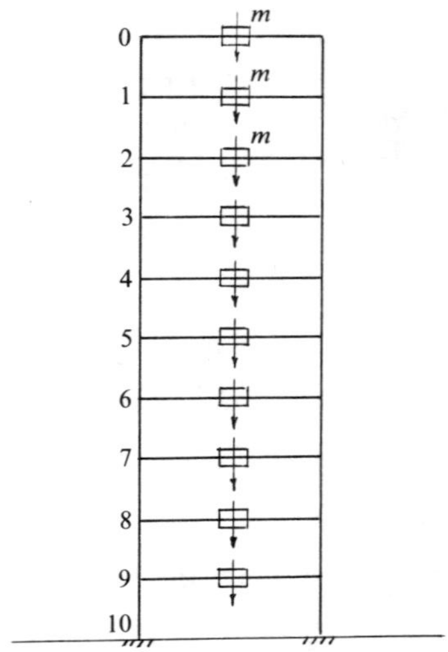

Fig. 7.7 Example

The corresponding modes of vibrations can now be evaluated as in Example 7.1.

It should be noted that, using Rayleigh's method, the fundamental frequency obtained was 0·67 Hz, while using the Ritz method, this is 0·6 or 0·655 Hz, depending on the assumed distribution of masses. The more accurate the assumption with regard to the distribution of masses, the more accurate are the values of frequencies and modes of vibration.

7.5 Damping of Vibrations of Tall Buildings

Damping is a property of a structure by which a structure set in motion comes to rest. The measure of damping is the rate of dissipation of the input energy within a structure or the rate of decay of amplitudes of vibrations. Damping characteristics of a structure govern its response to dynamic loads. Aerodynamic damping appears to be proportional to the ratio of air density to the average density of the building and to the frequency of the building.

When the part of wind energy absorbed by the structure becomes larger than that dissipated by structural damping, then the amplitude of oscillation continues to increase until resonance takes place, and a structure can fail due to aerodynamic instability. It is, therefore, important to predict the critical wind speed at which a given structure may become aerodynamically unsound.

Modern structural forms possess greater flexibility and smaller mass than traditional forms and this makes them more vulnerable to this form of instability. However, so far, this has been observed only in very light structures, such as transmission lines, chimneys, etc. Most tall buildings are able to dissipate extremely large quantities of energy by elastic and inelastic deformations without detrimental effect. In tall buildings, the mechanism of energy dissipation is very complicated and is not yet fully understood. It appears to be a function of the internal hysteresis in materials, mass of a structure, its stiffness, crack formation, imperfect functioning of joints, etc.

Damping forces are often assumed to be proportional to the velocity of the moving body, i.e., considered as viscous damping.

Damping takes place mainly through the structural materials, joints, internal damping of soil, and dissipation of energy into the soil.

A tall building can have several mechanisms by which input energy can be dissipated, i.e., by:

(a) internal damping within the material of a structure,
(b) damping through the walls and partitions,
(c) damping through the foundations and soil mass,
(d) aerodynamic damping, and
(e) added damping.

The material damping is usually small because materials behave basically elastically unless cracks develop. In flexible frames, a larger proportion of damping is through the structural deformations than in rigid shear walls. Damping through non-structural walls, partitions, windows, etc., can be considerable, and frequent cracks are visible indications of this effect. Internal damping through the soil mass is probably the most significant. In this, vibration energy travels into the soil mass and does not return; e.g., concrete foundations resting on the surface of the soil give 6 per cent to 20 per cent critical damping, while with backfill of up to 2 m it may reach 80 per cent. At high wind speeds, however, it appears that the natural damping of a structure is supplemented by the aerodynamic forces arising from the motion of the structure relative to the mean wind. Aerodynamic damping depends on the shape of a building, its surface texture, and vortex excitation.

Induced or added damping is a system separate from the basic structure of a building, the purpose of which is to dissipate the energy absorbed by a structure. Hydraulic dampers could, theoretically, be used for this purpose but no practical and economical solution exists suitable for tall buildings. Viscoelastic dampers, such as used in the World Trade Centre, appear to dissipate energy through heat and use up energy by resisting forces in shear. In the latter, three-plate dampers are used, laminated by thin layers of viscoelastic material.[9]

There are a number of terms used in describing aerodynamic damping of a structure, e.g., the term 'critical damping' is used to define damping in which the mass released from a displaced position just returns to its equilibrium position without passing it. 'Lesser' damping, c, is defined as a percentage of critical damping or as the 'logarithmic decrement'.

The 'logarithmic decrement' is frequently expressed as:

$$\delta_s = \log_e \frac{y_{n-1}}{y_n} \tag{7.17}$$

in which y_{n-1} and y_n are amplitudes of any two successive cycles. For small damping, it is often assumed that:

$$\delta_s = 2\pi c/c_{cr} \tag{7.18}$$

If there is no damping present ($c=0$), the motion is that of harmonic oscillation.

Damping is the dissipation of energy from a vibrating system and is related to the relative motion between the elements of the system. The exact mechanism by which a given structure dissipates energy is often difficult to identify. Approximate analyses

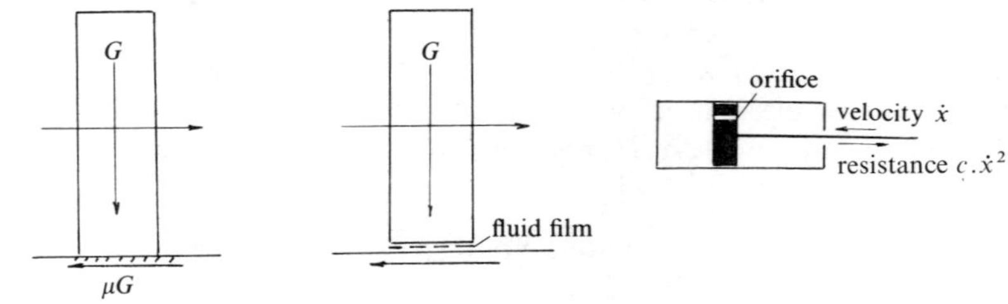

(a) Coulomb friction (b) Viscous friction (c) Orifice resistance

Fig. 7.8

of damping are based on idealized models such as shown in Fig. 7.8. A mechanism in Fig. 7.8(a) represents a Coulomb or dry friction which opposes the sliding of two members in contact. It is usually assumed that the friction force is directly proportional to the normal force G, in which the coefficient of proportionality is the coefficient of friction. This is assumed to be independent of the sliding velocity and acting in opposite direction to the velocity. The energy dissipated is equal to the work done against the friction forces, i.e., the product of $\mu.G$ and the relative displacement at the sliding interface.

Another damping mechanism used is shown in Fig. 7.8(b), which represents viscous friction. In this model, the energy dissipated is equal to the work done in shearing a fluid film contained between the sliding surfaces. The magnitude of the sliding force is directly proportional to the velocity of the relative displacement, and is opposite in direction to the motion.

Figure 7.8(c) shows a mechanism in which a piston with an orifice is moving inside a liquid-filled cylinder. In this model, the force which resists motion is approximately proportional to the square of the velocity of the piston.

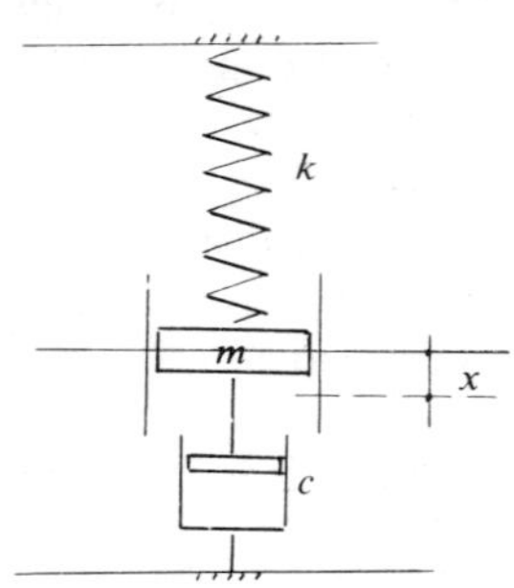

Fig. 7.9 Free vibration with viscous damping

Figure 7.9 shows a single-degree-of-freedom system with viscous damper. The differential equation of motion of mass m is:

$$m.x + c.x + k.x = 0 \tag{7.19}$$

The form of the solution of this equation depends on the relative value of the damping coefficient c to the critical damping coefficient c_{cr}, the latter being defined as:

$$c_{cr} = 2\sqrt{(k.m)} = 2m.\omega_n \tag{7.20}$$

The ratio $\lambda = c/c_{cr}$ is the fraction of critical damping.

(a) For $\lambda < 1$, i.e., less than critical damping, the solution of eq. (7.19) is:

$$x = e^{-ct/2m}.(A.\sin \omega_d.t + B.\cos \omega_d.t) =$$

$$= c.e^{-ct/2m}.\sin(\omega_d.t + \theta) \tag{7.21}$$

in which θ is the phase angle.

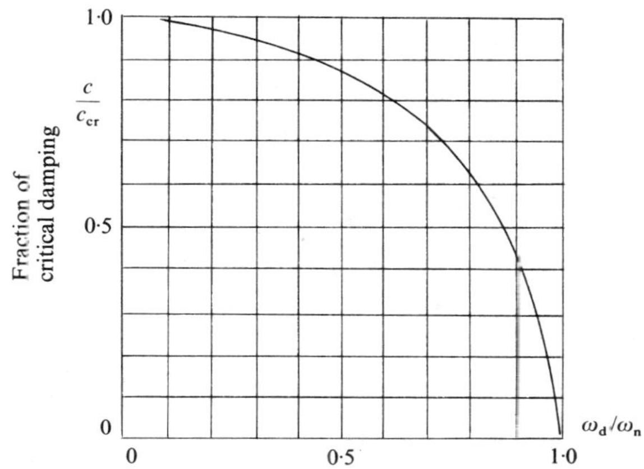

Fig. 7.10

Figure 7.10 shows the relationship between the damped natural frequency ω_d and the undamped natural frequency ω_n as a fraction of critical damping:

$$\omega_d = \omega_n(1 - \lambda^2)^{\frac{1}{2}} \tag{7.22}$$

(b) Critical damping occurs when $c = c_{cr}$, and then there is no oscillation. The solution of eq. (7.19) becomes:

$$x = (A + B.t).e^{-ct/2m} \tag{7.23}$$

(c) For greater than critical damping, $\lambda > 1$, eq. (7.23) has the solution:

$$x = e^{-ct/2m}(A.e^{\omega_n\sqrt{(\lambda^2-1)}\cdot t} + B.e^{-\omega_n\sqrt{(\lambda^2-1)}\cdot t}) \tag{7.24}$$

This motion is non-oscillatory and the system displaced tends to return gradually to its original position.

The rate of damping of a system when $\lambda < 1$ can be related to its successive amplitudes of free oscillations (Fig. 7.11), so that:

$$x = c.e^{-\lambda\omega_n t}.\sin(\omega_d + \theta)$$

Taking two maximum amplitudes separated by n cycles, we obtain:

$$\frac{x_n}{x_1} = e^{-2\pi n\lambda/(1-\lambda^2)^{\frac{1}{2}}} \tag{7.25}$$

The logarithmic decrement δ is the natural logarithm of the ratio of the amplitudes

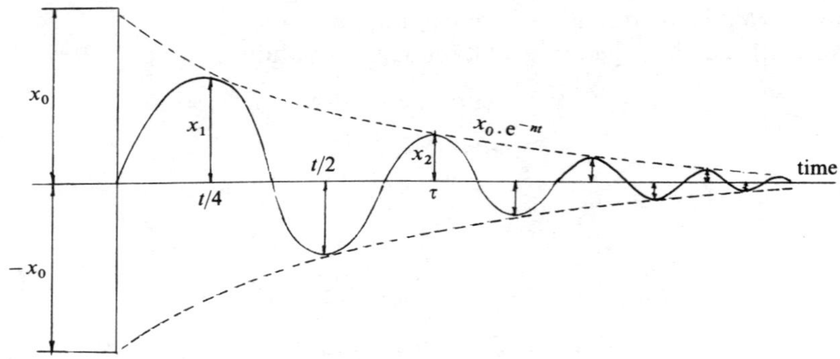

Fig. 7.11 Damped free vibration

of two successive cycles of the damped free vibration, i.e.,

$$\delta = \log\frac{x_1}{x_2} \quad \text{or} \quad \frac{x_2}{x_1} = e^{-\delta} \tag{7.26}$$

For $n=1$, eq. (7.25) gives:

$$\delta = \frac{2\pi\lambda}{(1-\lambda^2)^{\frac{1}{2}}}$$

For $\delta < 0{\cdot}2$, a good approximation can be obtained from:

$$\delta \approx \frac{x_1 - x_2}{x_1} \tag{7.27}$$

and for $\delta < 0{\cdot}1$:

$$\delta \approx 2\pi\lambda \tag{7.28}$$

In the discussion of motion of structures, it is usually assumed that the material is perfectly elastic and internal damping is neglected. However, observations show that all materials exhibit some damping so that instead of a straight line in a stress–strain diagram (see Fig. 7.12), there is a loop, the width of which depends on the material and on the level of stress applied. This diagram is called the hysteresis loop. The area of the loop gives the amount of energy dissipated per cycle due to the internal damping or hysteresis.

While mathematical expressions for damping forces described previously can be derived based on various models, the energy dissipation through the internal hysteresis of members is more complicated. The stress–strain relationship changes in cyclic manner and is affected by the past history of the strain cycles.

The effect of strain hysteresis can be taken into account by assuming that the stiffness of the system is related to the damping force and is inversely proportional to the frequency. When the viscous, damped, single-degree-of-freedom system undergoes vibration, defined by:

$$x = x_0 . \sin \omega t \tag{7.29}$$

the force exerted on the mass by the spring and damper is:

$$F = k . x_0 . \sin \omega t + c . \omega . x_0 . \cos \omega t \tag{7.30}$$

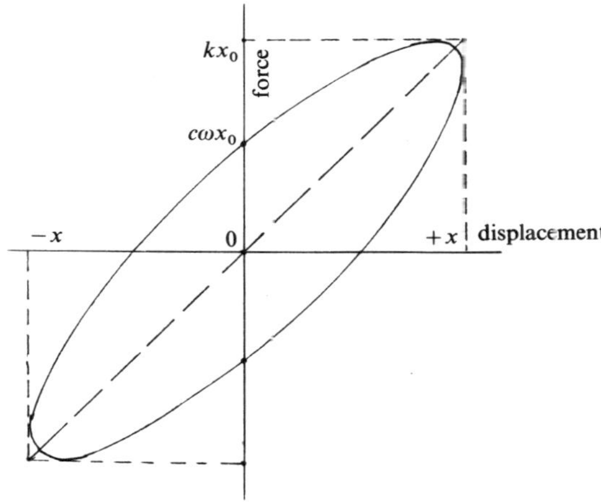

Fig. 7.12 Hysteresis loop

Equations (7.29) and (7.30) define the relation between F and x, which is in the form of an ellipse shown in Fig. 7.12. The energy dissipated in one cycle of oscillation is:

$$W = \int_{t}^{t+2\pi\omega} F\frac{dx}{dt}\cdot dt = \pi.c.\omega.x_0{}^2$$

Further references and discussion on this problem can be found in the references.[1–5]

References

1. W. C. Hurty and M. F. Rubinstein. *Dynamics of Structures*, Prentice Hall (1964).
2. S. Timoshenko, *Vibration Problems in Engineering*, Princeton N.Y., van Norstrand, 1955.
3. D. E. Goldman and H. E. Von Gierke. *The Effects of Shock and Vibration on Man*, Naval Medical Research Institute, Lecture and Review Series No. 60-3, Bethesda (1960).
4. 'Vibration in Buildings', *B.R.S. Digest*, No. 78, H.M.S.O. (June, 1955).
5. J. C. Guignard, *Journal of Sound and Vibration*, 11–16 (8 March, 1971).
6. F. Postlethwaite. 'Human Susceptibility to Vibration', *Engineering*, **157**, 61 (1944).
7. H. Reiher and F. J. Meister. (1931) *The Sensitivity of the Human Body to Vibration*, (USAF Air Material Command Translation No. F-TS-616-RE, (1946)).
8. B. J. Lazan and L. E. Goodman. 'Material and Interface Damping', Chapter 36 of *Shock and Vibration Handbook*, Vol. 2 (1965).
9. *Architectural Records* (September, 1971).
10. P. C. Jennings. 'Equivalent Viscous Damping for Yielding Structures', *J. Mech. Div., ASCE*, 103 (1968).
11. J. H. Wilkinson. *The Algebraic Eigenvalue Problem*, chapter 5, 282–355, Clarendon Press, Oxford (1965).
12. G. Temple and W. G. Brickley. *Rayleigh's Principle*, Oxford University Press, N.Y. (1933).

8 Secondary Effects in the Analysis of Tall Buildings

8.1 Introduction

The term 'primary analysis' is used to describe an analysis based on the assumption of linearly elastic relationship between stresses and strains, on validity of the principle of superposition, on the original geometry of a structure, on the action of loads considered to be static, and assuming that only the strain energy of bending need be taken into account.

The term 'secondary analysis' is used to describe the effect of axial forces and deformations, shearing forces, torsion, creep, shrinkage, temperature changes, flexible connections, differential settlement of supports, dynamic effect of loads, stress concentrations and cracks, inelastic response of materials, instability, buckling, etc. Many of these effects are interdependent in some complex way, and to simplify the analysis, one effect is treated separately from another.

The magnitude of these secondary stresses can vary from negligible to very high, and can even govern the response of members: cracks can appear, there may be excessive deflections and vibrations and, in the limit, a structure may collapse.

The following analysis is based on simplifying assumptions that (a) each effect can be treated separately, and (b) the principle of superposition is valid. The following effects are considered:

(a) the effect of axial forces on the stiffness or flexibility of members,
(b) the effect of axial deformations,
(c) the effect of shearing forces and deformations,
(d) the effect of torsion,
(e) the effect of flexible connections, and
(f) the effect of stress concentrations.

The effect of elastic instability is considered separately.

The effect of axial forces and deformations may become particularly important in the case of tall and slender frames subjected to high gravity and lateral forces, while shear forces and deformations may affect rigid shear walls. Torsional stresses are present in all buildings, although they are seldom considered because of difficulties of three-dimensional analysis, while two-dimensional analysis usually ignores torsion altogether. However, perforated service cores are often designed to resist torsional as well as flexural moments, particularly when the resultant wind force does not coincide with the centre of rotation of the supporting system.

Similarly, stress concentrations are present in all structures and can cause redistribution of stresses and deformations, both local and overall.

Flexible connections exist mainly in structural steel buildings when joints are bolted or riveted, but concrete buildings may behave in a similar manner if cracks develop at joints. Some joints in precast concrete buildings can also be considered as flexible.

Elastic instability can affect individual members as well as the whole structure. Slender steel buildings of small dead weight, and precast buildings with inadequate connections are particularly vulnerable to this effect. These problems will now be briefly reviewed.

8.2 The Effect of Axial Forces on Flexibility Coefficients

The effect of axial forces on secondary stresses and deformations can be divided into:

(a) the effect of stiffness or flexibility coefficients of members, particularly those in compression,
(b) the effect of differential elastic or plastic axial deformations on the change of geometry of the system, and therefore on the action of loads, and
(c) the effect of axial forces on instability and ultimate strength of a structure.

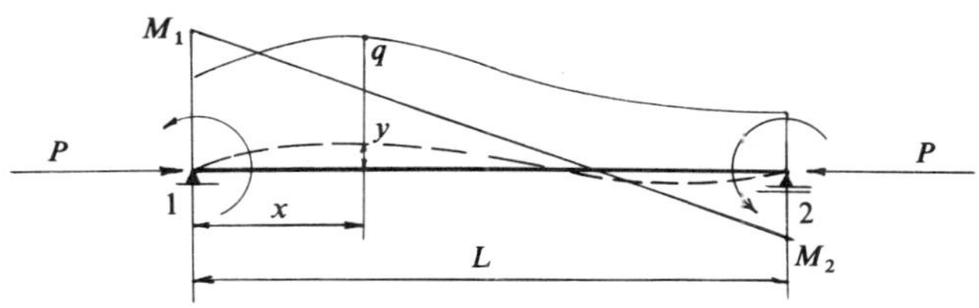

Fig. 8.1

The effect of axial forces on stiffness or flexibility coefficients of members can be studied on the basis of the well-known differential equation:

$$EI\frac{d^4y}{dx^4} + P\frac{d^2y}{dx^2} = q \tag{8.1}$$

in which y is the deflection at a distance x, and q is the loading (Fig. 8.1).

The solution of eq. (8.1) can be found in most textbooks on the subject for all standard loading conditions and will not be pursued here except for the effect of axial forces on the flexibility coefficients.

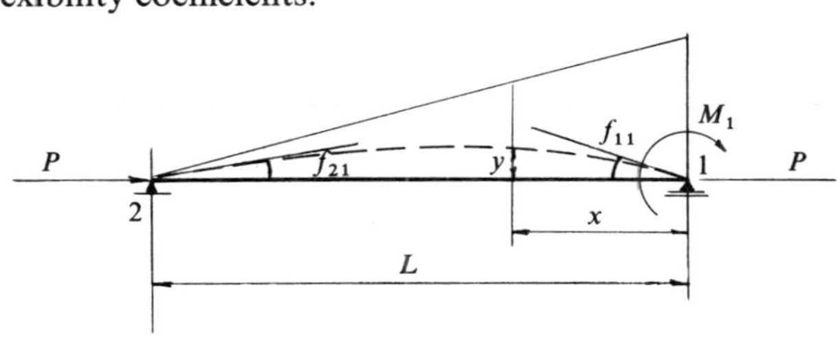

Fig. 8.2

For the beam in Fig. 8.2, subjected to axial forces P and bending moment M_1, at one end, we obtain from eq. (8.1):

$$y = \frac{M_1}{P}\left(\frac{\sin kx}{\sin kL} - \frac{x}{L}\right) \tag{8.2}$$

in which: $$k = \sqrt{(P/EI)}$$

Taking derivatives of eq. (8.2), with respect to x, we obtain the expression for the slope of the deflection line of the beam. The slopes at $x=0$ and $x=L$ are, by definition, equal to the flexibility coefficients. These are:

$$\begin{aligned} f_{11} &= -\left(\frac{\mathrm{d}y}{\mathrm{d}x}\right)_{x=0} = M_1\cdot\frac{L}{3EI}\cdot\frac{3}{2U}\left(\frac{1}{2U} - \frac{1}{\tan 2U}\right) = M_1\cdot\frac{L}{3EI}\cdot c_{11} \\ f_{21} &= \left(\frac{\mathrm{d}y}{\mathrm{d}x}\right)_{x=L} = M_1\cdot\frac{L}{6EI}\cdot\frac{3}{U}\left(\frac{1}{\sin 2U} - \frac{1}{2U}\right) = M_1\frac{L}{6EI}\cdot c_{21} \end{aligned} \tag{8.3}$$

in which $$U = \tfrac{1}{2}kL = \frac{L}{2}\sqrt{(P/EI)}$$

The functions c_{11} and c_{21} are shown in Table 8.1.

Table 8.1 Flexibility coefficients for a beam subjected to end bending moment and axial force (see eqs. (8.3))

$2U = kL$	c_{11}	c_{12}	$2U = kL$	c_{11}	c_{12}	$2U = kL$	c_{11}	c_{12}
0	1·0000	1·0000	2·05	1·4742	1·8706	3·10	23·5659	45·9234
0·1	1·0007	1·0008	2·10	1·5158	1·9493	π	$\pm\infty$	$\pm\infty$
0·2	1·0027	1·0047	2·15	1·5616	2·0366	3·15	−112·9747	−227·16
0·3	1·0061	1·0106	2·20	1·6124	2·1336	3·20	−15·7398	−32·7063
0·4	1·0108	1·0189	2·25	1·6690	2·2422	3·25	−8·1875	−17·6312
0·5	1·0171	1·0299	2·30	1·7325	2·3640	3·30	−5·4154	−12·0770
0·6	1·0249	1·0437	2·35	1·8041	2·5022	3·40	−3·0787	−7·4248
0·7	1·0343	1·0603	2·40	1·8854	2·6596	3·50	−2·0433	−5·3769
0·8	1·0460	1·0801	2·45	1·9786	2·8403	3·60	−1·4572	−4·2292
0·9	1·0625	1·1033	2·50	2·0864	3·0502	3·70	−1·0787	−3·4990
1·0	1·0765	1·1304	2·55	2·2124	3·2963	3·80	−0·8128	−2·9961
1·1	1·0912	1·1617	2·60	2·3618	3·5890	3·90	−0·6147	−2·6314
1·2	1·1114	1·1979	2·65	2·5415	3·9421	4·00	−0·4603	−2·3570
1·3	1·1345	1·2396	2·70	2·7619	4·3766	4·20	−0·2317	−1·9792
1·4	1·1610	1·2878	2·75	3·0386	4·9233	4·40	−0·0652	−1·7429
1·5	1·1915	1·3434	2·80	3·3963	5·6315	4·60	+0·0682	−1·5962
1·6	1·2266	1·4078	2·85	3·8765	6·5845	4·80	+0·1851	−1·5152
1·7	1·2673	1·4830	2·90	4·5550	7·9343	5·00	+0·2975	−1·4914
1·8	1·3147	1·5710	2·95	5·5875	9·9915	6·00	+1·8015	−3·7455
1·9	1·3704	1·6750	3·00	7·3486	13·5057	2π	$\pm\infty$	$\pm\infty$
2·0	1·4365	1·7993	3·05	11·0314	20·8629	6·50	−2·0242	+4·1490

8.3 Relationship between the Stability Functions and Flexibility Coefficients

The stiffness, s, and the carry-over factor, c, can be related to the flexibility coefficients f_{11} and f_{21} as follows. From Fig. 8.3,

$$M_1.f_{11} - M_2.f_{12} = \theta_1$$
$$-M_1.f_{21} + M_2.f_{22} = 0$$

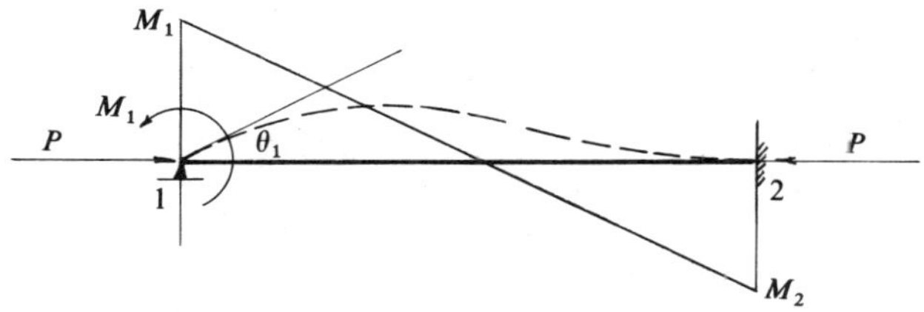

Fig. 8.3

from which:

$$\frac{M_2}{M_1} = c = \frac{f_{21}}{f_{11}} \tag{8.4}$$

and

$$\frac{M_1}{\theta_1} = s = \frac{f_{11}}{f_{11}^2 - f_{21}^2} \tag{8.5}$$

or solving for f_{11} and f_{21} we obtain:

$$f_{11} = \frac{1}{s(1 - c^2)} \quad \text{and} \quad f_{21} = \frac{c}{s(1 - c^2)} \tag{8.6}$$

Introducing in eqs. (8.4) and (8.5) the values for flexibility coefficients, we obtain:

$$s = \frac{kL(\sin kL - kL.\cos kL)}{2 - 2\cos kL - kL.\sin kL} \tag{8.7}$$

$$c = \frac{kL - \sin kL}{\sin kL - kL.\cos kL} \tag{8.8}$$

These coefficients are tabulated by Livesley and Chandler (see reference 8).

If the far end of the member is pinned, then the carry-over factor is zero and the stiffness, s, becomes:

$$M_1.f_{11} = \theta_1 \quad \ldots \quad M_1/\theta_1 = \bar{s} = \frac{1}{f_{11}}$$

Introducing $$s = f_{11}/(f_{11}^2 - f_{21}^2)$$

we obtain: $$\bar{s} = s[1 - (f_{21}/f_{11}^2)]$$

$$\bar{s} = s(1 - c^2)$$

8.4 Stiffness and Carry-over Factors, including Instability Functions

If the stiffness or distribution method is used, then the values of the stiffness and carry-over factors can be calculated based on Fig. 8.4 as follows:

$$M_1 \cdot \frac{c_{11}.L}{3EI} - M_2 \cdot \frac{c_{12}L}{6EI} = \theta_1$$

$$-M_1 \cdot \frac{c_{21}L}{6EI} + M_2 \cdot \frac{c_{22}.L}{3EI} = \theta^2$$

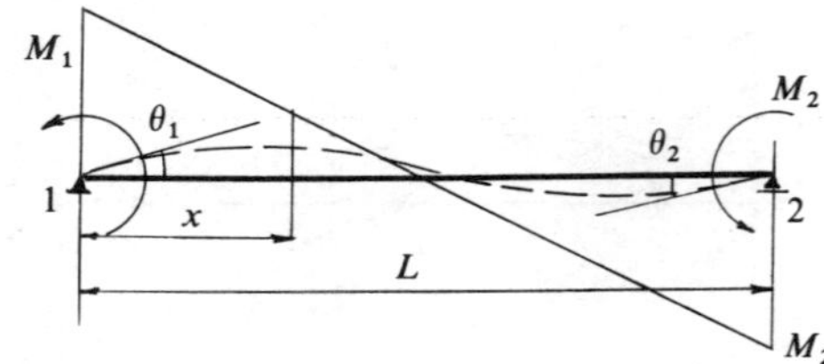

Fig. 8.4

By setting $\theta_2 = 0$, we obtain:

the stiffness
$$k = \frac{M_1}{\theta_1} = \frac{4EI}{L} \cdot \frac{c_{11}}{4c_{11} - c_{12}} \tag{8.9}$$

and carry-over factor
$$c = \frac{M_2}{M_1} = \frac{1}{2} \cdot \frac{c_{21}}{c_{22}} \tag{8.10}$$

If the end 2 is simply-supported, then the stiffness becomes:

$$k = \frac{3EI}{L} \cdot \frac{1}{c_{11}} \tag{8.11}$$

8.5 The Effect of Axial Deformations of Columns

The effect of axial deformations of columns and large lateral deflections is first discussed in relation to a portal frame and then extended to multi-storey frames.

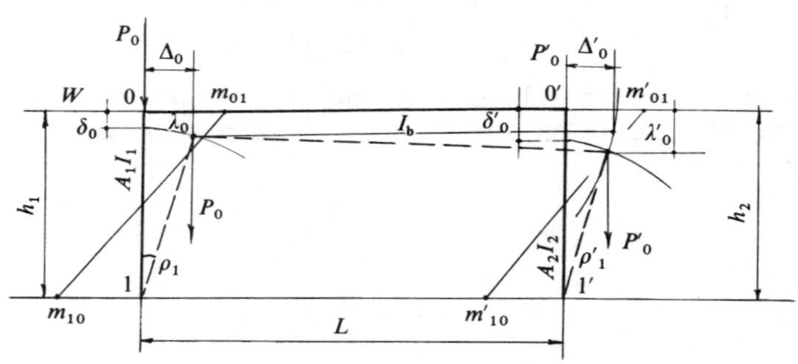

Fig. 8.5 The effect of axial deformations of columns

In Fig. 8.5 the frame is assumed to be subjected to the action of lateral force W and gravity loads P_0 and P'_0. The elastic axial deformations of columns are denoted by δ_0 and δ'_0 which can be expressed as follows:

$$\delta_0 = \frac{P_0 . h_1}{EA_1} + \frac{m_{01} + m'_{01}}{L . EA_1} \cdot h_1$$

$$\delta'_0 = \frac{P'_0 . h_2}{EA_2} + \frac{m_{01} + m'_{01}}{L . EA_2} \cdot h_2$$

While axial deformations δ_0 and δ'_0 depend on the unknown bending moments, the values of λ_0 and λ'_0 in Fig. 8.5 depend on the rotations of columns and on the condition that the beam length must remain at its original value L (neglecting small deformation from the axial forces in the beam).

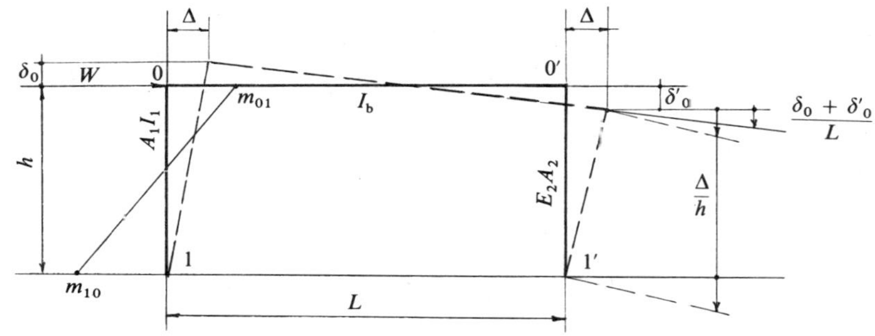

Fig. 8.6

For the portal frame in Fig. 8.6, taking only the effect of axial deformations resulting from the wind forces, the general form of the matrix is:

Table 8.2

$$\begin{bmatrix} \dfrac{h_1}{3EI_1}+\dfrac{L}{3EI_b} & -\dfrac{h_1}{6EI_1} & -\dfrac{L}{6EI_b} & 0 & -\dfrac{1}{h_1} \\ -\dfrac{L}{6EI_b} & 0 & \dfrac{h_2}{3EI_2}+\dfrac{L}{3EI_b} & -\dfrac{h_2}{6EI_2} & -\dfrac{1}{h_2} \\ 0 & 0 & -\dfrac{h_2}{6EI_2} & \dfrac{h_2}{3EI_2} & -\dfrac{1}{h_2} \\ -\dfrac{h_1}{6EI_1} & \dfrac{h_1}{3EI_1} & 0 & 0 & -\dfrac{1}{h_1} \\ \dfrac{1}{h_1} & \dfrac{1}{h_1} & \dfrac{1}{h_2} & \dfrac{1}{h_2} & 0 \end{bmatrix} \begin{Bmatrix} m_{01} \\ m_{10} \\ m'_{01} \\ m'_{10} \\ \Delta \end{Bmatrix} = \begin{Bmatrix} -\dfrac{\delta_0+\delta'_0}{L} \\ -\dfrac{\delta_0 \quad \delta'_0}{L} \\ 0 \\ 0 \\ W \end{Bmatrix}$$

in which $$\delta_0 = \frac{m_{01}+m_{10}}{L.EA_1}\cdot h_1; \qquad \delta'_0 = \frac{m'_{01}+m'_{10}}{L.EA_2}\cdot h_2$$

Example 8.1

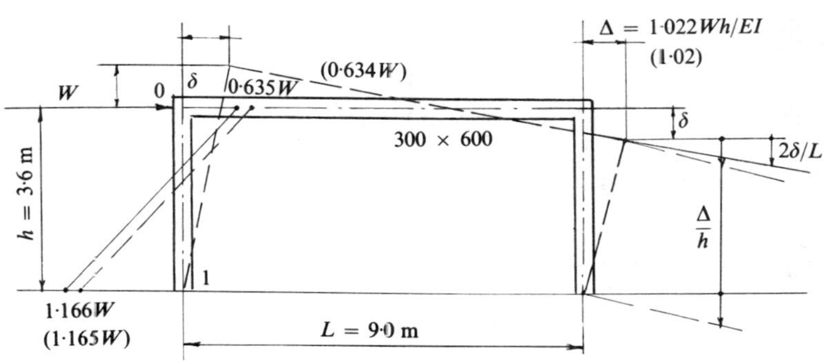

Fig. 8.7 Example

In this example, the effect of axial forces resulting from wind only is evaluated on the frame shown in Fig. 8.7. The required values of axial deformations are:

$$\delta = \frac{2m_{01}.h}{EA.L}$$

or

$$\frac{2\delta}{L}\cdot EI = 0{\cdot}005\,336 m_{01}$$

The matrix becomes now:

$$\begin{bmatrix} 2{\cdot}7054 & -0{\cdot}6 & -1 \\ -0{\cdot}6 & 1{\cdot}2 & -1 \\ 1 & 1 & 0 \end{bmatrix} \begin{Bmatrix} m_{01} \\ m_{10} \\ \Delta \end{Bmatrix} = \begin{Bmatrix} 0 \\ 0 \\ 1{\cdot}8W \end{Bmatrix}$$

from which the distribution of moments and deformations is as shown in Fig. 8.7. Figures in brackets show the values based on the primary analysis, i.e., in which the effect of axial deformations is neglected. It can be seen that, in this case, the effect of axial forces resulting from wind only is small. However, this may not always be the case particularly in tall and slender frames.

In multi-storey frames, axial forces and deformations are cumulative and their effect is felt most in the upper floors. Figure 8.8(a) shows diagrammatically the effect of axial forces arising from wind pressure, while Fig. 8.8(b) shows the effect of gravity forces on axial deformations. These effects can be taken into account in setting out of the force–displacement matrix without difficulty, e.g., from Fig. 8.8(a), axial forces can be expressed in terms of statically indeterminate bending moments as

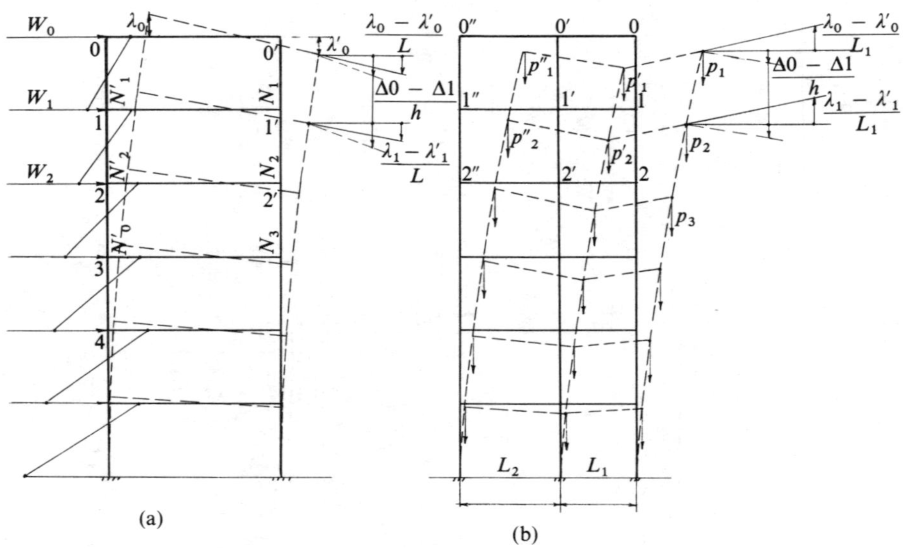

Fig. 8.8 Effect of axial deformations on frames due to lateral and gravity forces

follows:

$$N_{01} = -N'_{01} = \frac{m_{01} + m'_{01}}{L}$$

$$N_{12} = -N'_{12} = \frac{m_{01} + m_{10}}{h} + \frac{m'_{01} + m'_{10}}{h}$$

and the axial deformations of columns are then calculated based on the expression:

$$\lambda = \frac{N.h}{EA}$$

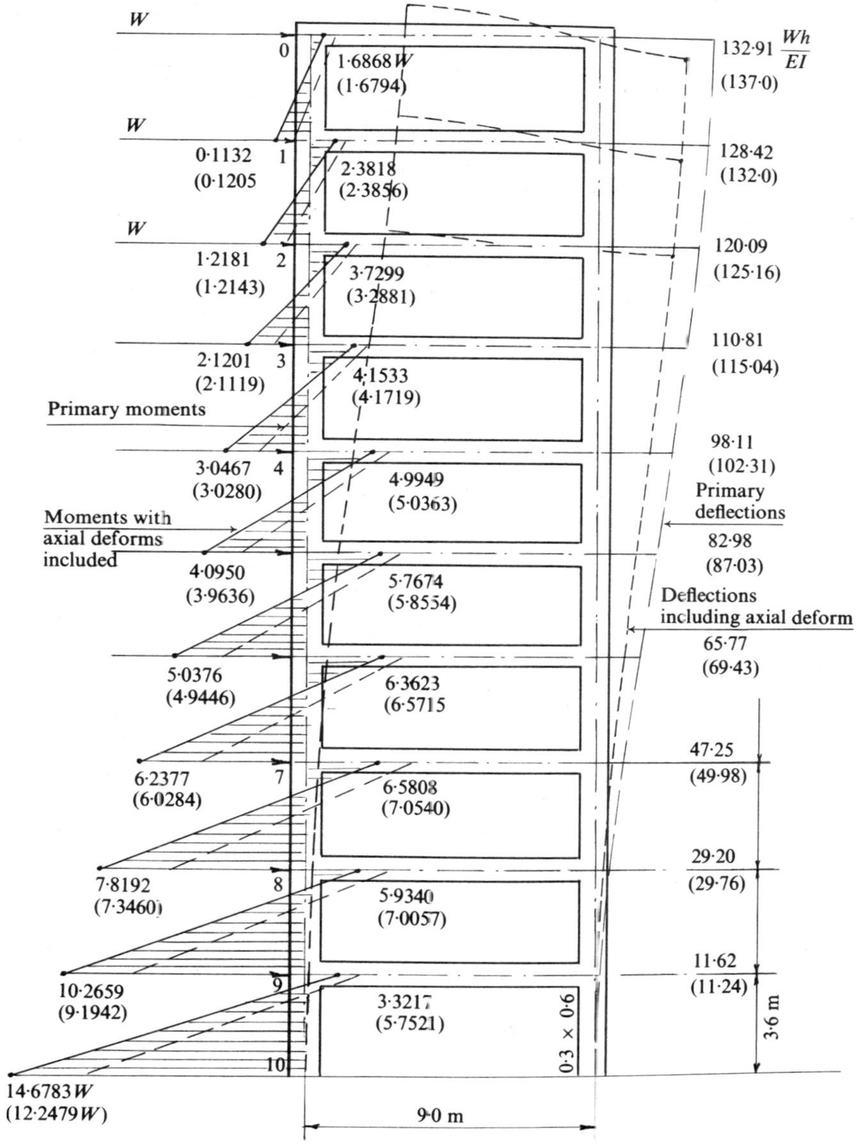

Fig. 8.9 Bending moments and deflections including the effect of axial deformations

Table 8.3 Matrix for the analysis of a frame in Fig. 8.9 including the effect of axial deformations

	0		1			2			3			4			5	
	0 1	Δ0	1 0	1 2	Δ1	2 1	2 3	Δ2	3 2	3 4	Δ3	4 3	4 5	Δ4	5 4	5 6
1 0 0′	2·7	−1	−0·599	0·646*	1	0·574*	0·574*	0	0·502*	0·502*	0	0·430*	0·430*	0	0·359*	0·359*
0–0′	1	0	1													
2 1 1′	1·292*	−1	1·501	2·701	1	−0·601	1·048*	0	1·048*	1·040*	0	0·860*	0·860*	0	0·718*	0·718*
0 1 2	−0·6	−1	1·2	−1·2	2·0	0·6	0	1								
1–1′	−1	0	−1	1	0	1	0	0								
3 2 2′	1·048*	0	1·048*	1·048*	−1	1·501	2·704	1	−0·599	1·004*	0	0·86*	0·86*	0	0·718*	0·718*
1 2 3				−0·6	−1	1·2	−1·2	2	0·6	0	1					
2–2′				−1	0	−1	1	0	1	0	0					
4 3 3′	1·004*	0	1·004*	1·004	0	1·004*	1·004*	−1	1·501	2·701	1	−0·599	0·86*	0	0·718*	0·718*
2 3 4							−0·6	−1	1·2	−1·2	2·0	0·6	0	1		
3–3′							−1	0	−1	1	0	1	0	0		
5 4 4′	0·86*	0	0·86*	0·86*	0	0·86*	0·86*	0	0·86*	0·86*	−1	1·5008	2·7008	1	−0·599	0·718*
3 4 5										−0·6	−1	1·2	−1·2	2	0·6	0
4–4′										−1	0	−1	1	0	1	0
6 5 5′	0·718*	0	0·718*	0·718*	0	0·86*	0·86*	0	0·86*	0·86*	0	0·718*	0·718*	−1	1·5007	2·7007
4 5 6													−0·6	−1	1·2	−1·2
5–5′													−1	0	−1	1
7 6 6′	0·574*	0	0·574*	0·574*	0	0·574*	0·574*	0	0·574*	0·574	0	0·574*	0·574*	0	0·574*	0·574*
5 6 7																−0·6
6–6′																−1
8 7 7′	0·43*	0	0·43*	0·43*	0	0·43*	0·43*	0	0·43*	0·43*	0	0·43*	0·43*	0	0·43*	0·43*
6 7 8																
7–7′																
9 8 8′	0·286*	0	0·286*	0·286*	0	0·286*	0·286*	0	0·286*	0·286*	0	0·286*	0·286*	0	0·286*	0·286*
7 8 9																
8–8′																
10 9 9′	0·143*	0	0·143*	0·143*	0	0·143*	0·143*	0	0·143*	0·143*	0	0·143*	0·143*	0	0·143*	0·143*
8 9 10																
9–9′																
10 9																

* All figures marked with asterisks to be multiplied by 10^{-3}.

6			7			8			9			10	$\frac{EI}{h}$
5	6 7	Δ6	7 6	7 8	Δ7	8 7	8 9	Δ8	9 8	9 10	Δ9	10 9	
86*	0·286*	0	0·215*	0·215*	0	0·143*	0·143*	0	0·070*	0·070*	0	0	0
													1·8W
74*	0·574*	0	0·430*	0·430*	0	0·286*	0·286*	0	0·143*	0·143*	0	0	0
													0
													1·8W
74*	0·574*	0	0·43*	0·43*	0	0·286*	0·286*	0	0·143*	0·143*	0	0	0
													0
													1·8W
74*	0·574	0	0·43*	0·43*	0	0·286*	0·286*	0	0·143*	0·143*	0	0	0
													0
													1·8W
74*	0·574*	0	0·43*	0·43*	0	0·286*	0·286*	0	0·143*	0·143*	0	0	0
													0
													1·8W
5994	0·574	0	0·43*	0·43*	0	0·286*	0·286*	0	0·143*	0·143*	0	0	0
6	0	−1											0
	0	0											1·8W
006	2·7005	1	−0·5993	0·43*	0	0·286*	0·286*	0	0·143*	0·143*	0	0	0
2	−1·2	2	0·6	0	−1								0
1	1	0	1	0	0								1·8W
3*	0·43*	−1	1·5004	2·7004	1	−0·5997	0·286*	0	0·143*	0·143*	0	0	0
	−0·6	−1	1·2	−1·2	2	0·6	0	−1					0
	−1	0	−1	1	0	1	0	0					1·8W
6*	0·286*	0	0·286*	0·286*	−1	1·5003	2·7003	1	−0·5996	0·143*	0	0	0
				−0·6	−1	1·2	−1·2	2	0·6	0	−1	0	0
				−1	0	−1	1	0	1	0	0		1·8W
3*	0·143*	0	0·143*	0·143*	0	0·143*	0·143*	−1	1·5001	2·7001	1	−0·6	0
							−0·6	−1	1·2	−1·2	2	0·6	0
							−1	0	−1	1	0	1	1·8W
										−0·6	−1	1·2	0

Example 8.2 Example of the analysis of frame for the effect of wind forces, including axial deformations

For the frame shown in Fig. 1.7, previously analysed for the primary effects of wind only, the effect of axial deformations can be included in the matrix as shown in Table 8.3. Figure 8.9 shows the distribution of bending moments and deformations. To facilitate comparison, the values from the primary analysis are shown by dotted lines and figures in brackets.

It can be seen from Fig. 8.9 that in this example bending moments are amplified, particularly in the lower floors, by approximately 20 per cent when compared with the primary analysis. Deflections are increased by approximately 4 per cent. The effect of axial forces on the flexibility coefficients was not included. A similar analysis can be made of the effect of the differential axial deformations of columns due to gravity loads.

8.6 The Effect of Shearing Forces in the Analysis of Shear Walls

Shearing forces have usually small effect on flexible beams and frames, and in most structures the major part of deformations is caused by the action of bending moments. However, there are structures in which shear effect can be significant and may need to be considered. This may occur when a heavily loaded member has a relatively low span to beam ratio, or when the material has low shear modulus of elasticity. In this category are many concrete shear walls and service cores in tall buildings.

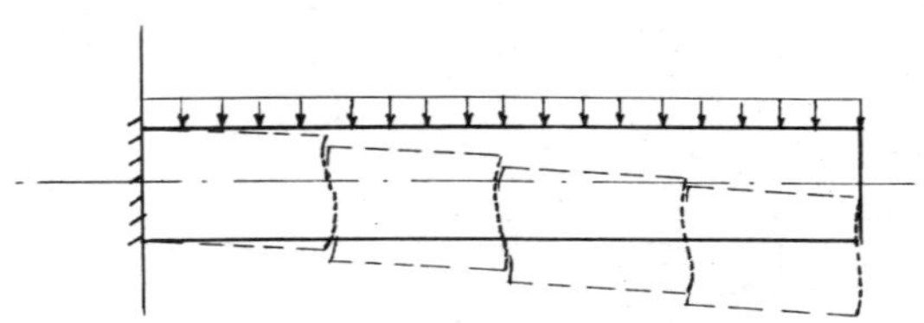

Fig. 8.10

The contribution of the shearing forces to deflections of beams and walls (Fig. 8.10) is in the form of a mutual sliding of adjacent cross-sections along each other. Since the distribution of shearing stresses is non-uniform, plane sections become curved and the slope of the deflection curve, due to the shear alone, is equal at each cross-section to the shearing strain at the centroid of the cross-section.

Shearing deflections are usually calculated by the Principle of Virtual Work or Unit Load Method. According to the Principle of Virtual Work, if an elastic body is in equilibrium under a set of forces and is given a small virtual displacement, the work done by the external forces during the virtual displacement is equal to the work done by the internal forces during the same displacement, i.e.,

$$U_{\text{ext}} = U_{\text{int}} \tag{8.12}$$

Equating the external and internal work, the following basic equation is obtained for the deflection of any point:

$$1. \quad \Delta = \int_l \frac{M.m.\mathrm{d}x}{EI} + \int_l \frac{V.v.\mathrm{d}x}{GI^2} \int \frac{Q^2}{b^2} \mathrm{d}A \tag{8.13}$$

The last integral in eq. (8.13) depends on the cross-sectional dimensions, and it is usual to denote this quantity by F:

$$F = \frac{A}{I^2}\int_A \frac{Q^2}{b^2}\,\mathrm{d}A \tag{8.14}$$

This value is called the shape factor and represents a non-dimensional parameter which must be evaluated for each particular cross-section.

Equation (8.13) can be written in the form:

$$\Delta = \int_l \frac{M.m.\mathrm{d}x}{EI} + F\int_l \frac{V.v.\mathrm{d}x}{GA}$$

The values of M and V are the bending moment and the shearing force in the beam due to the action of external loads, while m and v are the corresponding quantities resulting from the unit load applied in the direction of the deflection considered.

For a rectangular beam, the shape factor is calculated from eq. (8.14) by substituting the value:

$$Q = \frac{b}{2}\left(\frac{h^2}{4} - y^2\right)$$

so that:

$$F = \frac{A}{I^2}\cdot\int_{-h/2}^{+h/2} \frac{Q^2}{b^2}\cdot\mathrm{d}A = \frac{6}{5}$$

Similarly, for a thin tubular section, $F=2$, for a circular section $F=10/a$, and for I beams or closed rectangular boxes, $F=A/A_{\text{web}}$.

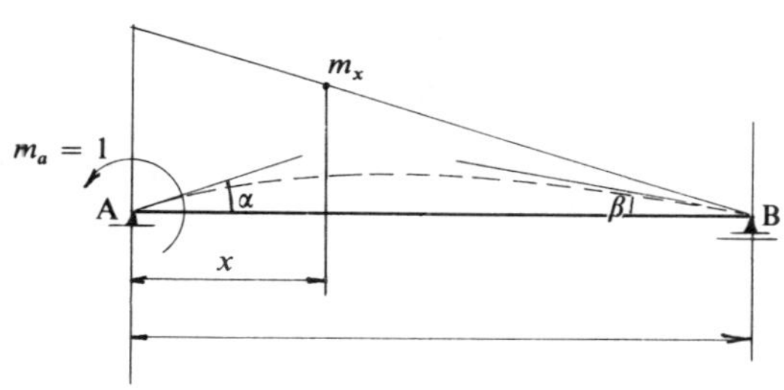

Fig. 8.11

The effect of shearing forces on the flexibility coefficients of members can be calculated from:

$$\alpha = f_{aa} = \int_0^L \frac{M.m.\mathrm{d}x}{EI} + F\int_0^L \frac{V.v.\mathrm{d}x}{GA} =$$

$$= \int_0^L \frac{(L - x/L)^2\,\mathrm{d}x}{EI} + F\int_0^L \frac{(1/L)^2}{GA}\,\mathrm{d}x = \frac{L}{3EI} + \frac{F}{GAL} \tag{8.15}$$

Similarly:

$$\beta = f_{ba} = \frac{L}{6EI} + \frac{F}{GAL} \tag{8.16}$$

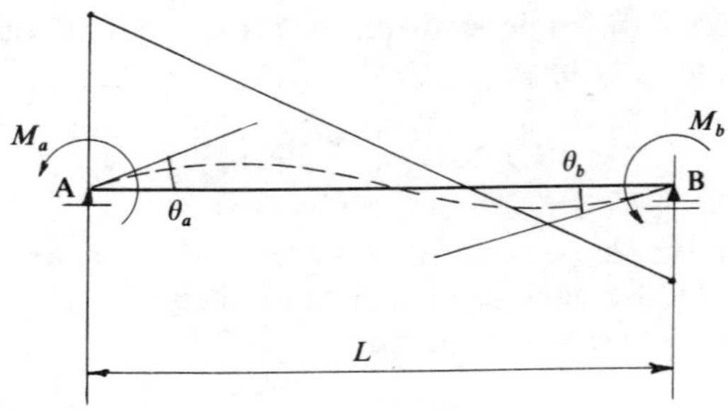

Fig. 8.12

The effect of shearing forces on the stiffness and carry-over factor can be derived from Fig. 8.12 by setting out the following conditions of compatibility of deformations:

$$M_a\left(\frac{L}{3EI}+\frac{F}{GAL}\right)-M_b\left(\frac{L}{6EI}+\frac{F}{GAL}\right)=\theta_a$$

and
$$-M_a\left(\frac{L}{6EI}+\frac{F}{GAL}\right)+M_b\left(\frac{L}{3EI}+\frac{F}{GAL}\right)=\theta_b$$

Assuming $\theta_b=0$, we obtain the following expressions for the stiffness and the carry-over factor which take into account shearing deformations:

$$k_{ab}=\frac{4EI}{L}\cdot\frac{2+g}{2(1+2g)} \tag{8.17}$$

$$c=\frac{1-g}{2+g} \tag{8.18}$$

in which $g=6F.EI/GAL^2$.

Equations (8.15) and (8.17) can be used in setting out the stiffness or force–displacement matrices.

Example 8.3

The following example illustrates the effect of shearing forces on the frame shown in Fig. 8.13, which has been previously analysed for the primary effect of bending deformations only (see Fig. 2.3(b), p. 59).

Assuming the Poisson's Ratio is $\mu=0{\cdot}18$, the flexibility coefficients for the wall $0{\cdot}3\times 2$ m become:

$$f_{00}=\frac{h}{3EI_c}+\frac{F}{GAh}=1{\cdot}218\frac{h}{3EI_c}$$

and
$$f_{01}=1{\cdot}436\frac{h}{6EI_c}$$

The element force–displacement matrix is shown in Table 8.4 and includes the effect of shearing deformations of column 0–10 only. Figure 8.14 shows the distribution of bending moments and deformations. Figures in brackets refer to the values obtained by considering bending deformations only. It can be seen that the redistribution of stresses and deformations is, in this case, of the order of 10 per cent.

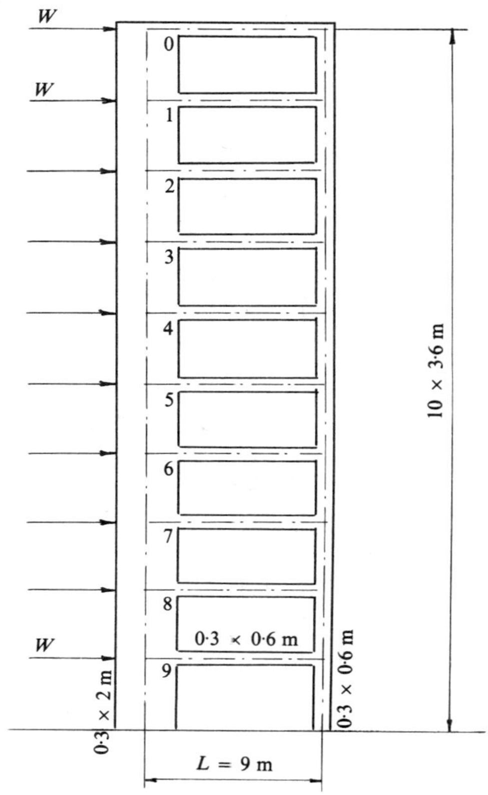

Fig. 8.13 Analysis of a shear wall, including bending and shearing deformations

Table 8.4

	1		1′		2			2′		3			3′
	1 2	Δ1	1′ 0′	1′ 2′	2 1	2 3	Δ2	2′ 1′	2′ 3′	3 2	3 4	Δ3	3′ 2′
3 2 2′	0	0	0	0	3·0	3·0394	−1	−1·5	−1·5	−0·0232	0	1	0
3′ 2′ 2	0	0	0	0	−1·5	−1·5	−1	3·0	4·2	0	0	1	−0·5
1 2 3	−0·0232	−1	0	0	3·0394	−3·0394	2	0	0	0·0232	0	−1	0
1′ 2′ 3′	0	−1	0	−0·6	0	0	2	1·2	−1·2	0	0	−1	0·6
2–2′	−1	0	0	−1	−1	1	0	−1	1	1	0	0	1

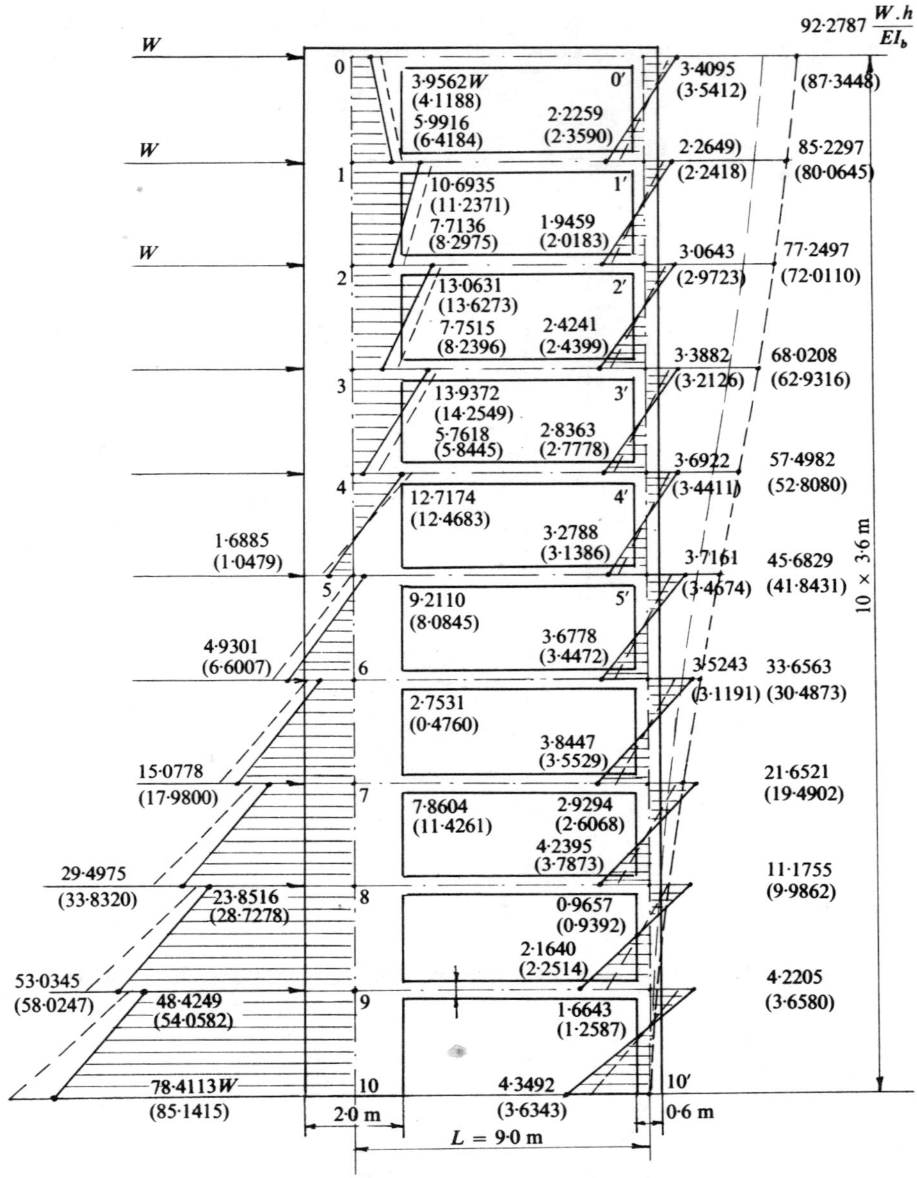

Fig. 8.14 Analysis of frame with shear deformations included

8.7 The Effect of Flexible or Semi-rigid Connections in the Analysis of Frames

The design of rigid frames is usually based on the assumption that all members meeting at a particular joint undergo the same amount of rotation, i.e., the members in a joint are assumed to be rigidly connected to one another, so that there is no relative rotation. Similarly, at the supports, members are assumed to be either ideally fixed or pinned. In reality, however, most steel as well as concrete connections permit a certain amount of relative rotation of members to take place. Also, most pinned connections show a certain amount of restraint against rotation. This property of joints is termed 'flexible' and is usually found by tests.

In tall building frames, flexible connections may result in redistribution of bending moments and in increased lateral deformations. In some cases, this source alone may become an important factor in instability of tall frames.

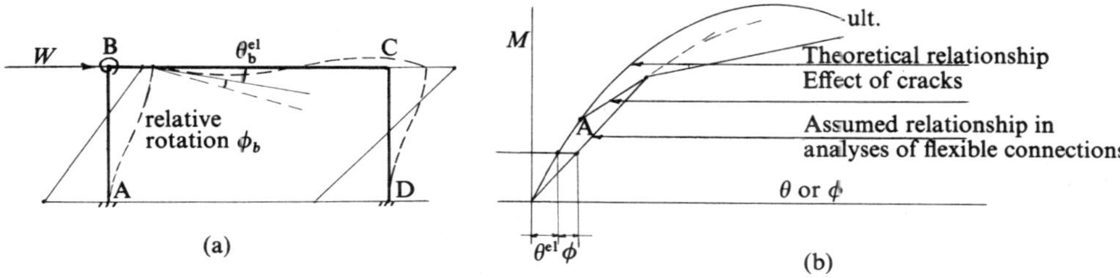

Fig. 8.15 The concept of the analysis of frames with flexible connections

In Fig. 8.15(a) the member BC at joint B is assumed to rotate through a larger angle than the member BA, under the effect of applied bending moment because of flexible connection at the joint. The significant characteristic of the flexible connection is the relationship between the applied moment and the relative angle of rotation ϕ. This can be expressed diagrammatically as shown in Fig. 8.15(b). Initially, the connection behaves nearly elastically and the curve OA is usually a straight line with a slope of:

$$J = M/\phi$$

or

$$\phi = M/J \tag{8.19}$$

The value of J represents the rotational spring constant of the connection, and is called the 'joint modulus'. At higher loads, the joint begins to deform inelastically and the angle of rotation ϕ increases rapidly. However, at normal working loads, the behaviour of the connections of most structures can be approximated by a straight line, such as OA in Fig. 8.15(b).

Experiments show that 'standard' types of steel connections possess moment-rotation characteristics with little scatter, and average values can be determined. Similarly, for a particular concrete joint, whether precast or in-situ, it is possible to determine an average relationship between the applied moment and the relative rotation.

Once the joint modulus has been found or assessed, the stiffness or flexibility coefficients of members can be determined which include the relative rotation of members. In an idealized framework, the action of flexible connections is represented by a hinge or a rotational spring. For clarity, the hinges are located at a small distance from the corresponding joint, this distance being taken as zero.

The stiffness and flexibility coefficients can be calculated from Fig. 8.16 as follows:

$$M_a(f_{aa} + \phi_a) - M_b\left(f_{ab} + \frac{\phi_b}{2}\right) = \theta_a$$

$$-M_a\left(f_{ba} + \frac{\phi_a}{2}\right) + M_b(f_{bb} + \phi_b) = \theta_b$$

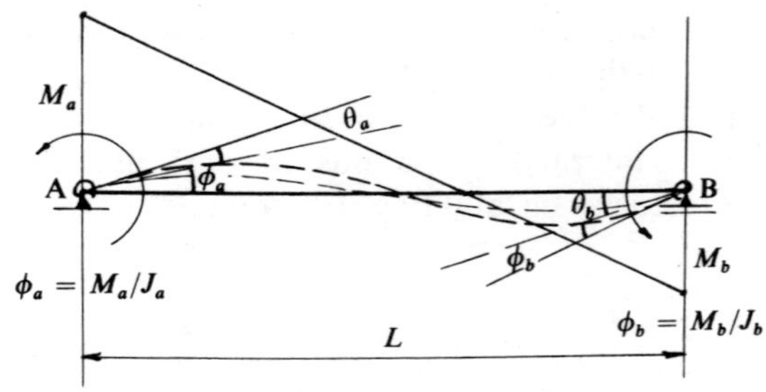

Fig. 8.16

from which, assuming $\theta_b = 0$, the stiffness of the beam becomes:

$$k = \frac{M_a}{\theta_a} = \frac{f_{bb} + \phi_b}{(f_{aa} + \phi_a)(f_{bb} + \phi_b) - (f_{ba} + \phi_a/2)(f_{ab} + \phi_b/2)} \tag{8.20}$$

Assuming $\phi_b = 0$, we obtain: for the stiffness:

$$k = \frac{4EI}{L} \cdot \frac{1}{1 + \phi_a/f_{aa}}$$

and the carry-over factor:

$$c = \frac{1}{2} \cdot \frac{2f_{ba} + \phi_a}{f_{bb}}$$

Figure 8.17(a) and (b) shows the diagram of the stiffness and carry-over factors versus the value of ϕ/f.

The values of the flexibility coefficients for a spring at the end A only are:

$$f_{aa} = \frac{L}{3EI} + \phi_a$$

$$f_{ba} = \frac{L}{6EI} + \frac{\phi_a}{2}$$

Example 8.4

If a beam AB in Fig. 8.18 has a flexible joint at C, then the moment at C is not affected but the deflection Δc increases and can be calculated from:

$$M_c(f_{cc}^{ca} + f_{cc}^{cb} + \phi_c) = \frac{\Delta c}{a} + \frac{\Delta c}{b}$$

from which:

$$\Delta c = M_c \cdot \frac{a . b}{3EI}\left(1 + \phi_c \cdot \frac{3EI}{L}\right)$$

If $a = b = L/2$, then

$$\Delta c = \frac{WL^3}{48EI}\left(1 + \phi_c \cdot \frac{3EI}{L}\right)$$

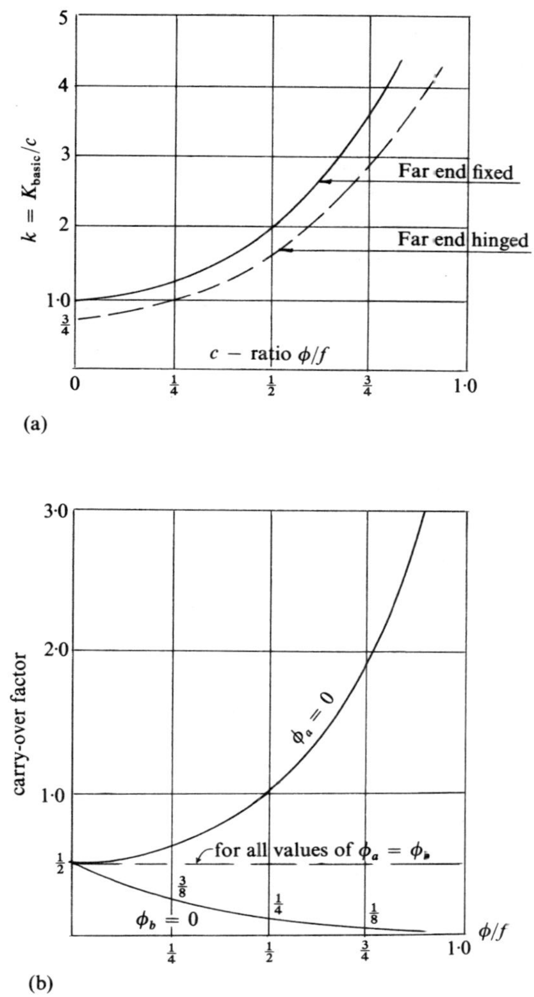

Fig. 8.17 (a) Beam stiffness vs. ratio of ϕ/f. (b) Diagram of carry-over factor vs. ratio of ϕ/f

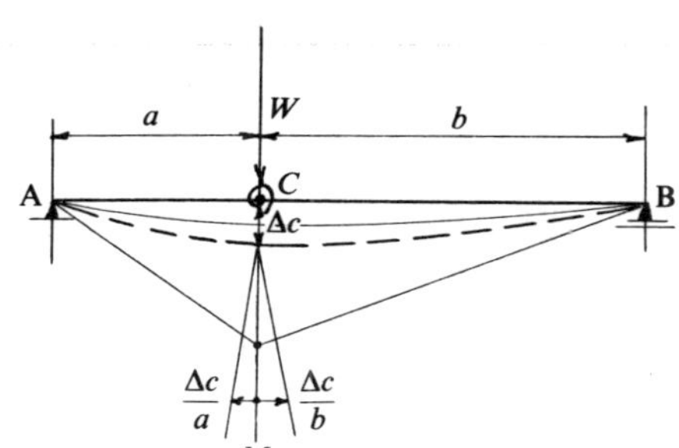

Fig. 8.18

and if $\phi_c=0$, then the deflection at point C is equal to its elastic value. If $\phi=f=L/3EI$, then the deflection at C is twice the elastic value. For $\phi_c \rightarrow \infty$, then also $\Delta c \rightarrow \infty$.

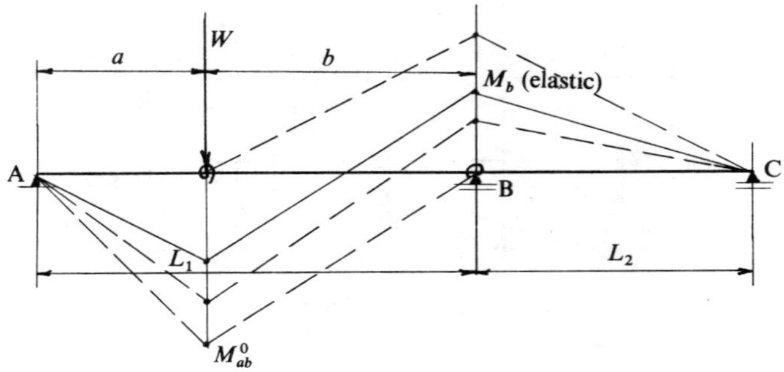

Fig. 8.19

For a two-span beam in Fig. 8.19, with flexible connections at B and D, the bending moment at B can be calculated from:

$$M_b \cdot \left(\frac{L_1}{3EI_1} + \frac{L_2}{3EI_2} + \phi_b \right) = \theta_{ba} + \theta_{bc}$$

from which:

$$M_b = \frac{3EI}{L_1 + L_2} \cdot (\theta_{ba} + \theta_{bc}) \cdot \frac{1}{1 + [3EI/(L_1 + L_2)] . \theta_b} =$$

$$= M_b^{e1} \cdot \frac{1}{1 + [3EI/(L_1 + L_2)] . \theta_b}$$

If $\theta_b=0$, then $M_b=M_b^{e1}$.
If $\theta_b=f_{ba}+f_{bc}$, then $M_b=\frac{1}{2}M_b^{e1}$ and if $\phi_b \rightarrow \infty$, then $M_b \rightarrow 0$.

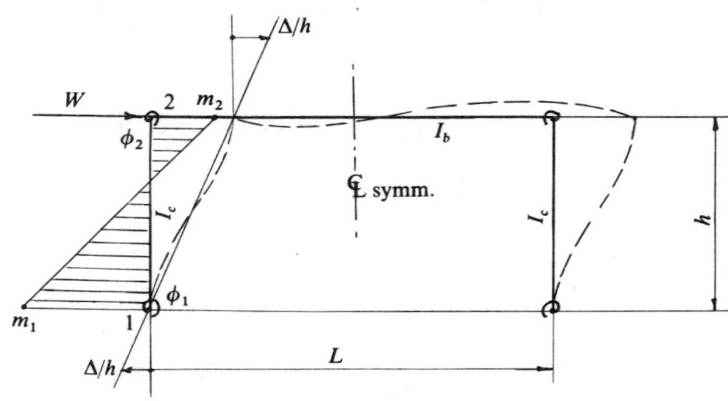

Fig. 8.20

Framed structures may have flexible joints resulting from bolted connections or from cracks. Such joints affect the distribution of primary stresses and deformations and may also adversely affect stability. If the values of ϕ for such joints are known, then the analysis can be carried out in the usual way; e.g., for a portal frame in Fig.

8.20, we have:

$$1.\quad m_1\left(\frac{h}{3EI_c}+\phi_1\right)-m_2\left(\frac{h}{6EI_c}+\frac{\phi_2}{2}\right)=\frac{\Delta}{h}$$

$$2.\quad -m\left(\frac{h}{6EI_c}+\frac{\phi_1}{2}\right)+m_2\left(\frac{h}{3EI_1}+\frac{L}{6EI_b}\right)=\frac{\Delta}{h}$$

$$\text{E.c.}\ \frac{m_1+m_2}{h}=\tfrac{1}{2}W$$

A similar matrix can be set out for multi-storey frames, taking into account the effect of any joint having a flexible connection or a crack.

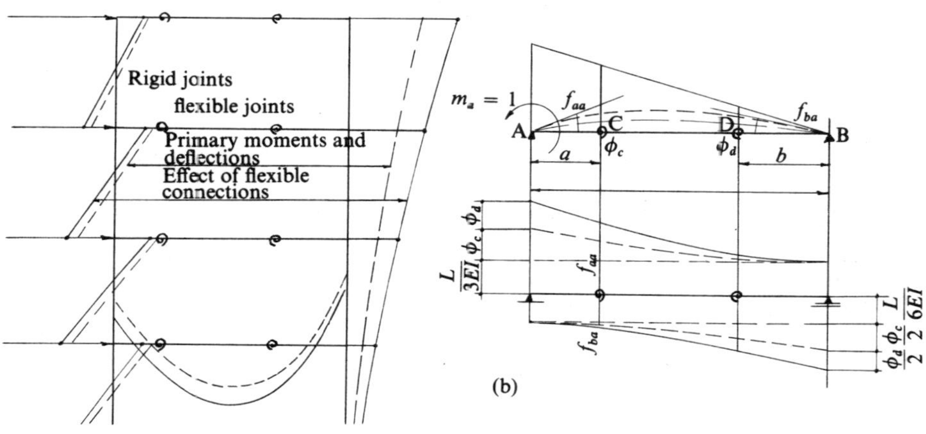

Fig. 8.21

In structural steel buildings, beam-column connections are sometimes made by welding in the fabricating shop, using short beam stubs to achieve full continuity of joints (Fig. 8.21(a)). Site connections are then moved away from the maximum bending moments and are made by the use of friction-grip, high-tensile bolts. Such joints are, therefore, flexible and can result in redistribution of bending moments and deflections. While bending moments from gravity forces may not be seriously affected, if such joints are near the points of contraflexure of beams, moments and deflections due to the effect of lateral forces may be affected to a much greater degree. These can be investigated by including in the matrix the effect of flexible joints on the stiffness of flexibility coefficients. For a beam with two flexible joints as shown in Fig. 8.21(b) the flexibility coefficients become:

$$\begin{aligned} f_{aa} &= \frac{L}{3EI}+\phi_c\cdot\left(\frac{L-a}{L}\right)^2+\phi_d\left(\frac{b}{L}\right)^2 \\ f_{ba} &= \frac{L}{6EI}+\frac{\phi_c}{2}\left(\frac{L-a}{L}\right)^2+\frac{\phi_d}{2}\left(\frac{b}{L}\right)^2 \end{aligned} \qquad (8.21)$$

The second powers used in eqs. (8.21) appear to correlate better with the experiments but have no theoretical justification.

Figure 8.21(b) shows the variation of the flexibility coefficients for various positions of flexible connections.

A similar effect can be produced by site connections in columns.

Example 8.5

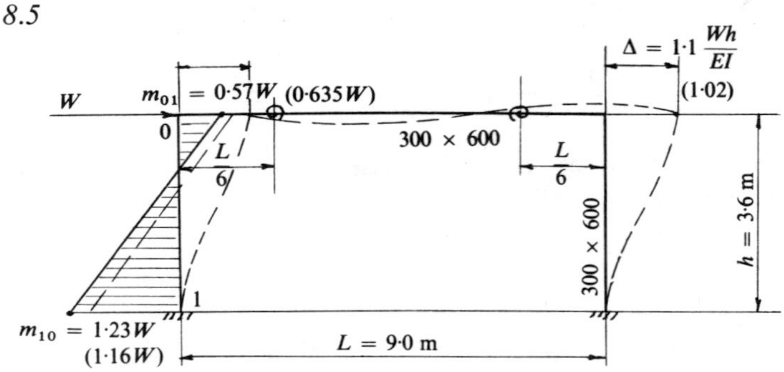

Fig. 8.22 Example of the analysis of a portal frame with flexible connections

Figure 8.22 shows a portal frame with two flexible connections and subjected to the effect of lateral force W. The following data are assumed: the two flexible joints are in the beam at a distance of $a=b=L/6$; the value of ϕ is taken to be equal to $\phi=\frac{1}{2}\theta^{el}$. The flexibility coefficients, based on eqs. (8.21), become:

$$f_{00} = 1{\cdot}36L/3EI \quad \text{and} \quad f_{10} = 1{\cdot}36L/6EI$$

The conditions of compatibility of deformations and equilibrium of forces can now be written as follows:

$$100', \quad -m_{10}\frac{h}{6EI} + m_{01}\left(\frac{h}{3EI} + 1{\cdot}36\frac{L}{6EI}\right) = \frac{\Delta}{h}$$

$$10, \quad m_{10}\cdot\frac{h}{3EI} - m_{01}\cdot\frac{h}{6EI} = \frac{\Delta}{h}$$

$$0\text{–}1, \quad \frac{m_{01} + m_{10}}{W} = \frac{W}{2}$$

from which bending moments and deflections can be calculated. These are shown in Fig. 8.22, together with the values in brackets which refer to the same frame but without flexible connections.

Example 8.6

Figure 8.23 shows a frame subjected to the effect of lateral forces. It is now assumed that flexible joints exist in all beams and in some columns as indicated, and that $\phi=\theta/2$, i.e., for beams, taking $a=b=L/6$, we have:

$$f_{00} = 1{\cdot}36L/3EI \quad \text{and} \quad f_{10'} = 1{\cdot}36L/6EI$$

and for columns, taking $a=0{\cdot}9$ m, we have from eq. (8.21)

$$f_{22} = 1{\cdot}5374h/3EI \quad \text{and} \quad f_{12} = 0{\cdot}7687h/6EI$$

and for column 10–9:

$$f_{10,10} = 1{\cdot}5h/3EI \quad \text{and} \quad f_{01,10} = 1{\cdot}5h/6EI$$

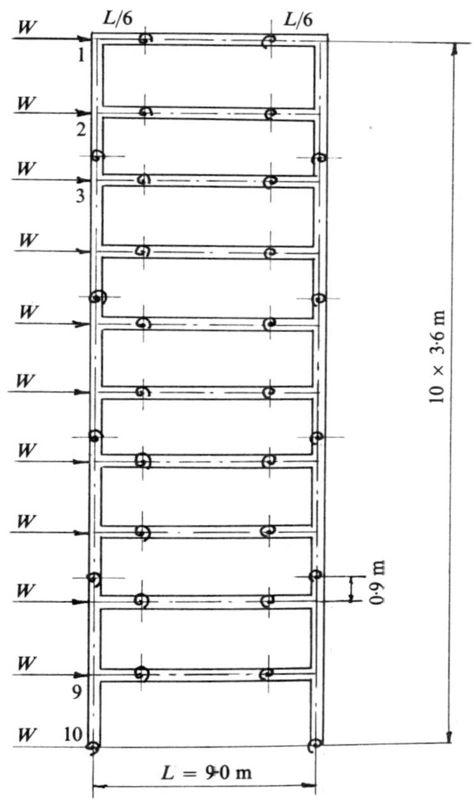

Fig. 8.23 Example of the analysis of a tall frame with flexible connections

The main matrix can now be written in the usual way and is shown in Table 8.5. It should be noted that the element matrices F_1^e and F_2^e repeat alternately throughout the main matrix. The difference between these element matrices is due to the fact that some columns have flexible connections and others do not.

Figure 8.24 shows the distribution of moments, forces, and deformations, together with the similar values shown in brackets or by dotted lines for the same frame but without flexible joints as analysed previously (see Fig. 2.36). It can be seen that the distribution of bending moments and particularly of deformations can be seriously affected by the presence of flexible joints.

Table 8.5 Force–displacement matrix for the analysis of frame in Fig. 8.23

	0		1			2			3			4			5			6			7			8			9			10	$\frac{EI}{h}$
	0 1	Δ0	1 0	1 2	Δ1	2 1	2 3	Δ2	3 2	3 4	Δ3	4 3	4 5	Δ4	5 4	5 6	Δ5	6 5	6 7	Δ6	7 6	7 8	Δ7	8 7	8 9	Δ8	9 8	9 10	Δ9	10 9	
1 0 0′	3·24	−1	−0·6	0	1																										0
0–0′	1	0	1	0	0																										1·8W
2 1 1′	0	0	2·04	3·5774	−1	−0·7687	0	1																							0
0 1 2	−0·6	−1	1·2	−1·5374	2	0·7687	0	−1	$-F_1^e$																						0
1–1′	−1	0	−1	1	0	1	0	0																							1·8W
3 2 2′				0	0	2·04	3·24	−1	−0·6	0	1																				0
1 2 3				−0·7687	−1	1·5374	−1·2	2	0·6	0	−1	$-F_2^e$																			0
2–2′				−1	0	−1	1	0	1	0	0																				1·8W
4 3 3′																															0
2 3 4											F_1^e																				0
3–3′																															1·8W
5 4 4′																															0
3 4 5														F_2^e																	0
4–4′																															1·8W
6 5 5′																															0
4 5 6																	F_1^e														0
5–5′																															1·8W
7 6 6′																															0
5 6 7																				F_2^e											0
6–6′																															1·8W
8 7 7′																															0
6 7 8																							F_1^e								0
7–7′																															1·8W
9 8 8′																															0
7 8 9																										F_2^e					0
8–8′																															1·8W
10 9 9′																															0
8 9 10																												F_1^e			0
9–9′																															1·8W
10 9																												−0·9	−1	1·8	0

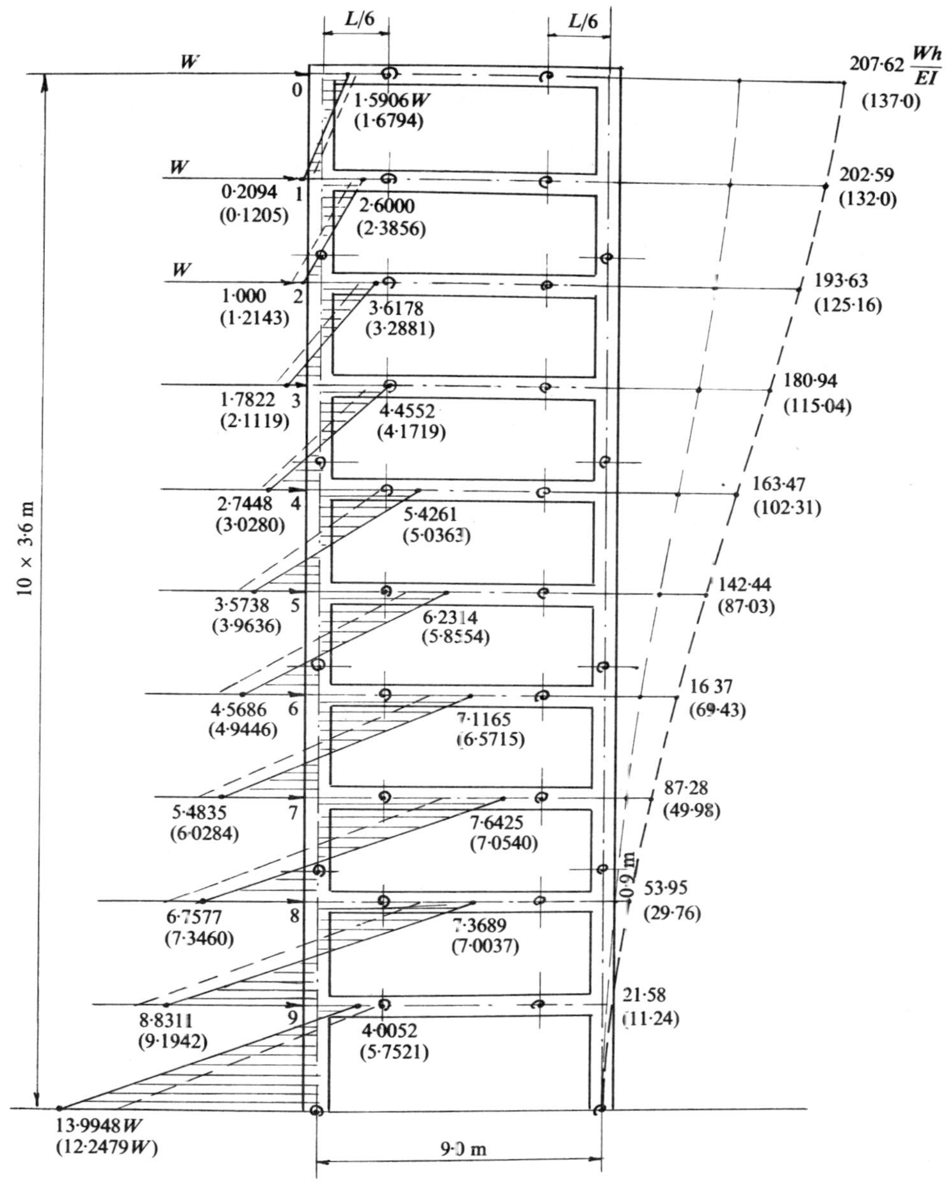

Fig. 8.24 Analysis of frame with flexible connections

References

1. V. I. Murashev. *Theory of Appearance and Opening of Cracks: Computation of Rigidity of RC members*, Stoitelnaya Promishlenost, Moscow (1940).
2. S. R. Lionberger and W. Weaver. 'Dynamic Response of Frames with non-rigid Connections', *Journal of the Eng. Mech. Div., ASCE*, 95–114 (February 1969).
3. I. Ohno. 'Stress distributed at the corner of a Stiff Frame', *Proc. 2nd Jap. Nat. Cong. for Applied Mech.* (1952).

4. I. Ohno. 'Stress Distribution in corner of Rigid Frame', *Proc. 4th Jap. Nat. Cong. for Applied Mech.* (1954).
5. D. R. Bayly. *An Investigation of the Stresses and Deformations at the Junction of Structural Members*, Ph.D. Thesis, The City University, London (1969).
6. R. A. Swann. *Flexural Strength of Corners of Reinforced Concrete Portal Frames*, Tech. Report TRA. 434, Cement and Concrete Association (1969).
7. H. E. Nilsson. 'Reinforced Concrete Frame Joints subjected to Positive Moments', *National Swedish Research Summaries* (in Swedish), (1969).
8. R. K. Livesley and D. B. Chandler. *Stability Functions for Structural Framework*, Manchester University Press, 1956.

9 Elastic Instability of Frames and Shear Walls

9.1 General Considerations

Instability of structural systems can be defined as a mode of elastic or inelastic deformation, buckling, or failure of the whole or part of a structure. At this stage, a structure (or a part of it), is in motion and it may or may not be in equilibrium with the forces causing instability.

Instability can arise from various effects, such as: excessive slenderness of members or excessive axial forces, eccentricities of loads, crookedness of columns, large axial or lateral deformations, cracks, excessive dynamic forces of wind or earthquakes, aerodynamic effect of shedding of vortices, etc. There are many other effects which can at least be of a contributory nature. These are: differential settlement of foundations, creep, shrinkage, temperature changes, etc. Any one or a combination of the above factors may lead to progressive movements and eventual failure. Instability can occur at high as well as at very low stresses, and can show itself in an excessive deformation of a structure, when equilibrium becomes restored, or in a local or overall collapse.

In a new equilibrium configuration after motion, internal forces must balance instability forces and such equilibrium may be temporary or even neutral, i.e., when no increase in the internal resisting forces can take place.

Instability of individual members is well established and is discussed in many textbooks on the subject. However, instability of frames and complete structures is not yet fully resolved, particularly with regard to critical loads and modes of failure in a form suitable for practical use in the office.

Exact analyses show that the overall stiffness of slender frames can reduce rapidly and to an alarming extent. Particularly tall and slender frames may buckle laterally under loads significantly smaller than those predicted by individual column buckling formulae. In its final stage of collapse, a structure behaves as a mechanism and deflections increase without increase of loads.

It is now recognized that, although no failures of tall buildings in service have been reported, the problem becomes increasingly more important as the structural weight is reduced, slenderness is increased, and joints may not be fully continuous.

In the analysis of instability of isolated members, energy methods have been used and have proved to be very useful. However, direct application of energy methods to full structures is often too complicated and other methods based on them are more suitable.

From a practical point of view, a system is considered to be stable if small disturbances, such as vibrations, eccentricities of loads, imperfections of materials, lack of homogeneity, etc., do not cause the system to deflect excessively from its original configuration. A system becomes unstable if a small amount of energy supplied from external source causes large deformations. If positive external forces are required to perform positive work to produce small displacement, the equilibrium is considered to be stable.

More accurate analysis is required to account for these secondary effects, particularly in tall and slender buildings. In such frames, stresses and deformations may not be proportional to the magnitude of loads, and become amplified by various instability effects.

In axially loaded members, the stiffness and flexibility coefficients depend on the magnitude of these forces. The stiffness or the force–displacement matrices are set out as a function of these coefficients. The critical loads can then be defined as those which correspond to the stiffness matrix becoming zero, or a force–displacement matrix becoming infinite. Theoretically, therefore, it should be possible to obtain critical loads based on this definition, as well as all buckling modes. However, in practice, only the most simple problems can be solved this way, for idealized frames, without imperfections, and for materials not subjected to yielding.

The main characteristics which distinguish instability problems from the conventional elastic analyses are as follows:

(a) At a certain stage of loading, deflections may become no longer proportional to loads (although the usual assumptions of the linear theory of elasticity may still be used). This is because bending moments become a function of deflections, while the simple bending theory is usually based on the equilibrium conditions being satisfied on the original undeformed configuration of the system. In other words, in instability analyses, the deformations of a structure are not uniquely determined nor are they proportional to loads.
(b) Buckling is an inherent problem of instability, and is the final stage of deflection of a structure. Buckling occurs because in instability problems small increments of loads may cause disproportional increase in deflections.
(c) Reduction of the stiffnesses of compression members or increase of their flexibility.
(d) The existence of critical loads is one of the main features of instability.
(e) Motion of a structure is an inherent feature of instability. A structure in motion may, or may not, be in equilibrium with the external forces, and finally:
(f) The natural frequency of vibrations is a characteristic of instability of a structure.

The analysis and mathematical formulation of instability problems is often based on the mechanical models of rigid bars and springs. These, although of idealized form, represent an equivalent stiffness and geometry of the system. In addition, such analyses require few approximate assumptions to be made and the solution based on such models is often sufficient for practical purposes. Some models can take into account even the inelastic response of materials. In these models, bars represent the geometry of the actual system, while the elastic springs represent its stiffness. In this way, a structure with infinite number of degrees of freedom can be represented by a model with finite number of degrees of freedom. Its response is then described by a set of simultaneous algebraic equations instead of by one or more differential equations.

9.2 Elastic Instability of Columns—General Review

The problem of the elastic instability can best be illustrated by a column eccentrically loaded as shown in Fig. 9.1, even if the assumptions of the elastic theories are accepted as valid, i.e., that the material is linearly elastic and obeys Hooke's Law.

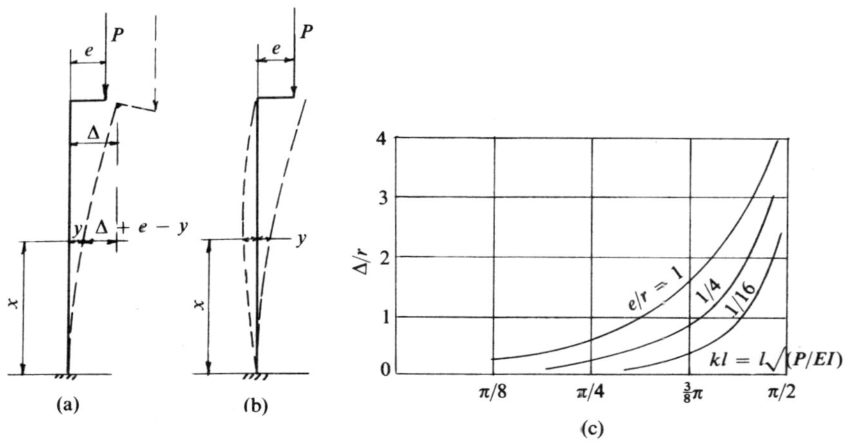

Fig. 9.1 Eccentrically loaded column

From Fig. 9.1 the bending moment at a section x is:

$$M = P(\Delta + e - y)$$

so that:

$$EI\frac{d^2y}{dx^2} = P(\Delta + e - y) \tag{9.1}$$

Denoting

$$k = \sqrt{\frac{P}{EI}}$$

the solution of eq. (9.1) is in the form:

$$y = A.\cos kx + B.\sin kx + \Delta + e \tag{9.2}$$

The constants of integration A and B can now be determined from the boundary conditions, i.e., for $x=0$, $y=0$, and $dy/dx=0$, and for $x=L$, $y=\Delta$.

Equation (9.2) becomes:

$$y = e\cdot\frac{1 - \cos kx}{\cos kl}$$

and

$$\Delta = e\left(\frac{1}{\cos kl} - 1\right) \tag{9.3}$$

Equation (9.3) shows a typical feature of the elastic instability, i.e., that the deflections are no longer proportional to the applied loads, as in the primary analysis. This is also indicated in Fig. 9.1(c), in which r is the radius of gyration of the cross-section. The deviation of proportionality of the load–deflection relationship takes place in spite of the assumptions of the linear-elastic theory, and is due to the fact that bending moments are a function of deflections.

From eq. (9.3), for $kl = \pi/2$ we obtain the critical value of the load:

$$P_{cr} = \frac{\pi^2 EI}{4l^2} \tag{9.4}$$

At this value of the load, deflections are infinitely large irrespective of the value of the eccentricity e. If it were possible to maintain zero eccentricity, the deflection would be zero up to the critical load at $kl = \pi/2$. However, in most cases, columns fail while deflections are very small, but the load thus calculated may still be considered as critical, and is referred to as the buckling load.

Similarly, for a hinged column with a load P acting at an eccentricity e (Fig. 9.1) we have:

$$EI\frac{d^2y}{dx^2} = -P.\left(y + \frac{e}{l}x\right)$$

from which:

$$y = e\left[\frac{\sin kx}{\sin kl} - \frac{x}{l}\right]$$

Since $y \to \infty$ for $kl \to \pi$ $(0 < x < l)$, the Euler buckling load is:

$$P_{cr} = \frac{\pi^2 EI}{l^2}$$

It follows from such analyses, that for a certain value of the load or static deflection, the equilibrium configuration becomes altered to such an extent that buckling may occur. This approach, however, becomes very complicated when applied to complex frames and whole structures.

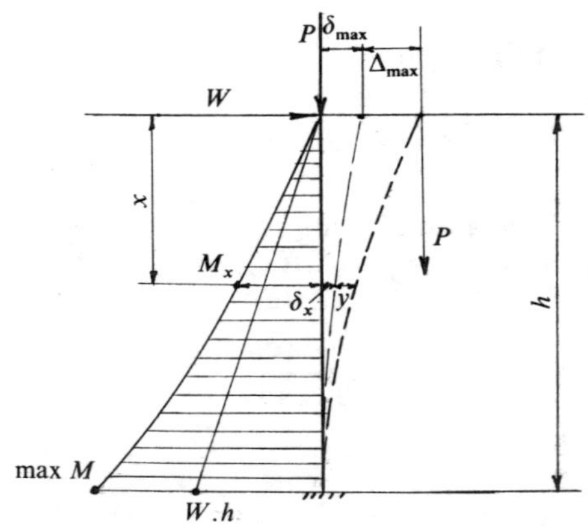

Fig. 9.2

Alternatively, the effect of large deformations of a cantilever column and therefore the effect of altered action of loads can be demonstrated by a cantilever as in Fig. 9.2. If δ_x corresponds to the primary deflection and y to the secondary effects of axial load P, then:

$$M_x = W.x + P(\delta_{\max} + \Delta_{\max} - \delta_x - y) \tag{9.5}$$

Introducing this value in the differential equation of the deflected line of a beam, we obtain:

$$y = \frac{P.W}{EI}\left[-\frac{x^5}{120} + \frac{h^2x^3}{12} - \frac{5h^4x}{12} + \frac{2h^5}{15}\right]$$

from which, for $x=0$,

$$\Delta_{\max} = \frac{Wh^3}{3EI} \cdot \frac{2Ph^2}{5EI}$$

In this equation, the value of $Wh^3/3EI$ is the primary deflection and the last value represents the non-dimensional magnification coefficient arising from the consideration of the altered geometry of the system.

The maximum bending moment is, from eq. (9.5):

$$\max M = W.h + P(\delta_{\max} + \Delta_{\max})$$

9.3 Elastic Instability of Frames

The elastic instability of a portal frame, such as shown in Fig. 9.3, can be investigated by the force–displacement method as follows: We assume that column flexibility coefficients are affected by the axial forces, and that the effect of the altered geometry of the frame is significant. The matrix required for determination of statically indeterminate bending moments, m_{12} and m_{21}, and deformation Δ, becomes now:

$$\begin{bmatrix} \left(\dfrac{c_{00}h}{3EI_c} + \dfrac{L}{2EI_b}\right) & -\dfrac{c_{10}h}{6EI_c} & -\dfrac{1}{h} \\ -\dfrac{c_{01}h}{6EI_c} & \dfrac{c_{11}.h}{3EI_c} & -\dfrac{1}{h} \\ \dfrac{1}{h} & \dfrac{1}{h} & -P.\tan\zeta \end{bmatrix} \begin{Bmatrix} m_{01} \\ m_{10} \\ \Delta \end{Bmatrix} = \begin{Bmatrix} 0 \\ 0 \\ W/2 \end{Bmatrix}$$

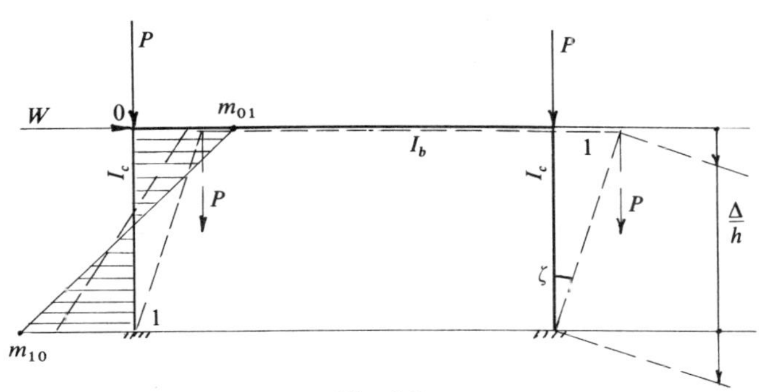

Fig. 9.3

It should be noted that the equilibrium condition must now be satisfied in the deformed configuration of the structure, and therefore includes the term $P.\tan\zeta$. The latter can be taken as $P.\Delta/h$, for loads not approaching critical. The solution of the above matrix can be obtained in a closed form. This is:

$$m_{01} = \frac{Wh}{2}\cdot\frac{A.B.h}{A.h(c_{10}+B)-(nL+A.h)(c_{00}-B)}$$
$$m_{10} = \frac{Wh}{2}\cdot\frac{B(nL+A.h)}{A.h(c_{10}+B)-(nL+A.h)(c_{11}-B)} \qquad (9.6)$$

in which

$$A = \frac{2c_{11}+c_{10}}{EI_c}, \qquad B = \frac{6EI_c}{P.h^2} \quad \text{and} \quad n = \frac{I_c}{I_b}$$

A similar analysis can be made to investigate the effect of out-of-plumb construction of frames. If a denotes the departure from plumb of the frame, as shown in Fig. 9.4, then the conditions of compatibility of deformations become:

$$\text{at 1.} \quad m_{12}\left(\frac{L}{6EI_b}+\frac{c_{11}.r}{3EI_c}\right)-m_{21}\cdot\frac{c_{12}.r}{6EI_c}=\frac{\Delta}{h}$$

$$\text{at 2.} \quad -m_{12}\frac{c_{10}r}{6EI_c}+m_{21}\cdot\frac{c_{22}.r}{3EI_c}=\frac{\Delta}{h}$$

The condition of equilibrium of forces becomes now a function of the large deformations as well as of the out-of-plumb error of construction, i.e.,

$$\frac{m_{12}+m_{21}}{r}\cdot\cos(\alpha+\zeta) = P.\tan(\alpha+\zeta)$$

If loads are not approaching critical, we can assume that:

$$\cos(\alpha+\zeta) \approx 1+\alpha+\zeta = 1+\frac{a+\Delta}{h}$$

and the equilibrium equation can be written as:

$$\frac{m_{12}+m_{21}}{r}\left(1+\frac{a+\Delta}{h}\right) = W + P\frac{a+\Delta}{h}$$

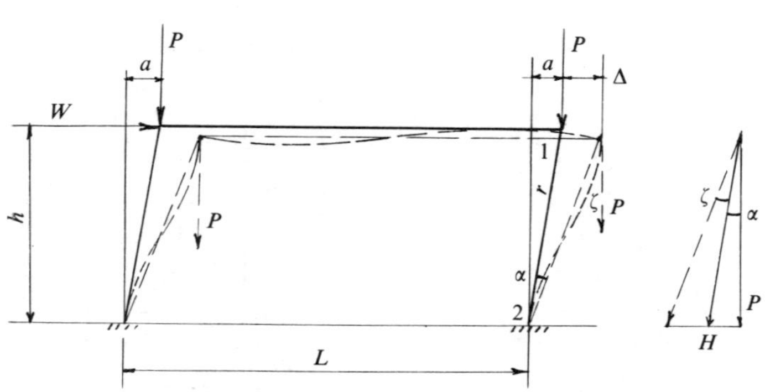

Fig. 9.4 Out-of-plumb construction

A similar analysis can be made to study the elastic instability of tall frames and shear walls, particularly the effect of altered configuration and therefore the action of loads, the effect of altered stiffnesses or flexibility coefficients of compression members, out-of-plumb construction, etc. The following example is used to illustrate the procedure.

Example 9.1

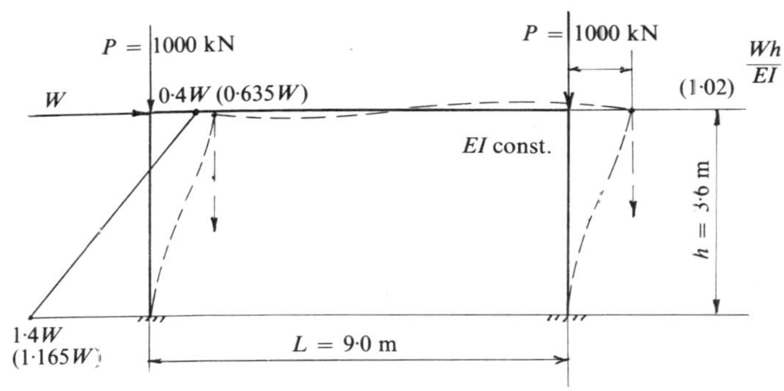

Fig. 9.5 Example

In this example a portal frame with all members 300 × 600 mm is analysed for the effect of elastic instability. It is assumed that $P = 1000$ kN and that $E_c = 28 \times 10^6$ kN/m^2. The values of the flexibility coefficients, based on Table 8.1, are: $c_{00} = 1{\cdot}0476$ and $c_{10} = 1{\cdot}0666$. The matrix becomes now:

$$\begin{bmatrix} 2{\cdot}7571 & -0{\cdot}6399 & -1 \\ -0{\cdot}6399 & 1{\cdot}2571 & -1 \\ 1 & 1 & -0{\cdot}0237 \end{bmatrix} \begin{Bmatrix} m_{01} \\ m_{10} \\ \Delta \end{Bmatrix} = \begin{Bmatrix} 0 \\ 0 \\ 1{\cdot}8W \end{Bmatrix}$$

from which the distribution of bending moments and deformations is as shown in Fig. 9.5.

It can be seen that the effect of elastic instability in this example is appreciable. The bending moment at point 1 is increased by 20 per cent and the deflection by 40 per cent.

Example 9.2

Figure 9.6 shows the frame analysed previously for the primary effect of lateral forces. Now it is assumed that gravity loads are present, of values as indicated in Fig. 9.6. These loads will increase the flexibility coefficients of columns and axial as well as lateral deformations. The flexibility coefficients can be obtained from Table 8.1 and the matrix can then be set out as shown in Table 9.1. It should be noted that the element matrices are no longer repetitive because of the effect of axial forces in columns, as well as the effect of gravity loads on the conditions of equilibrium of forces.

Table 9.1 Force–displacement matrix for the analysis of shear wall in Fig. 9.6 for the effect of elastic instability

	0		1			2			3			4		
	0 1	Δ0	1 0	1 2	Δ1	2 1	2 3	Δ2	3 2	3 4	Δ3	4 3	4 5	Δ4
1 0 0′	2·7022	−1	−0·6017	0	1									
0–0′	1	−0·0064	1	0	0·0064									
2 1 1′	0	0	1·5	2·7039	−1	−0·6033	0	1						
0 1 2	−0·6011	−1	1·2034	−1·2039	2	0·6033	0	−1						
1–1′	−1	0	−1	1	−0·0127	1	0	0·0127						
3 2 2′				0	0	1·5	2·7058	−1	−0·6048	0	1			
1 2 3				−0·6019	−1	1·2066	−1·2058	2	0·6050	0	−1			
2–2′				−1	0	−1	1	0·0181	1	0	0·0181			
4 3 3′							0	0	1·5	2·7075	−1	−0·6065	0	1
2 3 4							−0·6029	−1	1·2100	−1·2075	2	0·6048	0	−1
3–3′							−1	0	−1	1	0·0254	1	0	0·0254
5 4 4′										0	0	1·5	2·7096	−1
3 4 5										−0·6037	−1	1·2130	−1·2087	2
4–4′										−1	0	−1	1	0·0367
6 5 5′													0	0
4 5 6													−0·6043	−1
5–5′													−1	0
7 6 6′														
5 6 7														
6–6′														
8 7 7′														
6 7 8														
7–7′														
9 8 8′														
7 8 9														
8–8′														
10 9 9′														
8 9 10														
9–9′														
10 9														

5		6			7			8			9			10	$\frac{EI}{h}$
	Δ5	6 5	6 7	Δ6	7 6	7 8	Δ7	8 7	8 9	Δ8	9 8	9 10	Δ9	10 9	
															0
															1·8 *W*
															0
															0
															1·8 *W*
															0
															0
															1·8 *W*
															0
															0
															1·8 *W*
	1														0
	−1														0
	0·0367														1·8 *W*
0	−1	−0·6098	0	1											0
10	2	0·6098	0	−1											0
	0·0442	1	0	0·0442											1·8 *W*
	0	1·5	2·7130	−1	−0·6113	0	1								0
55	−1	1·2197	−1·2130	2	0·6113	0	−1								0
	0	−1	1	0·0495	1	0	0·0495								1·8 *W*
			0	0	1·5	2·7150	−1	−0·6121	0	1					0
			−0·6065	−1	1·2227	−1·2150	2	0·6121	0	−1					0
			−1	0	−1	1	0·0508	1	0	0·0508					1·8 *W*
						0	0	1·5	2·7175	−1	−0·6149	0	1		0
						−0·6075	−1	1·2241	−1·2175	2	0·6149	0	−1		0
						−1	0	−1	1	0·0621	1	0	0·0621		1·8 *W*
									0	0	1·5	2·7188	−1	−0·6165	0
									−0·6088	−1	1·23	−1·2189	2	0·6165	0
									−1	0	−1	1	0·0735	1	1·8 *W*
												−0·6094	−1	1·2330	0

Table 9.2 Force–displacement matrix for the analysis of out-of-plumb frame in Fig. 9.8

	0		1			2			3			4		
	0 1	Δ0	1 0	1 2	Δ1	2 1	2 3	Δ2	3 2	3 4	Δ3	4 3	4 5	Δ4
1 0 0′	2·7	−1	0·6	0	−1									
0–1	1	−0·00476	−1	0	0·00476									
2 1 1′	0	0	−1·5	2·7	−1	−0·6	0	−1						
0 1 2	0·6	1	1·2	1·2	−2	−0·6	0	−1						
1–2	0	0	0	1	−0·00952	1	0	−0·00952						
3 2 2′				0	0	−1·5	2·7	−1	+0·6	0	+1			
1 2 3				−0·6	−1	1·2	1·2	−2	0·6	0	+1			
2–3				0	0	0	1	−0·0143	−1	0	0·0143			
4 3 3′							0	0	−1·5	2·7	−1	−0·6	0	−1
2 3 4							0·6	1	1·2	1·2	−2	−0·6	0	−1
3–4							0	0	0	1	−0·019	1	0	−0·019
5 4 4′										0	0	−1·5	2·7	−1
3 4 5										−0·6	−1	1·2	1·2	−2
4–5										0	0	0	1	−0·0238
6 5 5′													0	0
4 5 6													0·6	1
5–6													0	0
7 6 6′														
5 6 7														
6–7														
8 7 7′														
6 7 8														
7–8														
9 8 8′														
7 8 9														
8–9														
10 9 9′														
8 9 10														
9–10														
10 9														

5	6			7			8			9			10	$\frac{EI}{h}$
Δ5	6 5	6 7	Δ6	7 6	7 8	Δ7	8 7	8 9	Δ8	9 8	9 10	Δ9	10 9	
														0
														9·0
														0
														0
														18·0
														0
														0
														−21·0
														0
														0
														28·0
+1														0
+1														0
0·0238														25·0
−1	−0·6	0	−1											0
−2	−0·6	0	−1											0
−0·0275	1	0	−0·0275											42·0
0	−1·5	2·7	−1	+0·6	0	+1								0
−1	1·2	1·2	−2	0·6	0	+1								0
0	0	1	−0·0333	−1	0	−0·0333								21·0
		0	0	−1·5	2·7	−1	−0·6	0	−1					0
		+0·6	1	1·2	1·2	−2	−0·6	0	−1					0
		0	0	0	1	−0·0381	1	0	−0·0381					24·0
					0	0	−1·5	2·7	−1	+0·6	0	+1		0
					−0·6	−1	1·2	1·2	−2	+0·6	0	+1		0
					0	0	0	1	−0·0428	−1	0	−0·0428		9·0
								0	0	−1·5	2·7	−1	−0·6	0
								+0·6	1	1·2	1·2	−2	−0·6	0
								0	0	0	1	−0·0476	1	10·0
											0·6	−1	1·2	0

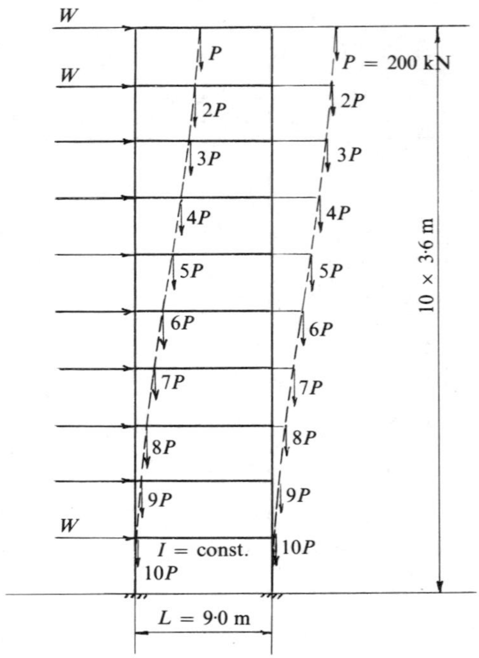

Fig. 9.6 Example

Figure 9.7 shows the distribution of moments and deformations. For ease of comparison, the values obtained from the primary analysis are indicated by dotted lines and figures in brackets.

This example shows, as could be expected, that tall and slender frames can be quite appreciably affected by elastic instability, e.g., some bending moments in this example are increased over $3\frac{1}{2}$ times, while deflections are almost doubled. As a consequence of large deformations, a proportion of gravity loads is carried by the frame in bending.

Example 9.3

In this example it is assumed that the frame shown in Fig. 9.8 is constructed out of plumb as shown. All members are of constant cross-sections 300 × 600 mm, $P = 200$ kN, and $E_c = 28$ kN/mm^2. This analysis takes into account only the eccentricities of gravity loads. The main matrix is shown in Table 9.2. It can be seen that the element force–displacement matrices are non-repetitive because of the variable eccentricities of loads at various junctions. Figure 9.9 shows the distribution of bending moments and deformations resulting from the crookedness of the frame alone.

It can be seen from Fig. 9.9 that maximum stress arising from out-of-plumb construction is 1·52 N/mm^2, and that the lateral deflection due to this cause alone is 5 mm.

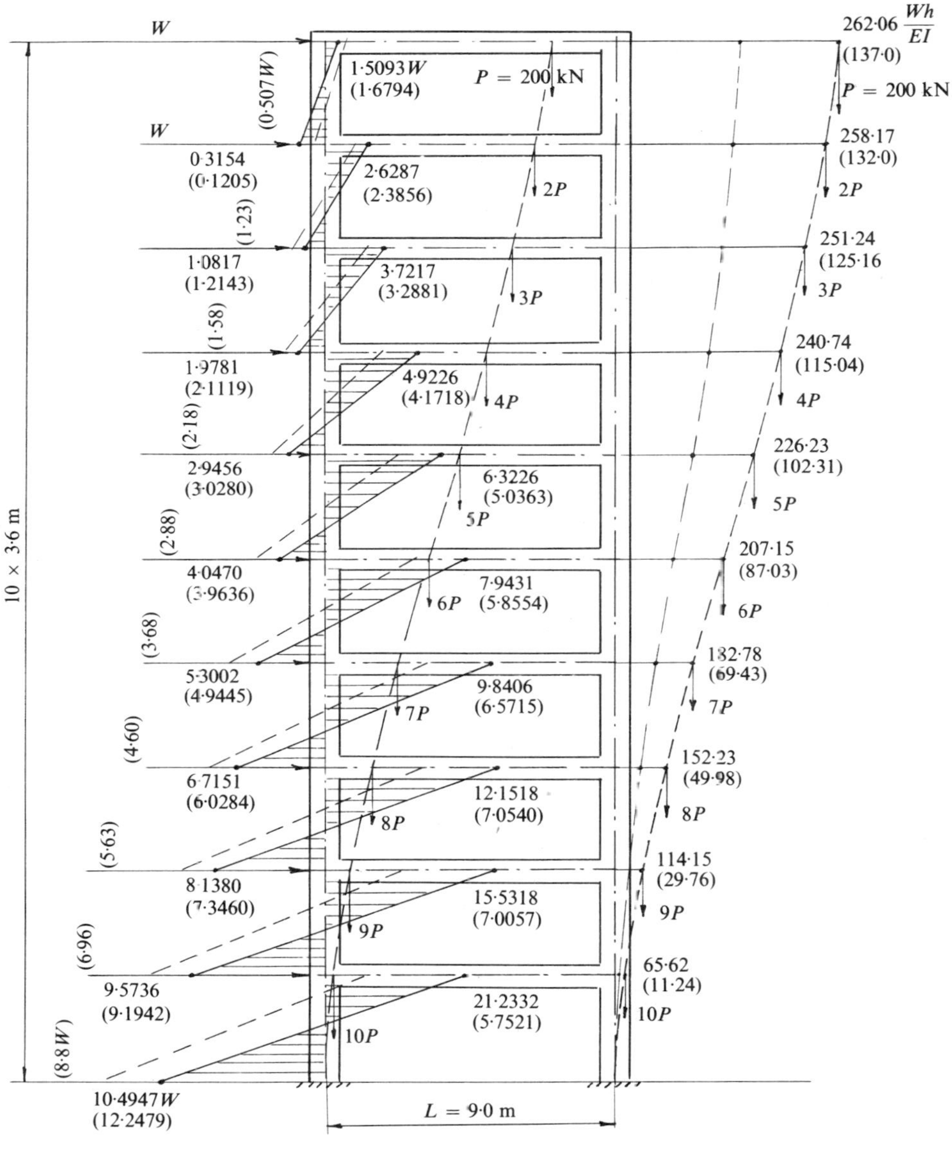

Fig. 9.7 Analysis of frame for the effect of elastic instability

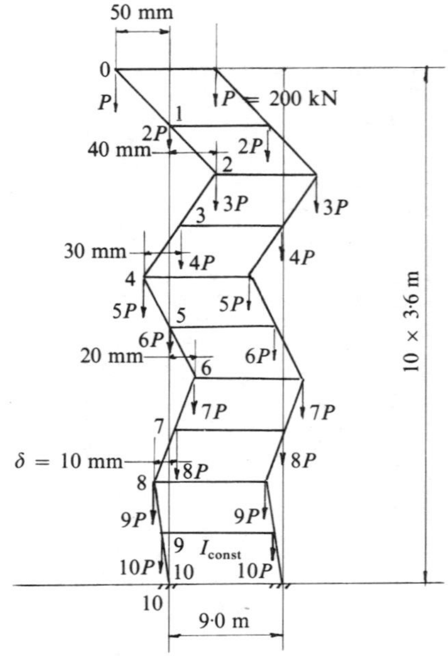

Fig. 9.8 Example of the analysis of a frame for the effect of out-of-plumb construction

References

1. S. Timoshenko and J. Gere. *Theory of Elastic Stability*, McGraw-Hill, N.Y. (1961).
2. F. Bleich. 'Die Knickfestigkeit Elastischer Stabverbindungen', *Der Eisenbau*, 10 (1919).
3. E. Chwalla. 'Die Stabilität Lotrecht Belasteter Rechteckrahmen', *Der Bauingenieur*, 19 (1938).
4. N. J. Hoff. 'Stable and Unstable Equilibrium in Plain Frameworks', *J. Aero. Sci.* (1941).
5. E. F. Masur. 'On the Lateral Stability of Multi-storey Bents', *Proc. A.S.C.E.*, **81**, separate No. 672 (1955).
6. E. F. Masur, I. C. Chang, and L. H. Donnell. 'Stability of Frames in the Presence of Primary Bending Moments', *Journal of the Engineering Mechanics Division, A.S.C.E.*, **87**, No. EM4 (1961).
7. F. Bleich. *Buckling Strength of Metal Structures*, McGraw-Hill, N.Y. (1952).
8. D. McGill. 'Column Instability under Weight and Follower Load', *Struct. Div. ASCE* (June 1971).

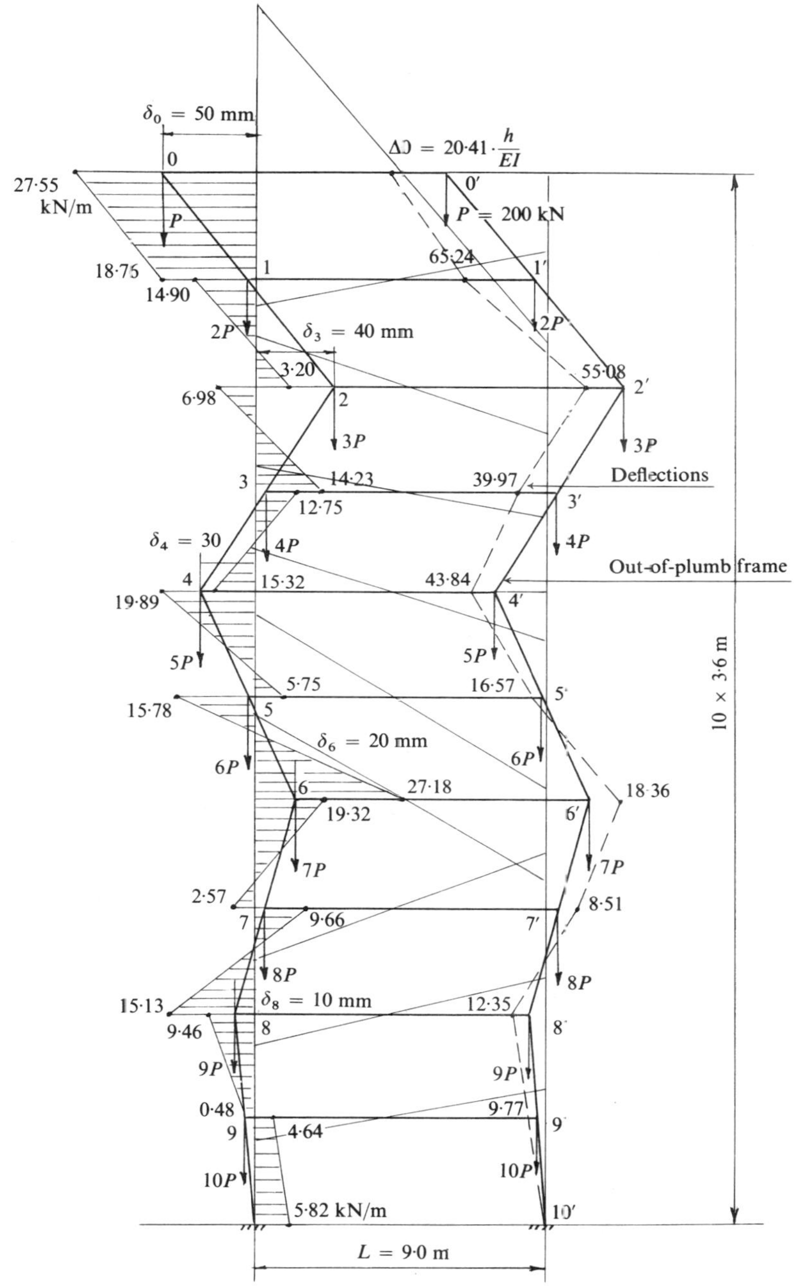

Fig. 9.9 Example of the analysis of frame for the effect of out-of-plumb construction

10 Volume Changes in Concrete and their Effect on Structures

10.1 Introduction

Volume changes in hardened concrete are caused by the hydration process and lack of hydrometric equilibrium of moisture content with the temperature and humidity of the surrounding atmosphere. The main causes of volume changes are shrinkage, creep, and temperature changes of concrete. As a result of volume changes, redistribution of internal stresses and strains takes place resulting in overall deformations of structures. Experiments show that the magnitude of these strains, expressed in terms of long-term ratio of creep strains to the elastic strains, can be as high as 1·5 or even 2·5.

The structure of the hardened cement paste and concrete has been extensively studied over the last thirty years and research still continues. The studies include macro- and micro-structure of the cement paste, its physical chemistry, and its microcrystalline properties. Electro-micrographs have been used to study its molecular and even atomic structure. As a result of these studies, a much clearer picture is beginning to emerge of the origin and the nature of the strength of concrete and of shrinkage and creep phenomena. However, the application of this research to practical design and exact prediction of the response of structures to loads with time still remains difficult and uncertain.

Conventional elastic theories predict quite well the overall behaviour of concrete members within the imposed limits, but cannot predict the response of structures strained beyond this limit and their true response as a function of time. This is because the properties of concrete depend on many parameters which are difficult to formulate exactly, particularly in explicit mathematical terms. Without such formulation, in turn, it is difficult to develop an exact analysis. It is mainly because of volumetric changes in concrete that the actual response of structures differs from the predicted when based on linearly-elastic theories. This discrepancy can be serious and even a reverse response to that predicted has been observed.

Creep and shrinkage are generally viewed as pathological properties of concrete which cause undesirable stresses and deformations. However, while shrinkage is always detrimental, creep can be in many cases extremely beneficial particularly in its ability to relieve stress concentrations.

10.2 Short Review of Mechanical Properties of Concrete

The mechanical properties of concrete are expressed in terms of stress–strain or load–deflection relationship, as shown in Fig. 10.1. The first part of the curve is relatively straight although not strictly linear. The maximum compressive strain in concrete is of the order of 0·003 and tensile strain is of the order of 0·000 17 for concrete of average strength. However, for concrete, considerable difficulties arise from the fact that the results of tests depend to a large extent on the method of testing, rate of loading, mass and shape of specimen tested, moisture conditions, temperature, restraints at the testing machine platens, etc.

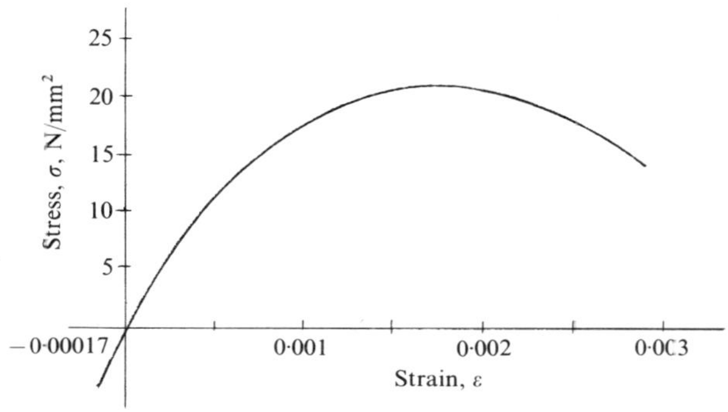

Fig. 10.1 Typical stress–strain diagram for concrete of compressive strength of 21 N/mm²

As in many other brittle materials, the apparent strength of concrete decreases with the increase of the size and volume of the tested specimen, and the failure is assumed to be caused by the propagation of micro-cracks. Fully saturated concrete appears to have reduced strength by 20 per cent to 30 per cent in comparison to air-dried specimens. Shrinkage, particularly if rapid and at early stage, can produce cracks and decrease of flexural strength and of modulus of elasticity.

The effect of temperature on strength of concrete is not yet fully resolved, but it is suggested that the ultimate strength in compression decreases by 2 per cent for each 1° C increase in temperature in the range of 0 to 55° C.

Dynamic loading, at the rate of up to 70 $N/mm^2/s$ can indicate an increase of the ultimate strength of concrete of 30 per cent to 80 per cent, as well as of its modulus of elasticity when compared with the standard rate of loading of 0·25 $N/mm^2/s$. However, at very slow loading rates, the creep and slow propagation of cracks can cause decrease of 20 per cent to 25 per cent of the ultimate strength of concrete.

The dynamic modulus of elasticity and the Poisson's ratio can be found by non-destructive testing using stress wave velocity and resonant frequency. The dynamic modulus is usually taken as equivalent to the tangent static modulus of elasticity at zero stress. It has been shown that at least 50 per cent of the ultimate static load can be applied at any rate of loading without seriously affecting the ultimate strength of concrete.

The 'elastic' constants, particularly the modulus of elasticity, are taken as tangent at zero stress, or at any other stress level, or as a secant modulus of elasticity. The value of the static modulus of elasticity is usually taken at a stress level not exceeding one-third of the ultimate strength, and for concrete of density of 1440 to 2300 kg/m^3 can be calculated from:

$$E_c = 0{\cdot}85w\sqrt{(u)} \times 10^{-6} \text{ N/mm}^2$$

and for concrete of density of 2300 kg/m^3, or more, from:

$$E_c = 4{\cdot}5\sqrt{u}$$

in which u is the cube strength in N/mm^2.

The dynamic modulus is calculated from:

$$E_d = 27{\cdot}5 + 2{\cdot}8\sqrt{u} \text{ kN/mm}^2$$

and static modulus from:

$$E_c = 1{\cdot}6E_d - 4{\cdot}4$$

The value of the Poisson's ratio for concrete depends mainly on the volume fraction of aggregate and the type of aggregate. Increase of the hard aggregate in the cement paste results in decrease of the Poisson's ratio. In normal conditions, the value of the Poisson's ratio for concrete varies between 0·15 and 0·25.

The coefficient of thermal expansion of concrete varies with the type of aggregate used and the proportion of the cement to aggregate. The generally accepted value of this coefficient is 0·000 012 per 1° C, and is assumed to be constant within the range of normal atmospheric temperatures. Limestone aggregates give concrete of low coefficients of thermal expansion of about one-tenth of the above value, while quartzite aggregates possess the highest coefficients of approximately six to eight times the average value shown above.

The modulus of elasticity of steel is usually taken as:

$$E_s = 200 \text{ kN/mm}^2$$

and the maximum tensile strain in steel is taken as 0·020 000. The maximum permissible strain in steel is usually taken as 0·0015 or $0{\cdot}75fy/E_s$, whichever is the lesser.

The coefficient of thermal expansion of steel is approximately 0·000 01 per 1° C.

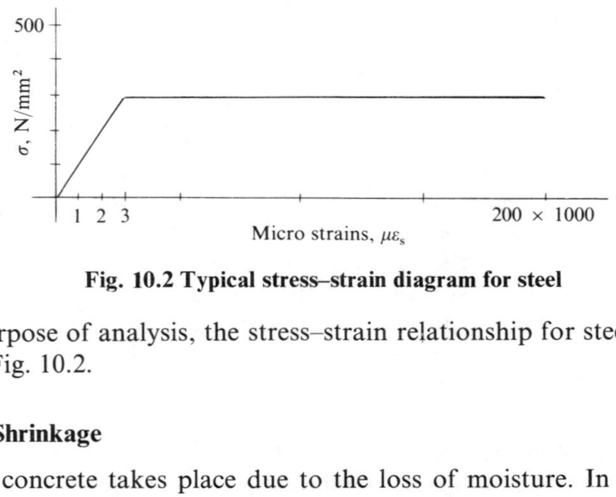

Fig. 10.2 Typical stress–strain diagram for steel

For the purpose of analysis, the stress–strain relationship for steel is often taken as shown in Fig. 10.2.

10.3 Drying Shrinkage

Shrinkage in concrete takes place due to the loss of moisture. In the presence of internal or external restraints, shrinkage produces high tensile stresses in concrete and even cracks frequently occur. These restraints to free volume changes arise from the presence of reinforcement, difference in masses, monolithic connections with other members, and even due to differential drying across the section.

The main factors affecting the nature and the magnitude of shrinkage of concrete are: the type of cement, the mix proportions, water–cement ratio of the paste, size, and composition of aggregate, the rate of evaporation of water, the length of drying period and curing conditions, the humidity of the surrounding air, and the size and shape of the concrete mass.

R. H. Evans[1] gives the following average values of the ultimate shrinkage suitable for practical use in design. The figures refer to standard concrete mixes made of Portland cement and crushed gravel aggregate, wet cured and exposed to 70 per cent relative humidity at $15\frac{1}{2}°$ C (60° F).

Table 10.1 Ultimate shrinkage of concrete of standard mixes (by R. H. Evans)

Mix proportion by weight	W/C	Shrinkage	Remarks
1:2:4	0·65	500×10^{-6}	Full size members
$1:1\frac{1}{2}:3$	0·55	600×10^{-6}	volume/surface
1:1:2	0·40	700×10^{-6}	ratio 100 mm

Table 10.2 Relation between shrinkage and duration of exposure

Duration of exposure	Percentage of long-term shrinkage
1 month	40
6 months	60
1 year	75
5 years	90
10 years	95
30 years	100

The effect of temperature on shrinkage is not yet fully resolved. R. H. Evans suggests that within the range of temperatures in the U.K., shrinkage can be assumed to increase at the rate of $\frac{1}{2}$ per cent of the 60° F shrinkage for each degree Fahrenheit rise in temperature.

The cement paste content has a predominant effect on shrinkage. Its content by volume varies from 28 to 40 per cent and within this range, shrinkage can be assumed to increase at the rate of 7 per cent for each per cent increase in the cement paste content by volume. The effect of water–cement ratio, the type of aggregate and the volume–surface ratio are summarized in the above paper quoted by R. H. Evans.

Long curing delays onset of shrinkage and permits concrete to attain higher strength in compression and in tension, but has small effect on the ultimate value of shrinkage. Prolonged cured concrete shrinks rapidly when subsequently exposed to evaporation of moisture and the beneficial effect of creep in relieving shrinkage stresses may be reduced. In addition, well-cured concrete has high strength and therefore small creep capacity, and shrinkage cracks may still appear even if long curing is used. Because of this, a contradictory opinion on curing exists and it is argued that, after initial curing, its further beneficial effect may be questioned.

The ultimate magnitude of shrinkage is almost independent of curing and the rate of drying, except that rapid drying may produce stresses which could lead to cracking.

In addition to the average drying shrinkage, concrete members are subjected to non-uniform shrinkage within the section itself. This is caused by non-uniform drying at the surface and inside the mass, so that a gradient is established. If this is unsymmetrical, warping may take place. The difference between shrinkage at the surface and the inside of beams can be very high, and may result in high internal restraint. High stresses and even surface cracks may then appear.

10.4 Creep in Concrete Structures

Creep is defined as time-dependent strain under sustained stress. Both steel and concrete are affected by creep, although creep in steel is negligible at working stresses and at room temperatures.

Creep and shrinkage are two major causes of time-dependent inelastic response of concrete. In concrete structures, a large proportion of non-elastic deformations, as well as relaxation of stresses, takes place due to these causes. It is only because of the complexity of the problem that these are not often included at the design stage.

The main factors affecting creep are: the cement content in the paste, water–cement ratio, properties of aggregate, sizes of members, the type and intensity of sustained stress, humidity, and temperature. The mechanism of creep is not yet fully resolved. Figure 10.3 shows that, on loading, concrete undergoes an instantaneous elastic strain, followed by a gradual increase in strain, which tends to a limiting value. On unloading, there is an instantaneous or elastic recovery, followed by a delayed elastic recovery and residual, irrecoverable strain.

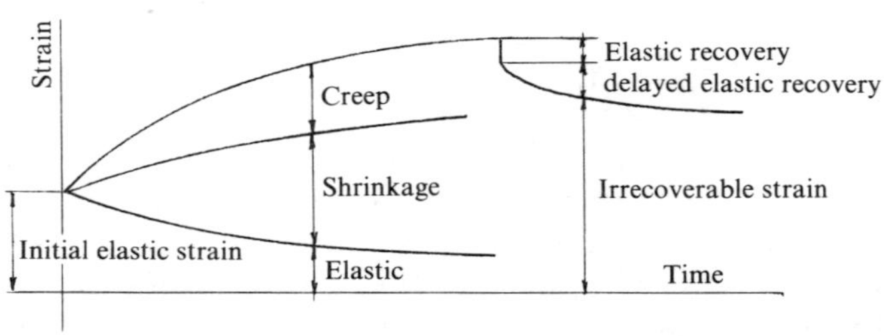

Fig. 10.3 Inelastic strains in concrete

At an early age, concrete under constant stress shows a very high rate of creep, which can have an important bearing on later response of a structure to loads. These early stresses can originate from shrinkage or temperature of hydration of cement, even if concrete is fully supported by formwork.

At a long period of sustained stress of 20 or 30 years, concrete shows a tendency to a limiting value of creep, and the greater the age or maturity of concrete, the lower is the rate of creep.

Most of the available creep data relate creep to time and constant stress, temperature, and humidity. Creep strain increases with increase of temperature of concrete and decreases with the increase in the size of members. Since creep is related to shrinkage, members with large surface-to-volume ratio undergo more shrinkage and therefore have larger creep strains.

Since the modulus of elasticity of concrete increases with time, the elastic strains decrease with time, as shown in Fig. 10.3, so that creep should be considered as strain in excess of the elastic strain, and strictly should not be related to the strain at the time of loading.

There are, therefore, two periods in concrete life which contribute most to the response of structures: an initial stage, prior to application of loads when concrete is rapidly gaining strength, and the following period 'at work'. There are indications that the initial period is of much greater importance than so far expected on deformations, relaxation of stresses, and formation of cracks. At this early stage, concrete has little strength but high rate of shrinkage and creep. Relatively high stresses can develop at this early stage, mainly from shrinkage, and concrete develops cracks, visible or micro-cracks. These are potential places for future cracks due to loads and other effects. Fortunately, these early stresses are partly relieved by the creep.

Shrinkage and creep are interdependent, although for simplicity these effects are often considered separately and as additive. Strictly, however, the principle of superposition is not valid for these inelastic strains. Shrinkage in fact increases creep, while creep tends to reduce shrinkage stresses.

In the analyses, it is convenient to use the term 'specific creep', which is defined as the magnitude of creep per unit stress, i.e.,

$$\text{Specific creep } c_s = c/\sigma \text{ in mm per N/mm}^2 \text{ of stress}$$

It is not yet entirely certain whether long-term creep tends to a limiting value, particularly in mass concrete structures. However, for tall buildings, which are built for a limited life, creep can be taken as having limiting value, with most creep strains taking place within the first four to five years.

A good quality concrete may have an ultimate specific creep value of 1·0 to 0·6 × 10 per N/mm^2. The difficulties in predicting creep strains arise from the fact that the change in stress causes also changes in the rate of creep, although most analyses assume that the change in creep strain is directly proportional to the instantaneous strain.

For practical purposes, creep is often related to the strength of concrete because

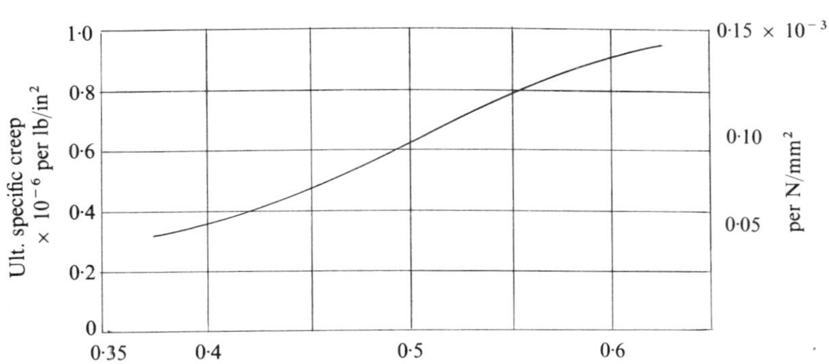

Fig. 10.4 Ultimate specific creep vs. water–cement ratio, by Hummel

Table 10.3 Ultimate creep of average concrete, by R. H. Evans

Mix by vol.	W/C ratio	Ult. creep per lb/in^2	N/mm^2
1:2:4	0·65	$1{\cdot}0 \times 10^{-6}$	$0{\cdot}145 \times 10^{-3}$
1:1½:3	0·55	$0{\cdot}85 \times 10^{-6}$	$0{\cdot}123 \times 10^{-3}$
1:1:2	0·40	$0{\cdot}40 \times 10^{-6}$	$0{\cdot}087 \times 10^{-3}$

Duration of loading	Percentage of long-term creep
28 days	40
6 months	60
1 year	75
5 years	90
30 years	100

Table 10.4 Average values of specific creep related to the strength of concrete and modulus of elasticity

Cube strength		Modulus of elasticity		Ultimate specific creep per	
lb/in^2	N/mm^2	lb/in^2	kN/mm^2	lb/in^2	N/mm^2
2000	14	2×10^6	14	$2{\cdot}0 \times 10^{-6}$	$0{\cdot}30 \times 10^{-3}$
4000	28	3×10^6	21	$1{\cdot}0 \times 10^{-6}$	$0{\cdot}15 \times 10^{-3}$
6000	42	4×10^6	28	$0{\cdot}5 \times 10^{-6}$	$0{\cdot}072 \times 10^{-3}$
8000	56	5×10^6	35	$0{\cdot}2 \times 10^{-6}$	$0{\cdot}03 \times 10^{-3}$
10 000	70	6×10^6	42	$0{\cdot}1 \times 10^{-6}$	$0{\cdot}01 \times 10^{-3}$

of the ease of approach and because creep, within a wide range of strength, is inversely proportional to the strength of concrete.

R. H. Evans gives average values of the ultimate creep for various types of concrete made of normal Portland cement to B.S.12 and aggregate to B.S.882, as well as the percentage of the long-term creep. These are shown in Table 10.3.

Since the strength and modulus of elasticity of concrete are related, creep and modulus of elasticity are, therefore, also related for a given concrete at any period of time. For the purpose of design, the average values of these parameters can be taken as shown in Table 10.4.

Experiments show that up to at least one-third of the cube strength, creep strains are proportional to stresses.

Most data on creep refer to creep in compression, because such tests are easier to perform in the laboratory. However, creep in tension, in flexure, and in bond are equally important and are not always identical with the creep in compression.

The available data on creep in tension are not adequate to permit of definite conclusions. However, experiments by Glanville and Thomas appear to indicate that creep in tension may not differ much from that in compression. Other research shows creep in tension as being smaller than in compression, particularly after a certain period of time.

Similarly, creep in flexure is not yet fully resolved, although it is likely that, in flexure, creep strains may differ on compression side and on tension, and this alone may affect the position of the neutral axis of strains. The latter may not coincide with the stress neutral axis.

Creep in bond has received little attention. This type of creep, however, may be particularly important in the performance of spliced reinforcement and reinforced concrete columns. In the latter, redistribution of stress takes place from concrete to steel, and bond creep may play a major part in this process. Based on limited tests performed. it appears that creep in bond can be very high but there is no general agreement on this.

10.5 Approximate Analysis of the Effect of Creep on Concrete Members

Creep in concrete structures shows itself in the redistribution of strains, stresses, and deformations, and in statically indeterminate structures in redistribution of restraint bending moments, as indicated diagrammatically in Fig. 10.5.

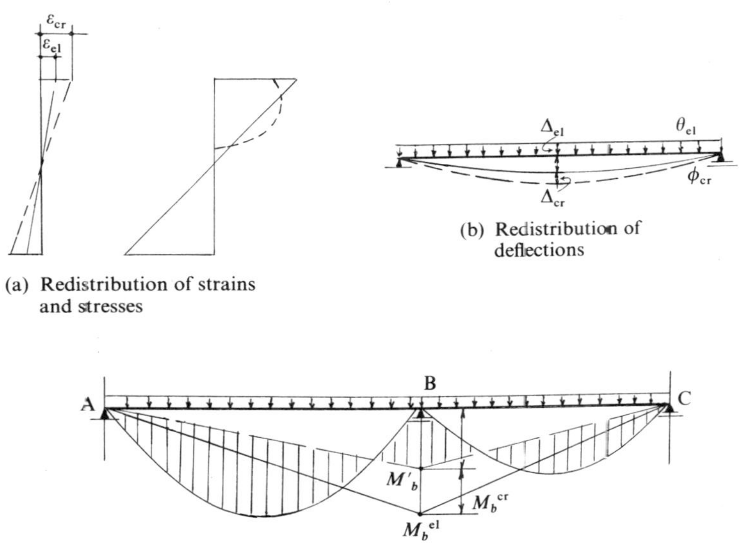

(a) Redistribution of strains and stresses

(b) Redistribution of deflections

(c) Relaxation of stresses

Fig. 10.5 Effect of creep on concrete members

It is important to note that the effect of creep depends on the source of stress: in statically determinate elements, creep cannot change the value of the static bending moment, although it can alter internal stress distribution as well as deformations, e.g., according to A. Neville 'a typical creep deformation after a year under load is three times the deformation on application of load'. However, if the source of stress is due to restraint, as in indeterminate structures, then even the magnitude of this stress can be affected. This is referred to as the relaxation of stresses. In statically indeterminate structures, both types of stress exist and both effects take place. Statically indeterminate structures are, therefore, more vulnerable to the effect of creep than determinate structures. Long and flexible members are particularly vulnerable to creep. Prestressed concrete beams are affected by creep because of high sustained stress imposed on concrete by tendons, resulting in the loss of prestressing force, and increased deformations.

Similarly, slender reinforced concrete columns, particularly in tall buildings, can show an increase in deformations and redistribution of stress from concrete to steel. However, once the steel yields, further increase in load can only be taken by concrete, so it is because of creep that the full strength of both steel and concrete tends to develop before failure takes place. The effect of creep on the magnitude of the ultimate strength is considered to be insignificant.

The adverse effect of creep of concrete lies, therefore, mainly in the increase of deformations and cracks, and, in the case of prestressed members, in the loss of force. It was mainly because of creep that early attempts to prestress beams by the use of mild steel resulted in failure, and only the introduction of high tensile steel, with elongations several times that of creep of concrete, made the concept of prestressing a practical proposition. Another case of possible adverse effect of creep is on the stability of some structures subjected to increasing deformations, e.g., cantilevers or tall, slender frames, particularly when eccentrically loaded.

However, an extremely valuable effect of creep lies in its ability to relieve stress concentrations from sustained loads, shrinkage, differential settlement of supports, etc.

For the purpose of design, a method of analysis is needed which would allow one to predict the effect of creep on redistribution of stresses and deformations in a structure after many years of service. However, even now only a general qualitative analysis of the effect of creep is possible in most practical cases. An exact analysis would require taking into account every case of variable stress, temperature, and moisture changes as well as all the past history of loading—an impossible task.

The object of this section is to consider a method of analysis of the effect of creep on structures, by necessity based on approximate assumptions. This analysis, therefore, makes no claim to being accurate for all practical purposes. At present, great refinement is neither justified nor practical in view of the incomplete understanding of creep and complexity of the parameters involved.

It should be mentioned that powerful methods of strain energy and complementary energy cannot be directly applied to transient creep but fortunately can be used for stationary steady state creep.

The following assumptions are, therefore, made in the analysis:

(a) that the materials are homogeneous and isotropic,
(b) that, within the working range of stresses, creep strains are proportional to the applied stresses,
(c) that the principle of superposition can be applied to this basically inelastic strain,
(d) that the effect of shrinkage and temperature can be considered separately and as additive,
(e) that originally plane sections remain plane after bending, and
(f) that the elements are subjected to a steady state of stress.

The above assumptions are not strictly true, and, therefore, the analysis cannot be exact. However, it has long been recognized that creep strains are almost proportional to stresses within working range, if concrete is well matured and at ambient temperatures. There is also a definite association of creep and elastic deformations, e.g., it can be shown that the governing equations of creep are of the same form as for the elastic problems. The proposed method of analysis is, therefore, closely linked to the elastic analysis. Furthermore, Hansen has shown that the principle of superposition of strains can be used for concrete, although it could not be used for materials with plastic deformations.

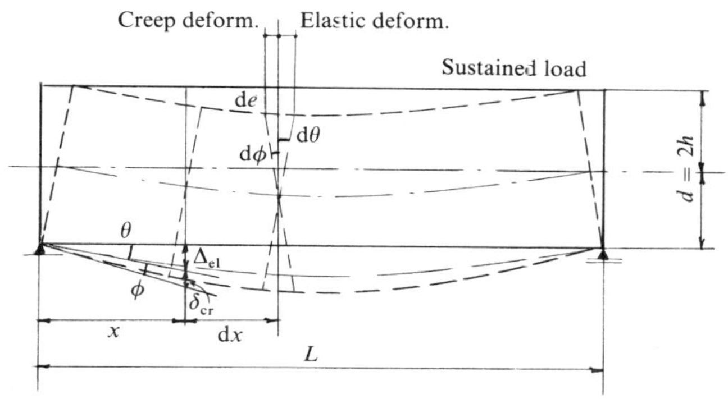

Fig. 10.6 Creep deformations in beams

Figure 10.6 shows a simply-supported beam subjected to a sustained load. An element dx is first affected by elastic deformations and then, after a given period of time, by creep deformation d$e = (c.\sigma x)$.dx at its outer fibres. The curvature at any section x is defined by:

$$\mathrm{d}\phi = \mathrm{d}e/h = M/EI$$

from which the total angular deformation of the beam is:

$$\phi = \frac{2}{d}\int_0^{L/2} c.\sigma_x.\mathrm{d}x \tag{10.1}$$

The creep deflection can then be calculated from: $\mathrm{d}\delta_c = x.\mathrm{d}\phi$ or, for symmetrical loading:

$$\delta_{cr} = \frac{2}{d}\int_0^{L/2} c.\sigma_x.x.\mathrm{d}x \tag{10.2}$$

A correlation coefficient could be introduced in these equations to account for the effect of parameters which are not included in the specific creep, such as: climatic and environmental conditions, temperature and moisture variations, past history of loading, sizes of members, volume–area ratio, percentage of steel, etc.

For example, for a cantilever beam in Fig. 10.7 the elastic deformations are:

$$\theta_{el} = \int_0^L \frac{M.m.\mathrm{d}x}{EI} = \frac{1}{EI}\int_0^L (\tfrac{1}{2}wx^2)(1)\,\mathrm{d}x = \frac{wL^3}{6EI}$$

and

$$\max \Delta_{el} = \frac{1}{EI}\int_0^L (\tfrac{1}{2}wx^2)(x)\,\mathrm{d}x = \frac{wL^4}{8EI}$$

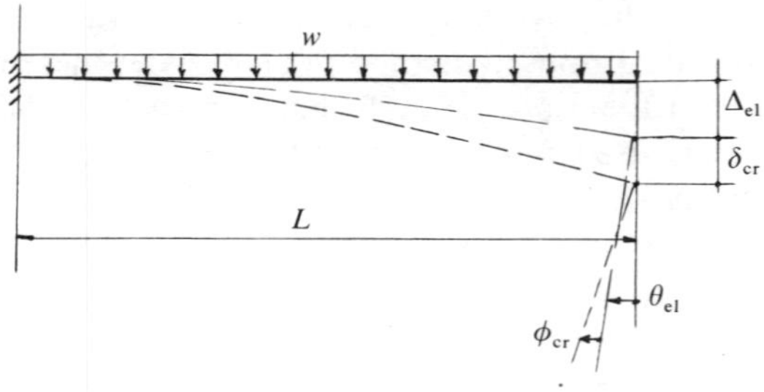

Fig. 10.7

The creep deformations, for an assumed ultimate value of specific creep c, are from eq. (10.2):

$$\sigma_x = \frac{d}{2I}(\tfrac{1}{2}wx^2)$$

$$\phi_{cr} = \frac{2}{d}\int_0^L c.\sigma_x.\mathrm{d}x = \frac{wL^3}{6EI} \times cE = \theta_{el}.cE$$

$$\delta_{cr} = \frac{2}{d}\int_0^L c.\sigma_x.x.\mathrm{d}x = \Delta_{el}.cE$$

The total deformations are therefore:

$$\theta_{el} + \phi_{cr} = \frac{wL^3}{6EI}\cdot(1 + cE)$$

$$\Delta_{el} + \Delta_{cr} = \frac{wL^4}{8EI}\cdot(1 + cE)$$

The value of parameter cE represents the magnification factor for creep deformations of concrete. Based on the average values of the ultimate specific creep and corresponding average values of the modulus of elasticity of concrete, the approximate variation of the coefficient cE, [as well as the value of $1/(1+cE)$] are shown in Table 10.5 and in Fig. 10.11.

Similarly, for a simply-supported beam subjected to a uniformly distributed load (Fig. 10.8), we obtain:

$$\theta_{el} + \phi_{cr} = \frac{M_0.L}{3EI}(1 + cE)$$

$$\Delta_{el} + \delta_{cr} = \frac{5M_0L_2}{48EI}(1 + cE)$$

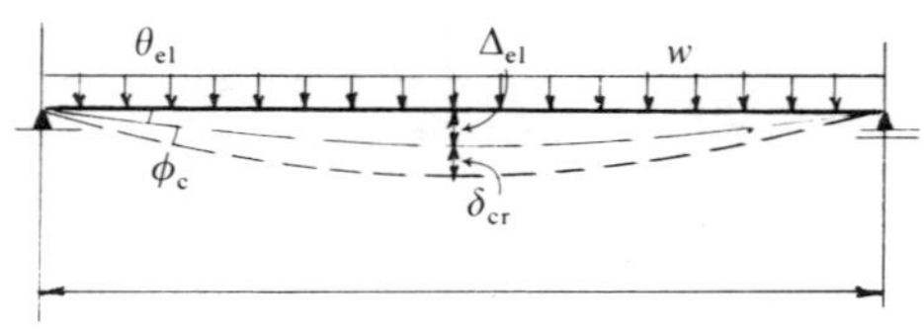

Fig. 10.8

Table 10.5 Average values of coefficients cE and $1/(1 + cE)$

U, N/mm^2	E, kN/mm^2	c, per N/mm^2	CE	$\frac{1}{1 + CE}$
14	14	$0{\cdot}3 \times 10^{-3}$	4·2	0·19
28	21	$0{\cdot}145 \times 10^{-3}$	3·0	0·25
42	28	$0{\cdot}072 \times 10^{-3}$	2·0	0·33
56	42	$0{\cdot}03 \times 10^{-3}$	1·2	0·45

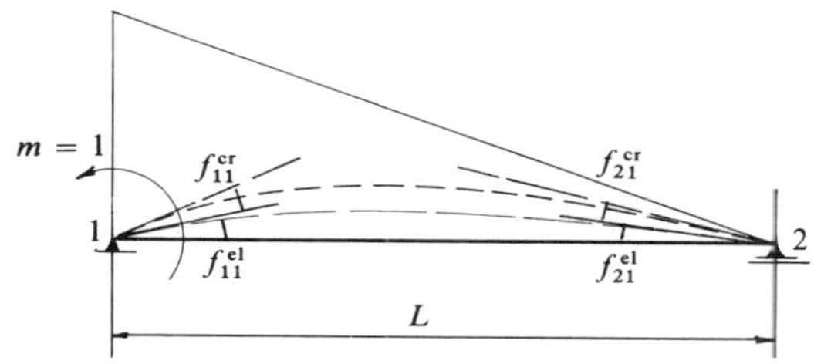

Fig. 10.9

The effect of creep strains on the flexibility coefficients can be approximately calculated in a similar way. From Fig. 10.9 we obtain:

$$f_{11}^{cr} = \frac{L}{3EI} \cdot cE \tag{10.3}$$

$$f_{21}^{cr} = \frac{L}{6EI} \cdot cE \tag{10.4}$$

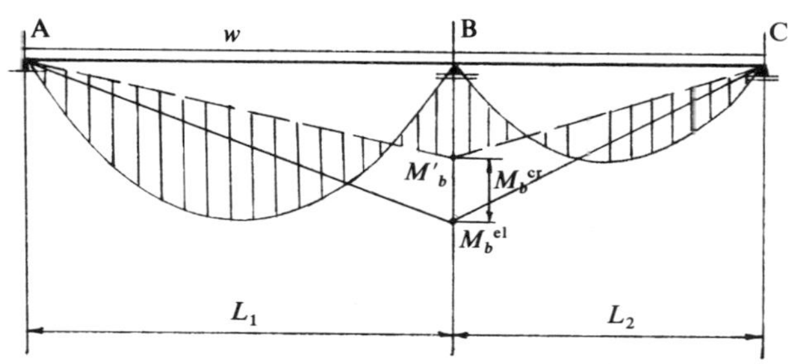

Fig. 10.10 Example of the analysis of the effect of creep on two-span beam

In statically indeterminate structures, creep strains result in redistribution of moments and deformations, and their effect can be approximately calculated by using the conditions of compatibility of deformations and the equilibrium of forces, e.g., for a two-span beam in Fig. 10.10 subjected to a sustained distributed load w, the residual bending moment M'_b retained over the support B at a time when the value of c is given, can be calculated from:

$$M'_b.\left(\frac{L_1}{3EI}+\frac{L_2}{3EI}\right)(1 + cE) = \frac{w(L_1^3 + L_2^3)}{24EI}$$

or

$$M'_b = M_b^{el}/(1 + cE)$$

or relaxation moment is: $M_b^{cr} = M_b^{el}.cE/(1 + cE)$

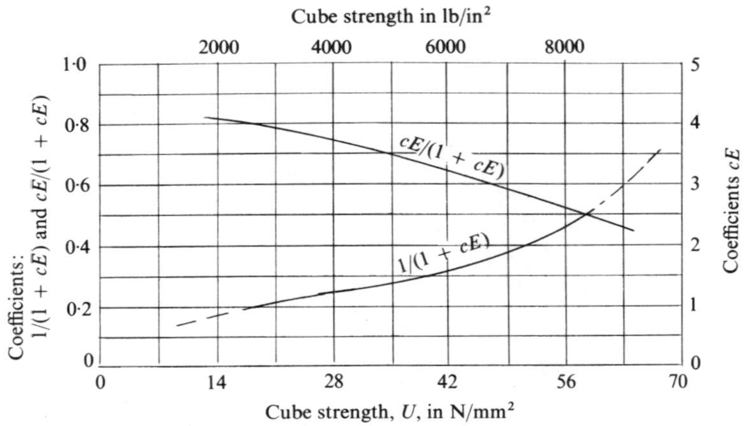

Fig. 10.11 Approximate values of the coefficients $1/(1+cE)$ and $cE/(1+cE)$

The approximate values of the coefficients $1/(1 + cE)$ and $cE/(1+cE)$ are shown in Fig. 10.11. These values are related to the cube strength of concrete only and do not take into account the dependence of creep on many other parameters, such as the temperature and humidity variation, shrinkage, past history of loading, etc., and should, therefore, be considered as very approximate only. The diagram, however, indicates the importance of using high quality concrete in avoiding undesirable effects of redistribution of stresses and deformations from creep.

It is usually assumed that the relaxation of stress due to creep results in decrease of restraint bending moments. However, at high creep strains, the formation of cracks may cause decrease as well as increase of restraint bending moments. When cracks appear, there is a sudden increase of the ratio of the creep moment to its elastic value, and the restraint bending moment can increase even above its elastic value. This process is a function of the relative amount of reinforcement at the span and at supports.

10.6 Effect of Creep on Columns

In columns, creep can have an important, although seldom evaluated, effect on redistribution of stresses and deformations. As mentioned earlier, creep in columns causes decrease of stresses in concrete and increase in steel stresses. However, at very high stresses, approaching yield stress in steel, the latter deforms without increase in stress and concrete alone carries increased load. This is because at ultimate the maximum strains in concrete are of the order of 0·0035, while the strain in steel at yield is approximately 0·02 (see Figs. 10.1 and 10.2).

It is recognized that, while creep has small effect on the ultimate strength, in slender columns creep can produce deformations which may affect the stability and load carrying capacity of a structure. However, very little quantitative information is available on this matter.

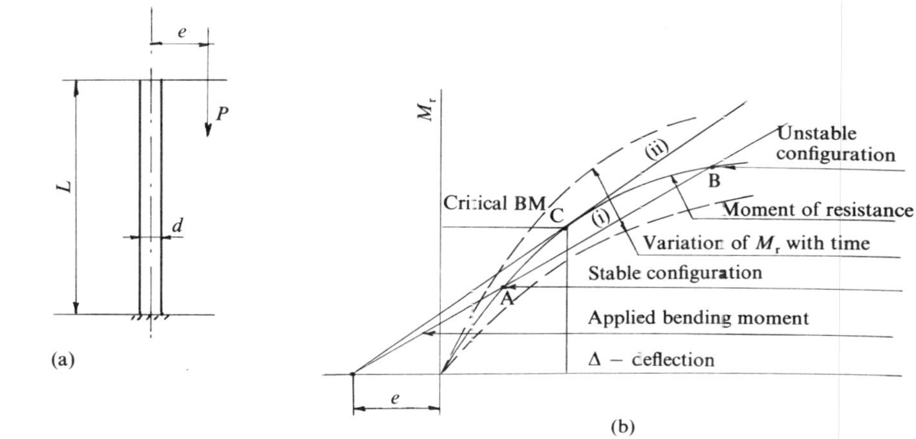

Fig. 10.12

Figure 10.12 shows diagrammatically a column subjected to an eccentric load and a typical moment–deflection relationship. The line (i) shows that, for an eccentricity e, two configurations can exist in which loads corresponding to the point A produce stable equilibrium, and at B unstable. A tangent at point C defines the maximum critical bending moment which a column can sustain. However, there is usually a considerable variation of the ultimate bending moment with time, as shown diagrammatically in Fig. 10.12 by dotted lines.

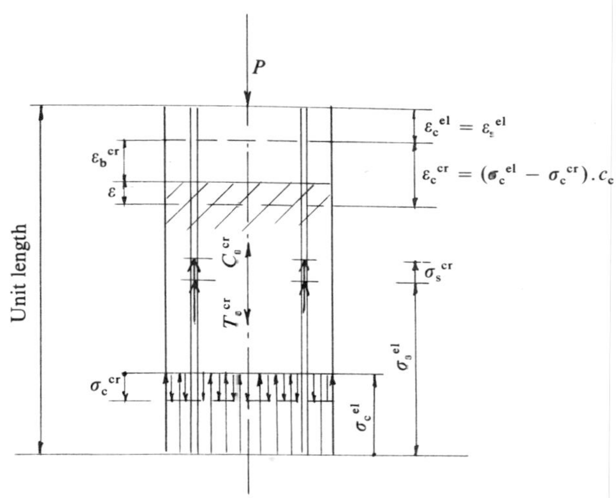

Fig. 10.13 Redistribution of strains and stresses in columns

An approximate analysis of the redistribution of stresses and deformations in reinforced concrete columns due to creep can be made as follows: Fig. 10.13 shows a unit length of a concrete column subjected to the axial force P which undergoes an elastic strain ε_c^{el} as well as creep strains ε_c^{cr}. If reinforcement is present, then it will partly prevent this creep from taking place and therefore redistribution of stresses will occur. Since the equilibrium must be maintained in the initial elastic range, as well as during creep strains, we can write that:

$$T_c = C_s \tag{10.5}$$

where T_c and C_s are respectively the decrease in the load carried by column (tension) and an increase of load carried by the reinforcement (compression) due to the effect of creep only.

Equation (10.5) can be written as:

$$\varepsilon . E . A_c = \varepsilon_s^{cr} . E_s A_s \tag{10.6}$$

in which, from Fig. 10.13,

$$\varepsilon = \varepsilon_c^{cr} - \varepsilon_b^{cr} \tag{10.7}$$

Referring to eq. (10.7) and Fig. 10.13, ε_c^{el} is the elastic instantaneous strain of concrete, ε_c^{cr} is creep strain of plain concrete, ε_s^{el} elastic instantaneous strain in steel, ε_b^{cr} is bond creep between concrete and steel, and ε is the residual strain in reinforced concrete column causing actual relaxation of concrete stress.

Introducing eq. (10.7) into eq. (10.6), we obtain:

$$\varepsilon = \frac{\sigma_c^{el} . c_c - \sigma_s^{el} . \varepsilon_b}{1 + K} \tag{10.8}$$

and

$$\varepsilon_s^{cr} = \sigma_c^{el} . c_c - \varepsilon \tag{10.9}$$

in which

$$K = \frac{E_c . A_c}{E_s . A_s} \tag{10.10}$$

The magnitudes of the redistributed stresses are therefore:

$$\sigma_c^{cr} = \varepsilon . E_c$$

and

$$\sigma_s^{cr} = \varepsilon_b^{cr} . E_s \tag{10.11}$$

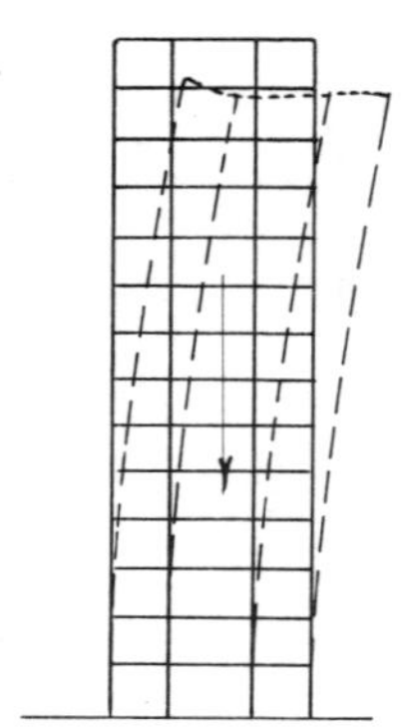

Fig. 10.14 Effect of creep in tall buildings

In tall and slender concrete buildings, creep can cause appreciable differential shortening of columns as shown diagrammatically in Fig. 10.14, and this in turn must result in redistribution of overall bending moments and deformations in addition to the transfer of stress from concrete to steel. This is a long-term effect on which there are no available data based on actual observations. However, based on the above discussed analysis of redistribution of stresses and deformations due to creep, an analysis of the overall response of tall building frames or shear walls can be made without difficulty using the force–displacement or other method.

Example 10.1

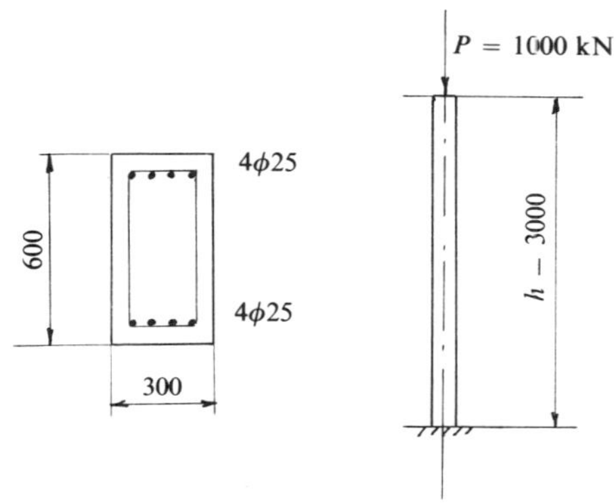

Fig. 10.15 Example of calculation of redistribution of stresses in columns due to creep

For a column in Fig. 10.15, assuming that $E_c = 14$ kN/mm², $E_s = 200$ kN/mm², and $c = 0{\cdot}15 \times 10^{-3}$ per N/mm², we have:

(a) in the elastic analysis:

$$\sigma_c^{el} = \frac{1\,000\,000}{300 \times 600 + 14 \times 3930} = 4{\cdot}25 \text{ N/mm}^2$$

$$\sigma_s^{el} = 15 \times 4{\cdot}25 = 63{\cdot}8 \text{ N/mm}^2$$

and

$$\Delta_{el} = \frac{1\,000\,000 \times 3000}{14\,000 \times 235\,020} = 0{\cdot}91 \text{ mm}$$

(b) in creep analysis:

$$k = \frac{14\,000 \times 176\,070}{200\,000 \times 3930} = 3{\cdot}14$$

neglecting the value of c_s, we have:

$$\varepsilon = \frac{4{\cdot}25 \times 0{\cdot}15 \times 10^{-3}}{4{\cdot}16} = 0{\cdot}153 \times 10^{-3}$$

$$\delta_c^{cr} = \varepsilon . E_c = 0{\cdot}153 \times 10^{-3} \times 14\,000 = 2{\cdot}15 \text{ N/mm}^2$$

$$\varepsilon_b^{cr} = 4{\cdot}25 \times 0{\cdot}15 \times 10^{-3} - 0{\cdot}153 \times 10^{-3} = 0{\cdot}483 \times 10^{-3}$$

$$\sigma_s^{cr} = 0{\cdot}482 \times 10^{-3} \times 200\,000 = 96{\cdot}4 \text{ N/mm}^2$$

so that the final stresses are:

$$\sigma_c = 4{\cdot}25 - 2{\cdot}15 = 2{\cdot}1 \text{ N/mm}^2$$

and

$$\sigma_s = 63{\cdot}8 + 96{\cdot}4 = 160{\cdot}2 \text{ N/mm}^2$$

Creep deformation is:

$$\Delta_{cr} = 0{\cdot}482 \times 10^{-3} \times 3000 = 1{\cdot}44 \text{ mm}$$

and total axial deformation of column is therefore 2·35 mm.

10.7 Effect of Creep on Frames and Shear Walls

It can be expected that a structure, after many years of service, will respond differently to loads, whether sustained or transient, than at the early stage after construction. This is mainly because of the effect of creep, but other causes no doubt also have some effect, e.g., the formation of cracks, deterioration of materials, loosening of bolts, settlement of foundations, etc. It is possible that creep will eventually reduce the stiffness of members (or increase their flexibility), increase axial as well as lateral deformations and redistribute stresses from concrete to steel.

At present, there is no exact way of predicting the response of complex structures to loads after a long period of time, say 30 or 50 years, when all the creep has taken place and stresses have reached a steady state. The following examples, therefore, are no more than an indication of what is needed and no claim is made that the procedure is rigorous. The analysis is based on the assumption that stresses have reached a steady state and on a given ultimate value of the specific creep and the corresponding value of E. Three effects are included: the effect of creep on the flexibility coefficients, on axial deformations of columns, and the effect of altered geometry of a structure.

Example 10.2

A portal frame in Fig. 10.16 is assumed to be subjected to the effect of the sustained gravity loads P=1000 kN, as well as to the transient lateral force W. It is assumed

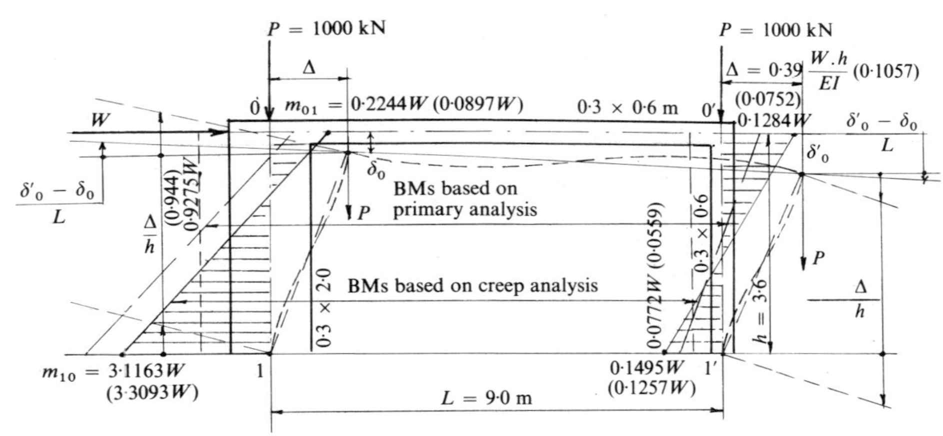

Fig. 10.16 Example

that $c=0{\cdot}15\times10^3$ per N/mm^2, $E=28$ kN/mm^2, and that the ratio of loads is $W/P=0{\cdot}1$. Axial deformations and the flexibility coefficients can now be calculated and the matrix can be set out in the usual way. For the above frame, this takes the form:

$$\begin{bmatrix} 0{\cdot}7825 & -0{\cdot}375 & -0{\cdot}0163 & 0 & -0{\cdot}25 \\ -0{\cdot}3750 & 1{\cdot}9954 & 0 & -0{\cdot}6384 & -0{\cdot}25 \\ -0{\cdot}0163 & 0 & 0{\cdot}0325 & 0 & -0{\cdot}25 \\ 0 & -0{\cdot}6384 & 0 & -1{\cdot}2014 & -0{\cdot}25 \\ 1 & 1 & 1 & 1 & -0{\cdot}0475 \end{bmatrix} \begin{Bmatrix} m_{01} \\ m'_{01} \\ m_{10} \\ m'_{10} \\ \Delta \end{Bmatrix} = \begin{Bmatrix} -0{\cdot}021W \\ -0{\cdot}021W \\ 0 \\ 0 \\ 3{\cdot}6W \end{Bmatrix}$$

It is interesting to note that the compatibility conditions become now a function of the magnitude of the external loads, while the equilibrium condition is a function of the deformation of a structure. The first is an inherent result of the creep problem, while the second is the effect of equilibrium being satisfied in deformed configuration of a structure.

Figure 10.16 shows the distribution of bending moments, forces, and deformations, including the effect of creep. Figures in brackets refer to similar values based on the primary analysis, i.e., without the effect of creep (see Fig. 1.7).

This example indicates that a considerable redistribution of stresses may take place in a structure after many years of service and particularly that the deformations can increase appreciably with time, even for transient loads.

Example 10.3

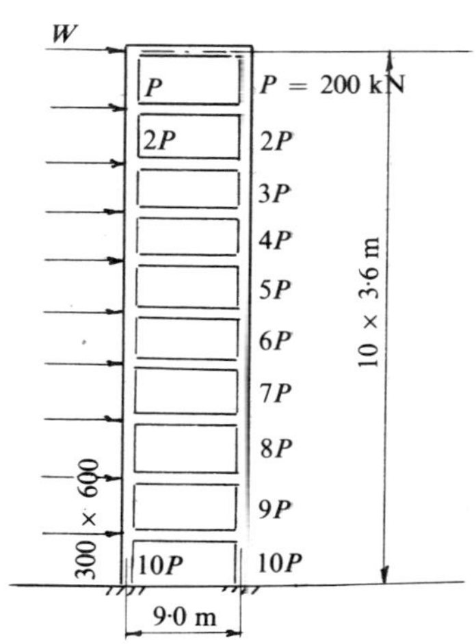

Fig. 10.17 Example

In this example, the effect of creep on the multi-storey frame shown in Fig. 10.17 is considered, based on similar assumptions as made in the previous example. Table 10.6 shows the setting out of the main matrix. In this case, the element matrices are non-repetitive because columns are affected by creep to a different extent at various floor levels.

Figure 10.18 shows the resulting moments, forces, and deformations, together with similar values based on the primary analysis shown in brackets. It can again be seen that a structure after many years of service may respond differently to imposed loads, even transient, than soon after construction. Both the stresses and the deformations may be affected. As a consequence of large deformations, now aggravated by creep, the frame carries in bending a proportion of gravity loads.

It is interesting to note, by comparing this analysis with the frame in Fig. 10.7 analysed for the effect of elastic instability, that the inclusion of creep results in

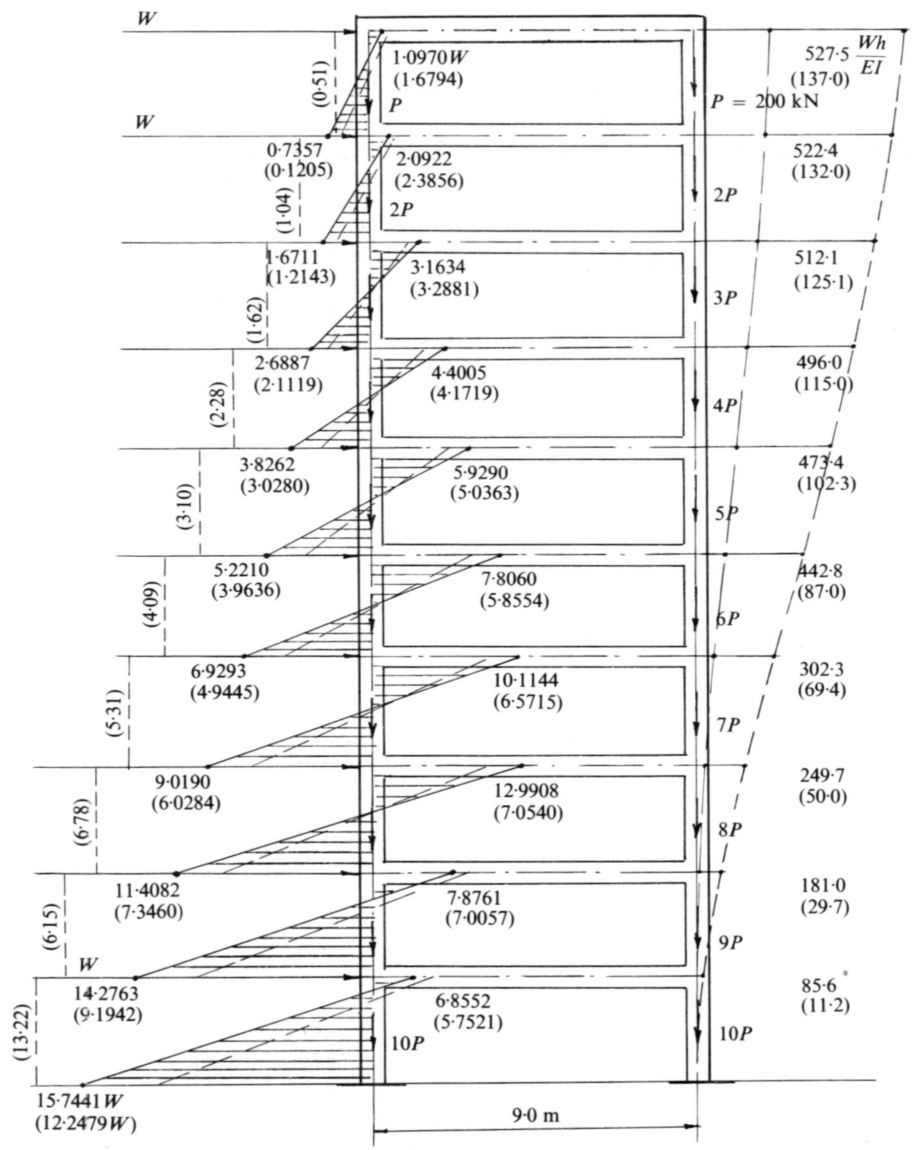

Fig. 10.18 Example of the analysis of a frame for the effect of creep

increased deformations and bending moments. A larger proportion of gravity loads is now carried by the frame in bending.

10.8 Approximate Analysis of the Effect of Shrinkage on Concrete Structures

It is well known that shrinkage of concrete, if not controlled, can produce high stresses and deformations. Cracks can develop in concrete, usually within the first three to six months, as well as excessive shrinkage deflections.

These stresses and deformations are caused by various restraints to free volume changes of concrete such as the presence of reinforcement (or tendons), difference in masses, thicknesses and shapes of members, monolithic connections with other members, and uneven drying shrinkage across the section.

The analysis shows, as is well known to practising engineers, that it is not possible to control effectively or to reduce the effect of shrinkage by reinforcement alone (except by post-tensioning). Such control can be made only by careful consideration of the main parameters of shrinkage. Even curing of concrete does not reduce the magnitude of shrinkage although it delays its effect until the concrete attains higher strength, and particularly a higher tensile strength. Creep, however, reduces concentration of stresses due to shrinkage.

In reinforced concrete members, shrinkage generally produces tension in concrete and compression in steel, and these are additive to the main stresses produced by gravity, wind loads, and prestressing tendons.

The effect of shrinkage is also additive to temperature changes and creep of concrete so that the actual stresses in concrete members may differ considerably from the design stresses. Theoretical analyses are complicated by the fact that the shrinkage strains, creep, and temperature are inter-related and change with time. The development of cracks further complicates the analyses.

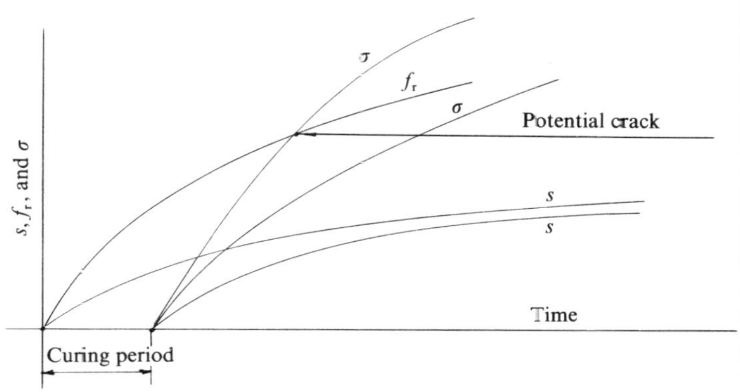

Fig. 10.19 Effect of shrinkage on concrete beams

Although shrinkage cracks in concrete members occur frequently, not all members crack or must inevitably crack. Cracks develop when shrinkage stresses (often added to other effects) exceed the tensile strength of concrete at a particular age of concrete. However, both the magnitude of shrinkage and the tensile strength of concrete vary with time and curing period. If the latter is sufficiently long, concrete strength, its modulus of elasticity, and particularly the tensile strength, may reach a

Table 10.6 Force–displacement matrix for the analysis of frame in Fig. 10.18 including the effect of creep

	0		1			2			3			4		
	0 1	Δ0	1 0	1 2	Δ1	2 1	2 3	Δ2	3 2	3 4	Δ3	4 3	4 5	
1 0 0′	1·5771	−0·25	−0·6017	0	0·25									
0–0′	1	−0·0063	1	0	0·0063									
2 1 1′	0	0	0·375	1·5788	−0·25	−0·6033	0	0·25						
0 1 2	−0·6011	−0·25	1·2033	−1·203	0·5	0·6033	0	−0·25						
1–1′	−1	0	−1	1	−0·0127	1	0	0·0127						
3 2 2′				0	0	0·375	1·5810	−0·25	−0·6049	0	0·25			
1 2 3				−0·6019	−0·25	1·2066	−1·2058	0·5	0·6050	0	−0·25			
2–2′				−1	0	−1	1	−0·0180	1	0	0·0180			
4 3 3′							0	0	0·375	1·5832	−0·25	−0·6065	0	[illegible]
2 3 4							−0·6029	−0·25	1·2100	−1·2074	0·5	0·6048	0	−
3–3′							−1	0	−1	1	−0·0254	1	0	0·
5 4 4′										0	0	0·375	1·5854	−
3 4 5										−0·6037	−0·25	1·2130	−1·2086	
4–4′										−1	0	−1	1	−
6 5 5′													0	
4 5 6													−0·6043	−
5–5′													−1	
7 6 6′														
5 6 7														
6–6′														
8 7 7′														
6 7 8														
7–7′														
9 8 8′														
7 8 9														
8–8′														
10 9 9′														
8 9 10														
9–9′														
10 9														

	5		6			7			8			9			10	
	5 6	Δ5	6 5	6 7	Δ6	7 6	7 8	Δ7	8 7	8 9	Δ8	9 8	9 10	Δ9	10 9	
																0
																1·8 *W*
																0
																0
																1·8 *W*
																0
																0
																1·8 *W*
																0
																0
																1·8 *W*
83	0	0·25														0
3	0	−0·25														0
	0	0·0367														1·8 *W*
	1·5865	−0·25	−0·6098	0	0·25											0
5	−1·211	0·5	0·6098	0	−0·25											0
	1	−0·0441	1	0	0·0441											1·8 *W*
	0	0	0·375	1·588	−0·25	−0·6113	0	0·25								0
	−0·6055	−0·25	1·2197	−1·2129	0·5	0·6113	0	−0·25								0
	−1	0	−1	1	−0·0494	1	0	0·0494								1·8 *W*
				0	0	0·375	1·590	−0·25	−0·6121	0	0·25					0
				−0·6065	−0·25	1·2227	−1·215	0·5	0·6121	0	−0·25					0
				−1	0	−1	1	−0·0508	1	0	0·0508					1·8 *W*
							0	0	0·375	1·5925	−0·25	−0·6149	0	0·25		0
							−0·6075	−0·25	1·2241	−1·2175	0·5	0·6149	0	−0·25		0
							−1	0	−1	1	−0·0621	1	0	0·0621		1·8 *W*
										0	0	0·375	1·5938	−0·25	−0·6165	0
										−0·6087	−0·25	1·2297	−1·219	+0·5	0·6165	0
										−1	0	−1	1	−0·0735	1	1·8 *W*
													−0·6094	0	1·233	0

higher value than the shrinkage stresses, and cracks may not form. This is shown diagrammatically in Fig. 10.19. If, however, shrinkage stresses increase at a higher rate than the increase of tensile strength, a time may be reached when the latter value is exceeded and cracks may then develop.

Because of difficulties in exact formulation of the many parameters involved, the following analysis is based on the approximate assumptions that:

(a) the effect of shrinkage can be considered independently from creep and temperature changes,
(b) the strain distribution is linear across the section or parts of the section,
(c) the stresses are proportional to strains,
(d) the potential shrinkage is uniform across uniform parts of sections, and that
(e) the sections are not cracked.

10.8.1 *The effect of shrinkage on rectangular reinforced concrete beams*

In the following, the distribution of stresses and deformations in concrete beams due to shrinkage is determined by consideration of the equilibrium and compatibility of strains of the system. In Fig. 10.20, s represents a potential shrinkage of a plain concrete beam of unit length, and ε_{ct} and ε_{cb} represent strains at the top and bottom of the beam as a result of the presence of reinforcement of the area A_{st} and A_{sb}. Strains in steel are indicated by σ_{ct} and σ_{cb}. The corresponding stresses and resulting forces in concrete and in steel are: T_c being tension in the concrete and C_{st} and C_{sb} are compression in the steel at the top and bottom respectively. For the equilibrium of forces and moments in the section we can write:

$$T_c - C_{sb} - C_{st} = 0 \tag{10.12}$$

$$(\tfrac{1}{2}\sigma_{cb} . A_c)\frac{d}{6} - (\tfrac{1}{2}\sigma_{ct} . A_c)\frac{d}{6} - C_{sb} . e_1 + C_{st} . e_2 = 0 \tag{10.13}$$

in which:

$$\begin{aligned}
T_c &= \tfrac{1}{2}(\sigma_{cb} + \sigma_{ct})A_c = \tfrac{1}{2}A_cE_c(\varepsilon_{cb} + \varepsilon_{ct})\\
C_{sb} &= A_{sb} . E_s . \varepsilon_{sb}\\
C_{st} &= A_{st} . E_s . \varepsilon_{st}\\
\varepsilon_{sb} &= s - \left(\frac{e_1}{d} + \frac{1}{2}\right)\varepsilon_{cb} + \left(\frac{e_1}{d} - \frac{1}{2}\right)\varepsilon_{ct}\\
\varepsilon_{st} &= s + \left(\frac{e_2}{d} - \frac{1}{2}\right)\varepsilon_{cb} - \left(\frac{e_2}{d} + \frac{1}{2}\right)\varepsilon_{ct}
\end{aligned} \tag{10.14}$$

Denoting:

$$\frac{A_c . E_c}{A_{sb} . E_s} = K; \qquad \frac{A_{st}}{A_{sb}} = K_s; \qquad \mu_1 = \frac{e_1}{d}; \qquad \mu_2 = \frac{e_2}{d}$$

and introducing the above values in eqs. (10.12) and (10.13), we obtain:

$$\left[\frac{K}{2} + \mu_1 + K_s(\tfrac{1}{2} - \mu_2) + \frac{1}{2}\right]\varepsilon_{cb} + \left[\frac{K}{2} - \mu_1 + K_s.(\mu_2 + \tfrac{1}{2})\right]\varepsilon_{cb} = s\,[K_s + 1]$$

$$\left[\frac{K}{12} + \mu_1(\mu_1 + \tfrac{1}{2}) + K_s.\mu_2(\mu_2 - \tfrac{1}{2})\right].\varepsilon_{cb} - \left[\frac{K}{12} + \mu_1(\mu_1 - \tfrac{1}{2}) + K_s.\mu_2(\mu_2 + \tfrac{1}{2})\right]\varepsilon_{ct}$$

$$= s[\mu_1 - K_s.\mu_2] \quad (10.15)$$

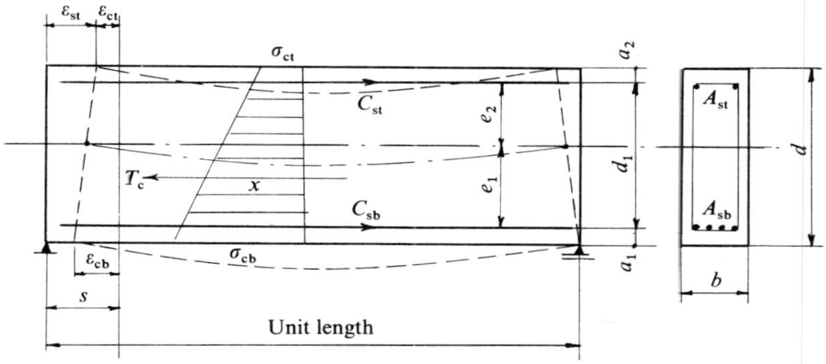

Fig. 10.20 Analysis of shrinkage stresses in reinforced concrete beams

Denoting dimensionless parameters in square brackets in eqs. (10.15) by:

$$A = \frac{K}{2} + \mu_1 + K_s(\tfrac{1}{2} - \mu_2) + \tfrac{1}{2}$$

$$B = \frac{K}{2} - \mu_1 + K_s(\tfrac{1}{2} + \mu_2) + \tfrac{1}{2}$$

$$C = 1 + K_s$$

$$D = \frac{K}{12} + \mu_1(\mu_1 + \tfrac{1}{2}) + K_s.\mu_2(\mu_2 - \tfrac{1}{2})$$

$$E = \frac{K}{12} + \mu_1(\mu_1 - \tfrac{1}{2}) + K_s.\mu_2(\mu_2 + \tfrac{1}{2})$$

$$F = \mu_1 - K_s.\mu_2,$$

eqs. (10.15) can be re-written in the form:

$$A.\varepsilon_{cb} + B.\varepsilon_{ct} = C.s$$

$$D.\varepsilon_{cb} - E.\varepsilon_{ct} = F.s$$

from which:

$$\varepsilon_{cb} = s\cdot\frac{BF + CE}{AE + BD} \quad (10.16)$$

$$\varepsilon_{ct} = s\cdot\frac{CD - AF}{AE + BD} \quad (10.17)$$

The strains in the steel can now be determined from eqs. (10.14).

For a beam with a symmetrical reinforcement at the top and bottom $A_{st}=A_{sb}$, and $e_1=e_2$, eqs. (10.16) and (10.17) reduce to:

$$\varepsilon_{cb} = \varepsilon_{ct} = \frac{s}{A} \tag{10.18}$$

and

$$\varepsilon_{sb} = \varepsilon_{st} = s\left(1 - \frac{1}{A}\right)$$

Since in this case $K_s=1$, the value of A reduces to

$$A = \frac{K}{2} + 1$$

i.e., the strains in concrete and in steel in symmetrically reinforced members are independent of the position of the steel, but depend on the magnitude of steel area.

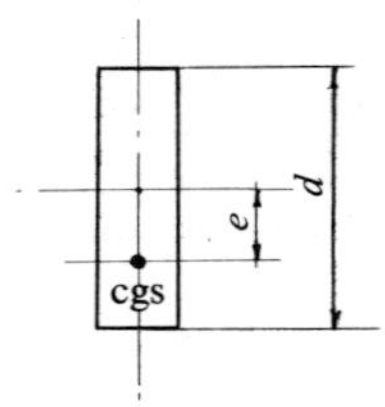

Fig. 10.21

If the centre of gravity of steel is denoted by e (Fig. 10.21) we obtain:

$$\varepsilon_{cb} = s\cdot\frac{BF + E}{AE + BD} = s\frac{6\mu + 1}{K + 12\mu^2 + 1} \tag{10.19}$$

$$\varepsilon_{ct} = s\cdot\frac{D - AF}{AE + BD} = s\cdot\frac{1 - 6\mu}{K + 12\mu^2 + 1} \tag{10.20}$$

and

$$\varepsilon_s = s\cdot\frac{K + 12\mu^2 - 8\mu + 1}{K + 12\mu^2 + 1}; \qquad \mu = \frac{e}{d} \tag{10.21}$$

which, for $e=0$ reduce to eq. (10.18). The diagram in Fig. 10.23 is based on eqs. (10.16) and (10.17).

Example 10.4

Taking: $E_s=200$ kN/mm^2 and $E_c=14$ kN/mm^2 we have:

for case (a): $\sigma_{ct} = 0{\cdot}332$ N/mm^2 $\quad \sigma_{cb} = 2{\cdot}2$ N/mm^2

$\sigma_{st} = 57{\cdot}3$ and $\sigma_{sb} = 341{\cdot}0$ N/mm^2

for case (b): $\sigma_{ct} = \sigma_{cb} = 1{\cdot}685$ and $\sigma_{st} = \sigma_{sb} = 392{\cdot}0$ N/mm^2

for case (c): $\sigma_{ct} = 2{\cdot}135$ and $\sigma_{cb} = 3{\cdot}15$ N/mm^2

$\sigma_{sb} = 248{\cdot}0$ N/mm^2

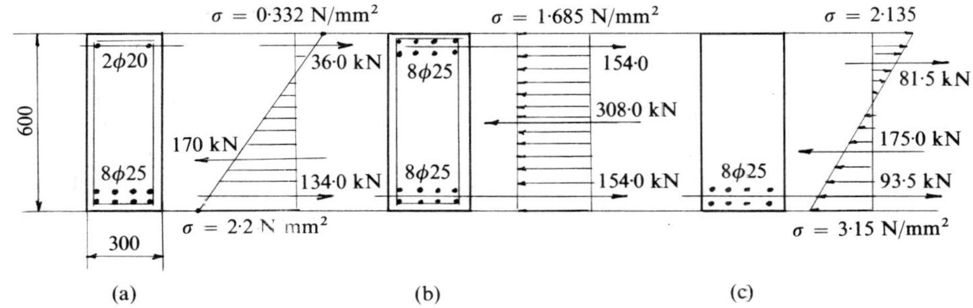

Fig. 10.22 Example

10.8.2 *Shrinkage stresses in 'T' beams*

In reinforced concrete 'T' beams webs are sometimes cast prior to the slab, and high shrinkage stresses may develop, particularly along the construction joint and in the slab. At the time of casting the slab, the beam may have already been subjected to shrinkage strains s with the corresponding stresses as indicated in Fig. 10.24. These could be calculated from eqs. (10.16) and (10.17) or from Fig. 10.23.

The subsequent stresses will occur due to further shrinkage in the web s_1 (Fig. 10.24) and in the slab s_2. These stresses can be calculated in a similar way, resulting in the following values:

$$\varepsilon_1 = [(C_1E - F_1B)s_1 + (C_2E - F_2B)s_2] \div (AE - BD)$$

$$\varepsilon_4 = [(C_1D - F_1A)s_1 + (C_2D - F_2A)s_2] \div (BD - AE) \qquad (10.22)$$

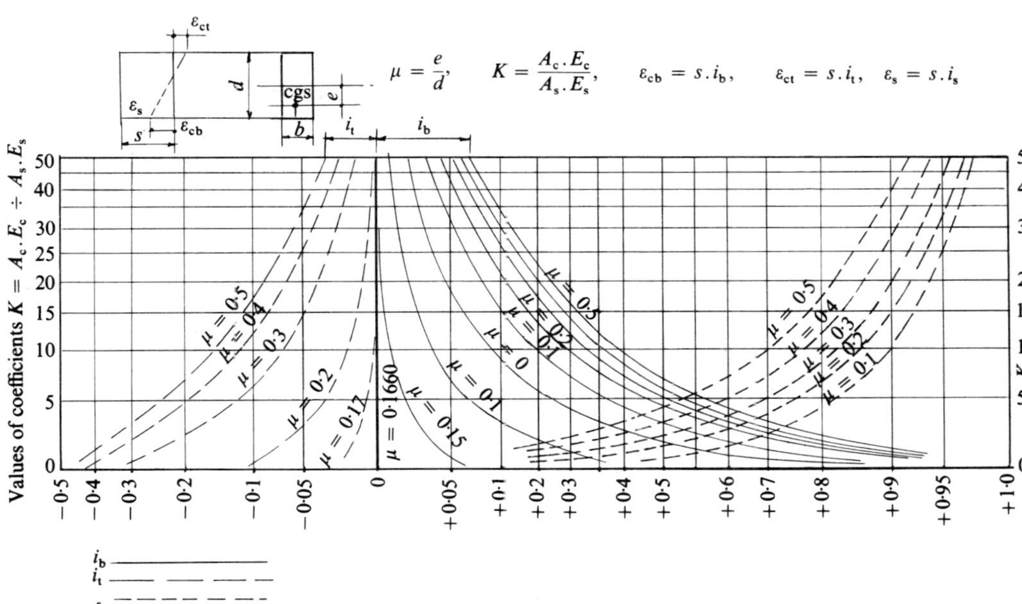

Fig. 10.23 Diagram of shrinkage strain coefficients *i* (see eqs. 10.13 and 10.14)

and

$$\varepsilon_2 = (1 - m_2)(s_2 - s_1) - m_2.\varepsilon_1 - (1 - m_2)\varepsilon_4$$

$$\varepsilon_3 = m_2(s_2 - s_1) + m_2.\varepsilon_1 + (1 - m_2)\varepsilon_4$$

$$\varepsilon_{s1} = [1 - (1 - \mu_1)(1 - m_2)]s_1 + (1 - \mu_1)(1 - m_2)s_2 - [\mu_1 + m_2(1 - \mu_1)]\varepsilon_1 - (1 - \mu_1)(1 - m_2)\varepsilon_4$$

$$\varepsilon_{s2} = m_2(1 - \mu_2)s_1 + [1 - m_2(1 - \mu_2)]s_2 - m_2(1 - \mu_2)\varepsilon_1 - [1 - m_2(1 - \mu_2)].\varepsilon_4 \quad (10.23)$$

in which:

$$K_1 = \frac{A_{c1}.E_{c1}}{A_{s1}.E_s}, \qquad K_2 = \frac{A_{c2}.E_{c2}}{A_{s1}.E_s}, \qquad K_s = \frac{A_{s2}}{A_{s1}}$$

$$m_1 = \frac{d_1}{d}, \qquad m_2 = \frac{d_2}{d}, \qquad \mu_1 = \frac{e_1}{d_1}, \qquad \mu_2 = \frac{e_2}{d_2}$$

$$A = \frac{K_1}{2}(1 + m_2) + \frac{K_2}{2}\cdot m_2 + \mu_1 + m_2(1 - \mu_1) + K_s.m_2(1 - \mu_2)$$

$$B = \frac{K_1}{2}(1 - m_2) + \frac{K_2}{2}\cdot(2 - m_2) - (1 - \mu_1)(1 - m_2) + K_s(1 - m_2 + m_2.\mu_2)$$

$$C_1 = -\frac{K_1}{2}(1 - m_2) + \frac{K_2}{2}m_2 - (1 - \mu_1)(1 - m_2) + K_s.m_2(1 - \mu_2) + 1$$

$$C_2 = \frac{K_1}{2}(1 - m_2) - \frac{K_2}{2}m_2 + (1 - \mu_1)(1 - m_2) + K_s.(1 - m_2 + m_2.\mu_2)$$

$$D = \frac{K_1}{6}(2 + m_2) + \mu_1^2 + \mu_1.m_2(1 - \mu_1) - \frac{m_2^{\,2}}{m_1}\left[\frac{K_2}{6} + K_s.\mu_1(1 - \mu_2)\right]$$

$$E = \frac{K_1}{6}(1 - m_2) - \frac{K_2}{6}\left[2 + \frac{m_2}{m_1}(1 - m_2)\right] - K_s.\mu_2\cdot\frac{m_2}{m_1}\cdot(1 - m_2 + m_2.\mu_2) + \mu_1(1 - \mu_1)(1 - m_2)$$

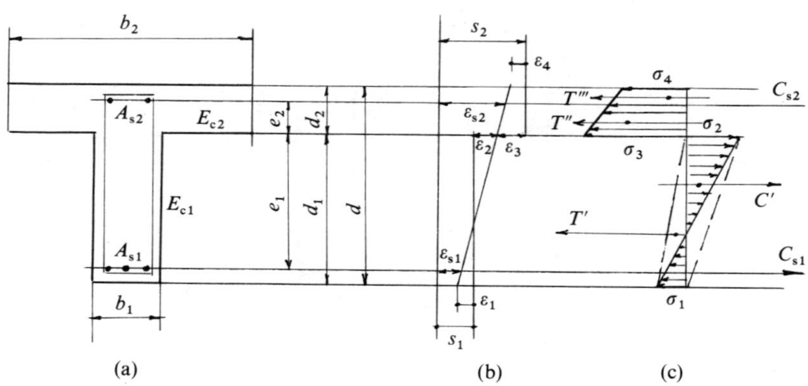

Fig. 10.24 Shrinkage stresses in T beams

$$F_1 = -\frac{K_1}{6}\cdot(1 - m_2) + \mu_1[1 - (1 - m_2)(1 - \mu_1)] - \frac{m_2^2}{m_1}\cdot\left[\frac{K_2}{6} - K_s.\mu_2(1 - \mu_2)\right]$$

$$F_2 = \frac{K_1}{6}(1 - m_2) + \frac{K_2}{6}\cdot\frac{m_2^2}{m_1} + \mu_1(1 - m_2)(1 - \mu_1) -$$

$$- K_s.\mu_2\frac{m_2}{m_1}(1 - m_2 + m_2.\mu_2)$$

10.9 Effect of Shrinkage on Parapet Beams

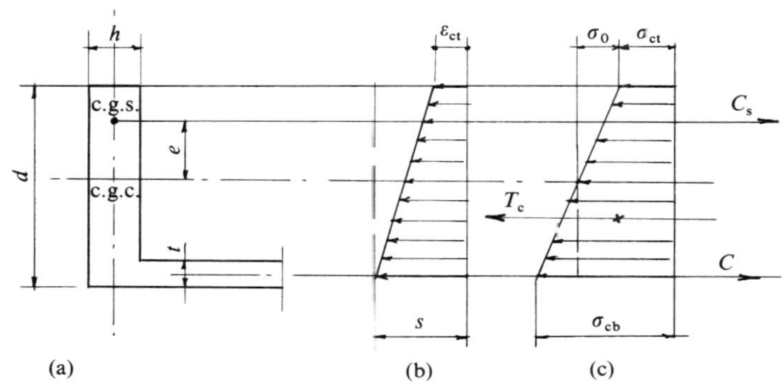

Fig. 10.25 Shrinkage stresses in parapet beams

A similar problem occurs in parapet beams restrained by the reinforcement as well as by the floor slab as shown in Fig. 10.25. If we assume that the slab is infinitely rigid in its own plane, then the conditions of equilibrium can be written in the form:

$$T_c - C - C_s = 0$$

$$\frac{\sigma_0}{2}\left(\frac{bd}{2}\right) + \tfrac{2}{3}d - C\left(\frac{d}{2} - \frac{t}{2}\right) + C_s.e = 0 \qquad (10.24)$$

in which:

$$T_c = \tfrac{1}{2}(\sigma_{cb} + \sigma_{ct})A_c - \tfrac{1}{2}.A_c.E_c(\varepsilon_{cb} + \varepsilon_{ct})$$

$$C_s = A_sE_s.\varepsilon_s$$

$$\sigma_0 = \tfrac{1}{2}(\sigma_{cb} - \sigma_{ct}) = \tfrac{1}{2}E_c(\varepsilon_{cb} - \varepsilon_{ct}) = \tfrac{1}{2}E_c(s - \varepsilon_{ct})$$

and

$$\varepsilon_s = (s - \varepsilon_{st})\frac{d/2 + e}{d} = (s - \varepsilon_{ct})(K_3 + \tfrac{1}{2}) \qquad (10.25)$$

Substituting the above values in eqs. (10.24), we obtain:

$$\left[\frac{K}{2} + \mu + \frac{1}{2}\right]\varepsilon_{ct} - \left[\frac{1}{A_s.E_s}\right].C = s\left[\mu - \frac{K}{2} + \frac{1}{2}\right] \qquad (10.26)$$

$$\left[\frac{K}{12} + \mu(\mu + \tfrac{1}{2})\right]\varepsilon_{cb} + \left[\frac{1 - \rho}{2A_s.E_s}\right].C = s\left[\frac{K}{12} + \mu(\mu + \tfrac{1}{2})\right] \qquad (10.27)$$

in which:

$$K = \frac{A_c E_c}{A_s E_s}; \qquad \mu = \frac{e}{d}; \qquad \rho = \frac{t}{d}$$

or by denoting the values in the square brackets by:

$$p_1 = \frac{K}{2} + \mu + \frac{1}{2}, \qquad p_2 = \frac{1}{A_s . E_s}, \qquad p_3 = \mu - \frac{K}{2} + \frac{1}{2}$$

$$p_4 = p_6 = \frac{K}{12} + \mu(\mu + \tfrac{1}{2}), \qquad p_5 = \frac{1-\rho}{2A_s . E_s}$$

we obtain:

$$\varepsilon_{ct} = s \cdot \frac{p_2 \cdot p_6 + p_3 \cdot p_5}{p_1 \cdot p_5 + p_2 \cdot p_4} \tag{10.28}$$

and

$$C = s \cdot \frac{p_1 \cdot p_6 - p_3 \cdot p_4}{p_1 \cdot p_5 + p_2 \cdot p_4} \tag{10.29}$$

Example 10.5

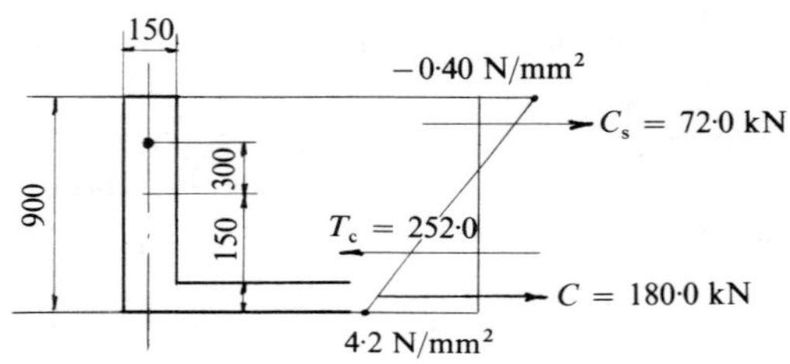

Fig. 10.26 Example

For the parapet beam shown in Fig. 10.26, assuming $s = 0{\cdot}0003$, $E_c = 14$ kN/mm², and $A = 1260$ mm², we have: $K = 7{\cdot}13$, $\mu = 0{\cdot}33$, $\rho = 0{\cdot}16$, $p_1 = 4{\cdot}4$, $p_2 = 0{\cdot}166 \times 10^{-7}$, $p_3 = -2{\cdot}73$, $p_4 = p_6 = 0{\cdot}87$, and $p_5 = 0{\cdot}695 \times 10^{-8}$. Introducing these values in eq. (10.28), we obtain:

$$\varepsilon_{ct} = -0{\cdot}000\ 029\ 2$$

Therefore: $\sigma_{ct} = -0{\cdot}40$ N/mm² (compression)

and $\sigma_{cb} = 4{\cdot}2$ N/mm² (tension)

From eq. (10.25), $\varepsilon_s = 0{\cdot}000\ 274$ and $\sigma_s = 57{\cdot}5$ N/mm².
The resultant forces are:

$$T_c = \tfrac{1}{2}(\sigma_{cb} + \sigma_{ct})A_c = 252 \text{ kN} \quad \text{and} \quad C = 180 \text{ kN}$$

10.10 Analysis of Shrinkage Deformations

The curvature $d\phi$ of the beam element dx (Fig. 10.27) can be expressed as:

$$d\phi = \frac{\varepsilon_{cb} - \varepsilon_{ct}}{d} \cdot dx \tag{10.30}$$

For a beam with constant reinforcement throughout the length of the beam, the total angular deformation at the support is:

$$\phi = \int \frac{l}{d} d\phi = \frac{l}{d}(\varepsilon_{cb} - \varepsilon_{ct}) \tag{10.31}$$

or in terms of the dimensionless parameters A, B, C, D, E, and F:

$$\phi = \frac{l.s}{d} \cdot \frac{(A + B)F + C(E - D)}{AE + BD} \tag{10.32}$$

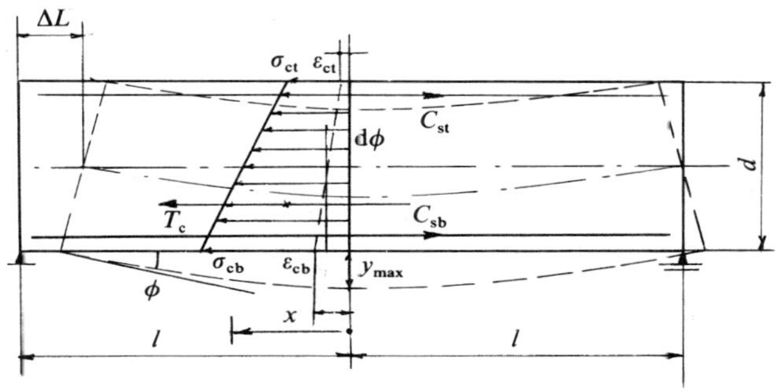

Fig. 10.27 Analysis of shrinkage deformations

The magnitude of the bending moment produced by shrinkage can be calculated from:

$$\frac{\varepsilon_{cb} - \varepsilon_{ct}}{d} = \frac{M}{EI}$$

from which

$$M = \frac{EI}{d}(\varepsilon_{cb} - \varepsilon_{ct}) \tag{10.33}$$

and the maximum shrinkage deflection at the centre of the beam

$$\max y = \phi \cdot \frac{l}{2} = \frac{l^2}{2d}(\varepsilon_{cb} - \varepsilon_{ct}) \tag{10.34}$$

The shortening of the centre line of the beam is

$$\Delta L = \frac{\varepsilon_{cb} + \varepsilon_{ct}}{2} \cdot L \tag{10.35}$$

For beams with variable reinforcement eq. (10.34) could be solved by the method of summation.

The diagram in Fig. 10.30 is based on eqs. (10.32) and (10.34).

Example 10.6

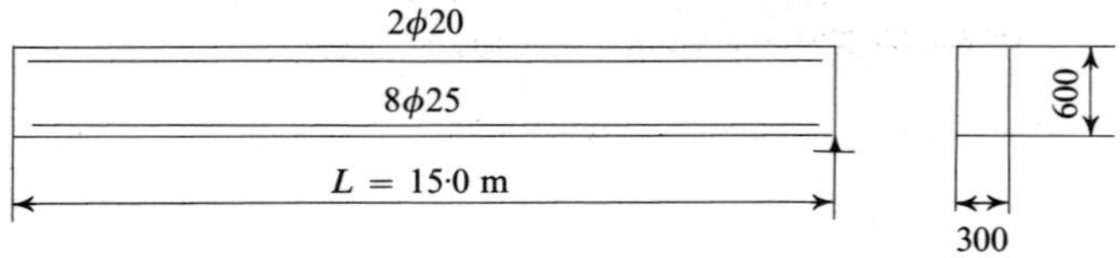

Fig. 10.28 Example

For the beam shown in Fig. 10.28, and assuming all parameters as in Example 10.4, we have:

$$M = \frac{EI}{d}(\varepsilon_{cb} - \varepsilon_{ct}) = 36{\cdot}0 \text{ kN/m}$$

$$\phi = \frac{L}{2d}(\varepsilon_{cb} - \varepsilon_{ct}) = 0{\cdot}003\,54 \text{ rad}$$

$$\max y = \phi . L/2 = 1{\cdot}35 \text{ mm} \quad \text{and} \quad \Delta L = L(\varepsilon_{cb} + \varepsilon_{ct}) = 1{\cdot}05 \text{ mm}$$

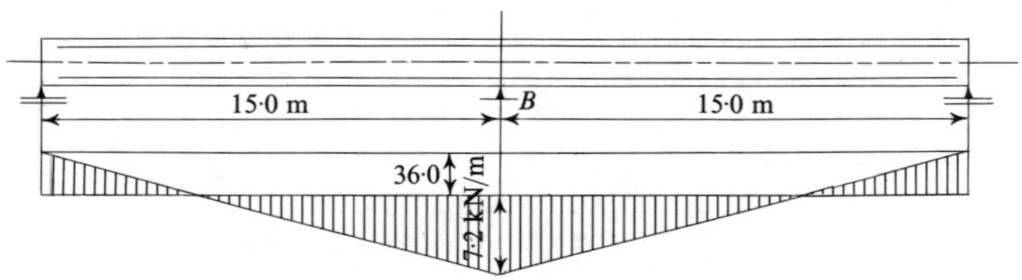

Fig. 10.29 Example

If the same beam is two-span, then the statically indeterminate bending moment at the support B can be calculated from:

$$M'_b \cdot \frac{L}{3EI} = \phi$$

from which:

$$M'_b = \phi \cdot \frac{3EI}{L}$$

Substituting the value for ϕ, we obtain:

$$M'_b = 43{\cdot}2 \text{ kN/m}$$

The actual bending moment at the support is therefore:

$$M_b = 7{\cdot}2 \text{ kN/m}$$

10.11 The Effect of Shrinkage on Frames

The effect of shrinkage on complex frames and shear walls is difficult to evaluate with certainty. Various parts of concrete structures are cast and cured at different times and there is a considerable time lag between shrinkage of various parts. In addition,

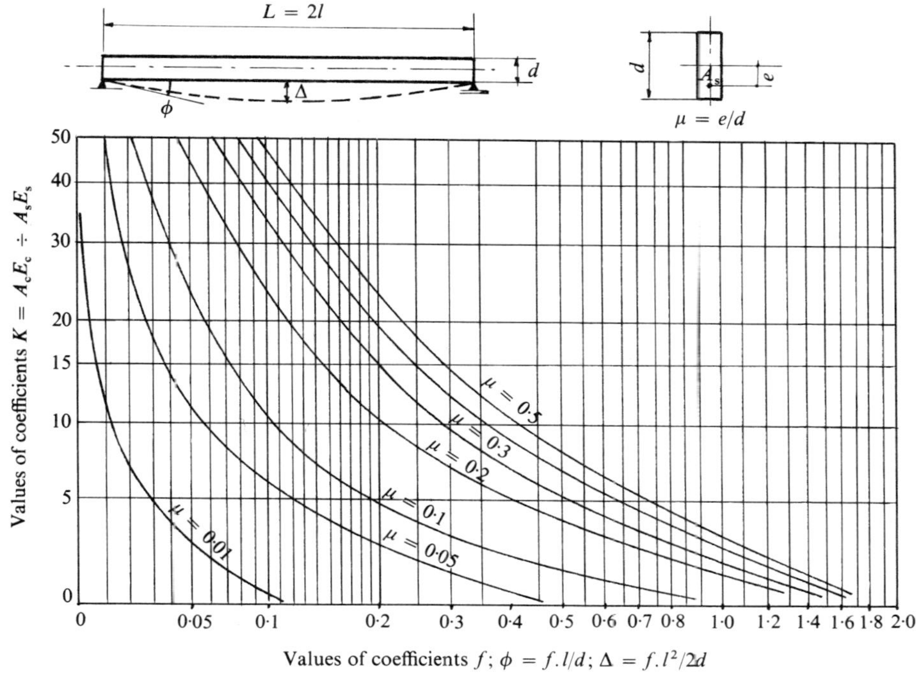

Fig. 10.30 Diagram of coefficients f for shrinkage deflection and rotation of beams

creep and temperature modify shrinkage stresses considerably, particularly at early stages. Generally, complex arrangements of shear walls are not designed for shrinkage, and rather an effort should be made to avoid these high and parasitic stresses by the control of mix.

However, the following example is chosen to illustrate the magnitude of shrinkage stresses which, theoretically, may develop even in a relatively simple frame.

Example 10.7

In this example, it is assumed that concrete in the portal frame shown in Fig. 10.31 has, at one stage, a shrinkage coefficient of 0·0003 and the modulus of elasticity $E_c = 28 \times 10^6$ kN/m². Based on Example 10.4, we obtain for the beam $\varepsilon_{cb} = 0{\cdot}000\ 150$ and $\varepsilon_{ct} = -0{\cdot}000\ 024$, so that $\phi = l/d(\varepsilon_{cb} + \varepsilon_{ct}) = 0{\cdot}002\ 360$ and $\Delta L = l/2(\varepsilon_{cb} + \varepsilon_{ct}) = 0{\cdot}000\ 565$. The following values are also required for the matrix:

$$\phi . EI = 387{\cdot}7, \qquad \frac{\Delta L}{h} EI = 23{\cdot}73$$

$$M_0 = \frac{EI}{d}(\varepsilon_{cb} - \varepsilon_{ct}) = 46{\cdot}4 \text{ kN/m}$$

The setting out of the force–displacement matrix does not present any difficulties

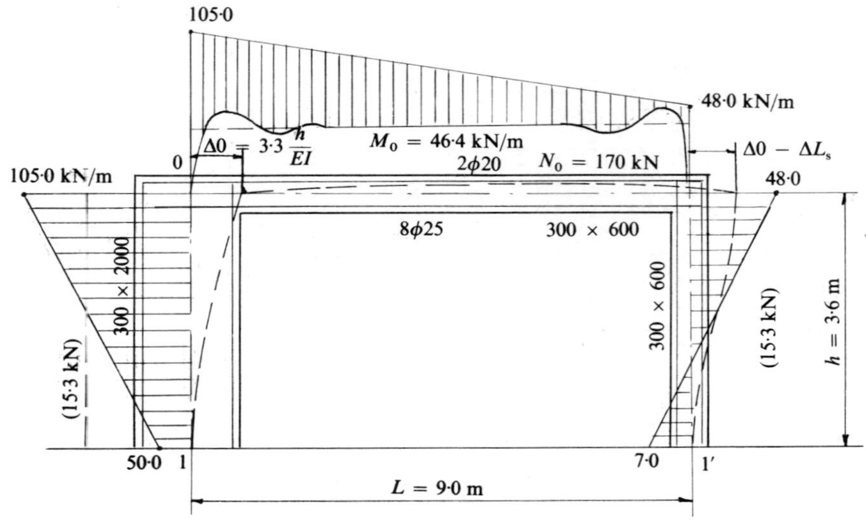

Fig. 10.31 Example

and is of the usual form:

$$\begin{bmatrix} \dfrac{h}{3EI_1} + \dfrac{L}{3EI} & \dfrac{h}{6EI_1} & \dfrac{L}{6EI} & 0 & +1 \\ \dfrac{L}{6EI} & 0 & \dfrac{L}{3EI} + \dfrac{h}{3EI} & -\dfrac{h}{6EI} & -1 \\ \dfrac{h}{3EI_1} & \dfrac{h}{6EI_1} & 0 & 0 & -1 \\ 0 & 0 & -\dfrac{h}{6EI} & \dfrac{h}{3EI} & -1 \\ +1 & -1 & -1 & -1 & 0 \end{bmatrix} \begin{Bmatrix} m_{01} \\ m_{10} \\ m'_{01} \\ m'_{10} \\ \Delta_0 \end{Bmatrix} = \begin{Bmatrix} \phi \\ \phi - \dfrac{\Delta L_s}{h} \\ 0 \\ -\dfrac{\Delta L_s}{h} \\ 0 \end{Bmatrix}$$

or, introducing numerical values, we obtain:

$$\begin{bmatrix} 3{\cdot}0324 & 0{\cdot}0162 & 1{\cdot}5 & 0 & +1 \\ 1{\cdot}5 & 0 & 4{\cdot}2 & -0{\cdot}6 & -1 \\ 0{\cdot}0162 & 0{\cdot}0324 & 0 & 0 & -1 \\ 0 & 0 & -0{\cdot}6 & 1{\cdot}2 & -1 \\ 1 & -1 & -1 & -1 & 0 \end{bmatrix} \begin{Bmatrix} m_{01} \\ m_{10} \\ m'_{01} \\ m'_{10} \\ \Delta_0 \end{Bmatrix} = \begin{Bmatrix} 387{\cdot}7 \\ 363{\cdot}34 \\ 0 \\ -23{\cdot}73 \\ 0 \end{Bmatrix}$$

Figure 10.31 shows the distribution of bending moments, forces, and deformations. In addition to statically indeterminate bending moments, there is a statically determinate moment $M_0 = 46{\cdot}4$ kN/m and an axial force $N_0 = 170$ kN in the beam, which are self-equilibrating.

The problem of shrinkage stresses in complex structures is very difficult: while it would be unreasonable to design structures for shrinkage (the more steel provided, the higher is the degree of restraint), these stresses can be very high and can drastically alter stresses and deformations from primary loads. The provision of high quality concrete and curing is one obvious solution, but even this may have an adverse effect by reducing creep capacity of members and, at stress concentrations, cracks may still appear. The problem is not yet resolved and the formation of cracks from this source will continue to present some risk.

References

1. R. H. Evans and F. K. Kong. 'Estimation of Shrinkage of Concrete in Reinforcement and Prestressed Concrete Design', *Civil Eng. and P.W.R.*, 559–561 (May 1967).
2. R. H. Evans and E. W. Bennett. *Prestressed Concrete*, p. 180–183, Calculations of Shrinkage Stresses, Chapman & Hall (1962).
3. A. M. Neville. 'Shrinkage and Creep in Concrete', *Structural Concrete*, **1** (March/April, 1962).
4. A. Neville. *Creep of Concrete: plain, reinforced and prestressed*, North Holland Publ. Co. (1970).
5. G. E. Troxell, J. M. Raphael, and R. E. Davis. 'Long-time creep and shrinkage tests of plain and reinforced concrete', *Proc. ASTM*, **58**, 1101–1200 (1958).
6. A. M. Neville. *Properties of Concrete*, Sir Isaac Pitman and Son, London (1963).
7. A. M. Neville. 'Non-elastic deformations in concrete structures', *J. New Zealand Inst. Eng.* (April, 1957).
8. T. C. Hansen. 'Estimating Stress Relaxation from Creep Data', *ASTM Mat. Research and Standards*, No. 1 (1964).
9. T. C. Hansen. 'Creep and Stress Relaxation of Concrete', *Swedish Cement and Concrete Research Inst., Stockholm, Proc.*, **31** (1960).

11 Analysis of Trusses by the Force–displacement Method

11.1 Introduction

Trusses are made of straight elements which are, theoretically, pin-connected and form two- or three-dimensional structures as shown in Fig. 11.1. In tall buildings, trusses are used as floor joists, roofs, wind bracings, frame tube structures, and also for temporary structures such as scaffolding and shoring.

Trusses, in contrast to frames, are subjected mainly to axial forces and deformations. Bending stresses are nearly always present in trusses due to various factors which will be discussed later, but these are seldom evaluated in practice and are considered to be small.

From the analysis point of view, trusses can be statically determinate or indeterminate internally or externally, or both. In this chapter some simple forms of trusses are analysed using the force–displacement method.

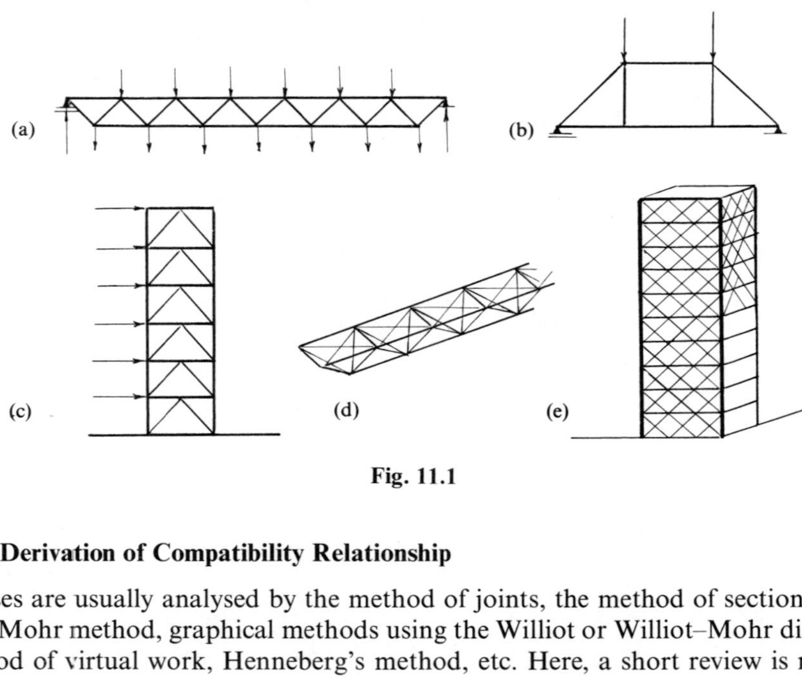

Fig. 11.1

11.2 Derivation of Compatibility Relationship

Trusses are usually analysed by the method of joints, the method of sections, Maxwell–Mohr method, graphical methods using the Williot or Williot–Mohr diagrams, method of virtual work, Henneberg's method, etc. Here, a short review is made of the application of the force–displacement method to the analysis of forces and displacements in trusses, both statically determinate and indeterminate. In this method matrices are set out in a similar way to those in the analysis of frames, by expressing the conditions of compatibility of deformations and equilibrium of forces at joints. The load–deflection characteristics of members are used in the form of the linear flexibility coefficients and these provide an analytical link between the geometry of a structure, system of loading, compatibility of deformations, and equilibrium of forces.

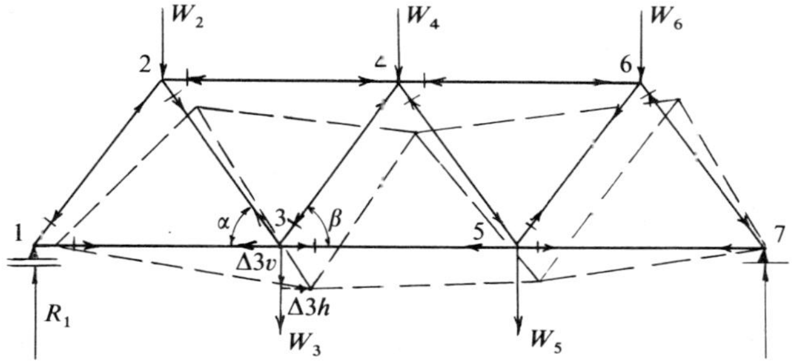

Fig. 11.2

For the purpose of setting out the matrix, a structure is assumed to deflect to its final position of equilibrium defined by the linear components of deflections Δv and Δh in the vertical and horizontal directions respectively as indicated in Fig. 11.2. Next, releases are introduced in all members, but at one end only in any preferred sequence, e.g., in Fig. 11.2 at joint 1 both members (1–2 and 1–3) are assumed to be released, at joint 2 members 2–3 and 2–4 are released, while at joint 3 the members 3–4 and 3–5 are released, etc. The direction of the forces and displacements must also be assumed at this stage and these must be used in setting out the matrix. The sign 'minus' obtained from the solution will then mean that the action of the force or the direction of the displacement are to be reversed. The number of the compatibility conditions is always equal to the number of the unknown forces, while the number of the equilibrium conditions corresponds to the number of the unknown deformations.

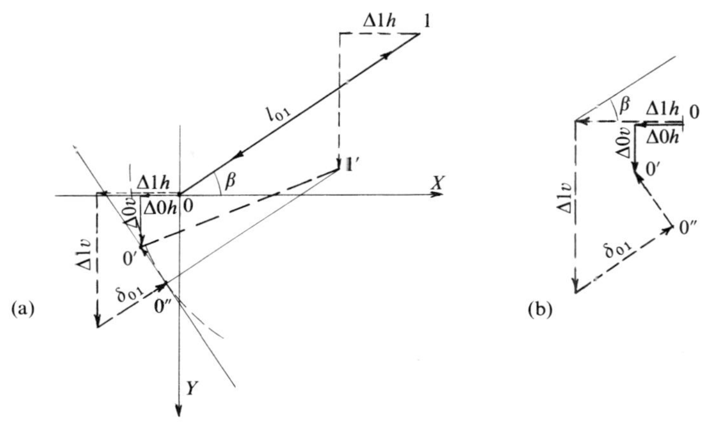

Fig. 11.3

However, before the compatibility conditions can be set out, it is necessary to develop the relationship between the displacements of a joint, such as $\Delta 0h$ and $\Delta 0v$ of the joint 0 in Fig. 11.3 and the deformations of one member such as 0–1 at its other end given by the values of $\Delta 1v$, $\Delta 1h$, and δ_{01}. This can be obtained by assuming that the member 0–1 (see Fig. 11.3a) first moves bodily by the values $\Delta 1v$ and $\Delta 1h$, then is subjected to the axial deformation δ_{01} and that finally, it rotates with respect to point 1′ until the other end coincides with the final position of the whole junction at 0′.

Assuming that the deformations are small, then in lieu of circular rotation of the member, the tangent at 0″ can be used as shown in Fig. 11.3(a) and this greatly simplifies the relationship required. To write the compatibility relationship, it is necessary first to assume the direction of the forces (compression or tension) and the corresponding axial deformation δ. The displacement diagram can then be drawn as shown in Fig. 11.3(b) for the joint 0. From this diagram we can write that:

$$\delta_{01} = \Delta 1h.\cos\beta + \Delta 1v.\sin\beta - \Delta 0h.\cos\beta - \Delta 0v.\sin\beta$$

or

$$(\Delta 0h - \Delta 1h).\cos\alpha + (\Delta 0v - \Delta 1v)\sin\beta + \delta_{01} = 0$$

By multiplying the whole equation by the value EA_n/l_n and substituting for δ_{01} the value of $N_{01}.l_{01}/EA_{01}$ we obtain the required relationship in the form:

$$\frac{EA_n}{l_n}(\Delta 0h - \Delta 1h)\cos\beta + \frac{EA_n}{l_n}(\Delta 0v - \Delta 1v)\sin\beta + N_{01}\cdot\frac{l_{01}}{A_{01}}\cdot\frac{A_n}{l_n} = 0$$

or C_{01}

$$r(\Delta 0h - \Delta 1h)\cos\beta + r(\Delta 0v - \Delta 1v)\sin\beta + N_{01}.s_{01} = 0 \qquad (11.1)$$

in which

$$r = \frac{EA_n}{l_n}, \qquad s_{01} = \frac{l_{01}}{A_{01}}\cdot\frac{A_n}{l_n}$$

l_n and A_n are the length and the cross-sectional area of any chosen reference member (not necessarily the member actually used in the truss analysed). The symbol, such as C_{01}, is used to denote the linear compatibility condition for a particular member, while the symbols E_{0h} or E_{0v} will refer to the equilibrium conditions at a particular joint in a horizontal and vertical direction respectively.

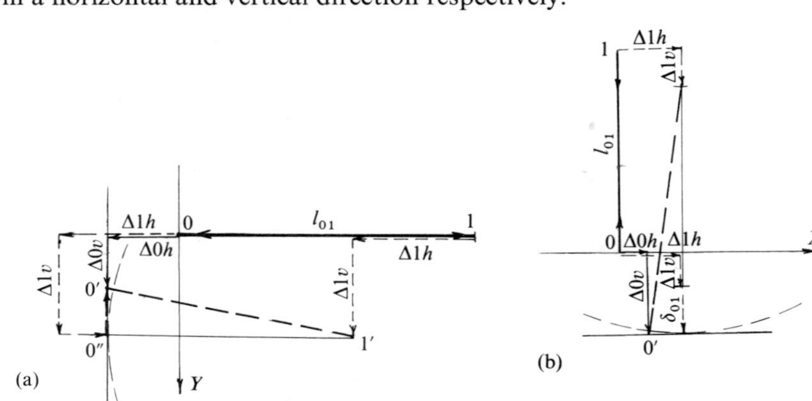

Fig. 11.4

The following two extreme cases are also useful, i.e., when the member 0–1 is either horizontal or vertical. In the first case, we obtain either from eq. (11.1) or directly from Fig. 11.4(a):

for $\beta = 0$:

$$C_{01}, \quad r(\Delta 0h - \Delta 1h) + N_{01} . S_{01} = 0 \tag{11.2}$$

For the vertical member 0–1 as shown in Fig. 11.4(b) we obtain:
for $\beta = 90°$:

$$C_{01}, \quad r(\Delta 0v - \Delta 1v) + N_{01} . S_{01} = 0 \tag{11.3}$$

More exact relationships can be obtained based on the actual circular rotation of the member 01 with respect to point 1′ as shown in Fig. 11.3 from which we can write that:

$$[x - (l_{01} . \cos\beta + \Delta 1h)]^2 + [y - (l_{01} . \sin\beta - \Delta 1v)]^2 = [l_{01} + \delta_{01}]^2$$

Introducing the boundary conditions for the displacements of the point 0″, i.e., for $x = \Delta 0h$, $y = \Delta 0v$ we obtain:

$$[\Delta 0h - l_{01} . \cos\beta - \Delta 1h]^2 + [\Delta 0v - l_{01} . \sin\beta - \Delta 1v]^2 = [l_{01} + \delta_{01}]^2 \tag{11.4}$$

The expression (11.4) is more accurate than (11.1), but it leads to more complicated non-linear analysis.

Example 11.1

As a simple example, a three-bar bracket shown in Fig. 11.5 is used. The compatibility conditions at point 1 based on eqs. (11.1) and (11.2) are:

$$C_{12}, \quad r(\Delta 1h) + N_{12} . S_{12} = 0$$

$$C_{13}, \quad r(\Delta 1h) \cos\alpha_3 + r(\Delta 1v) \sin\alpha_3 + N_{13} . S_{13} = 0$$

$$C_{14}, \quad r(\Delta 1h) \cos\alpha_4 + r(\Delta 1v) \sin\alpha_4 + N_{14} . S_{14} = 0$$

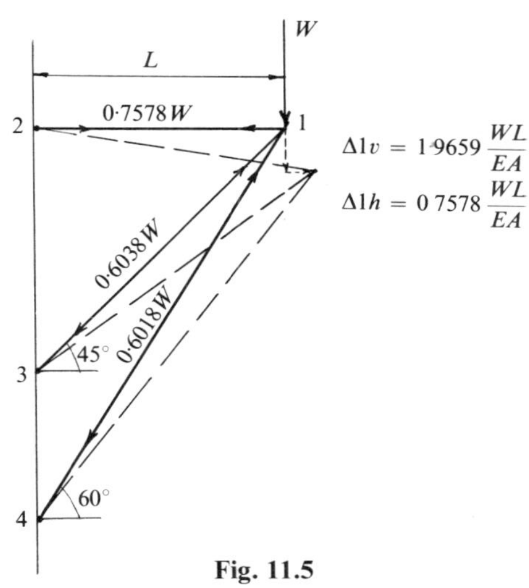

Fig. 11.5

The equilibrium equations at point 1 are:

$$E_{1v}, \quad N_{13}.\sin\alpha_3 + N_{14}.\sin\alpha_4 = W$$

$$E_{1h}, \quad N_{12} - N_{13}.\cos\alpha_3 - N_{14}.\cos\alpha_4 = 0$$

These equations are now presented in a matrix form as follows

	12	13	14	$r.\Delta 1v$	$r.\Delta 1h$	
C_{12}	$+S_{12}$	0	0	0	$+1$	0
C_{13}	0	$+S_{13}$	0	$\sin\alpha_3$	$\cos\alpha_3$	0
C_{14}	0	0	$+S_{14}$	$\sin\alpha_4$	$\cos\alpha_4$	0
E_{1v}	0	$\sin\alpha_3$	$\sin\alpha_4$	0	0	W
E_{1h}	$+1$	$-\cos\alpha_3$	$-\cos\alpha_4$	0	0	0

Taking $l_n = L$, A=constant throughout

$$S_{12} = 1$$

$$S_{13} = \frac{1}{\cos\alpha_2} = 1{\cdot}4144; \qquad S_{14} = \frac{1}{\cos\alpha_4} = 2{\cdot}0$$

so that the above matrix becomes

	12	13	14	$r.\Delta 1v$	$r.\Delta 1h$	
C_{12}	$+1$	0	0	0	$+1$	0
C_{13}	0	$+1{\cdot}4144$	0	$0{\cdot}707$	$0{\cdot}707$	0
C_{14}	0	0	$+2{\cdot}0$	$0{\cdot}866$	$0{\cdot}5$	0
E_{1v}	0	$0{\cdot}707$	$0{\cdot}866$	0	0	W
E_{1h}	$+1$	$-0{\cdot}707$	$-0{\cdot}5$	0	0	0

Figure 11.5 shows the resulting values of forces and displacements.

Example 11.2

Figure 11.6 shows an idealized arrangement of shoring using four members. The compatibility equations at point 1 are:

$$C_{12}, \quad r(\Delta 1v) + N_{12}.S_{12} = 0$$

$$C_{13}, \quad r(\Delta 1h)\cos\alpha_3 + r(\Delta 1v)\sin\alpha_3 + N_{13}.S_{13} = 0$$

$$C_{14}, \quad r(\Delta 1h)\cos\alpha_4 + r(\Delta 1v)\sin\alpha_4 + N_{14}.S_{14} = 0$$

$$C_{15}, \quad r(\Delta 1h)\cos\alpha_5 + r(\Delta 1v).\sin\alpha_5 + N_{15}.S_{15} = 0$$

and the equilibrium equations are:

$$E_{1v}, \quad N_{12} - N_{13}.\sin\alpha_3 - N_{14}.\sin\alpha_4 - N_{15}.\sin\alpha_5 = 0$$

$$E_{1h}, \quad N_{13}.\cos\alpha_3 + N_{14}.\cos\alpha_4 + N_{15}.\cos\alpha_5 = W$$

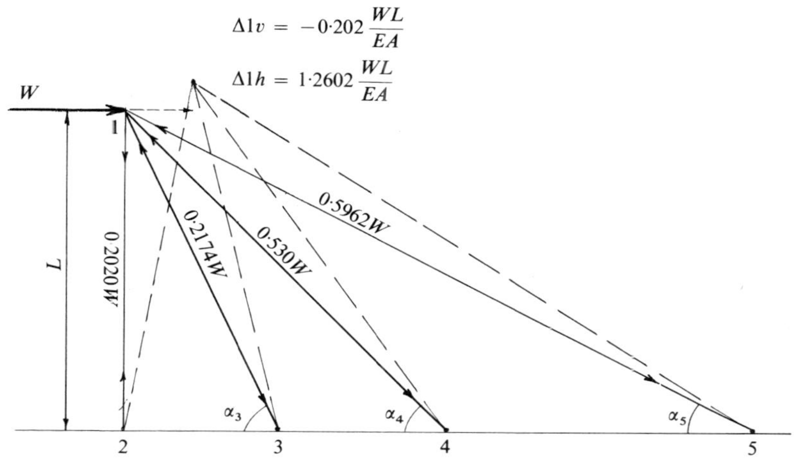

Fig. 11.6

These can be presented by the following general matrix:

	12	13	14	15	$r.\Delta 1v$	$r.\Delta 1h$	
C_{12}	$+S_{12}$	0	0	0	+1	0	0
C_{13}	0	$+S_{13}$	0	0	$\sin\alpha_3$	$\cos\alpha_3$	0
C_{14}	0	0	$+S_{14}$	0	$\sin\alpha_4$	$\cos\alpha_4$	0
C_{15}	0	0	0	$+S_{15}$	$\sin\alpha_5$	$\cos\alpha_5$	0
E_{1v}	+1	$-\sin\alpha_3$	$-\sin\alpha_4$	$-\sin\alpha_5$	0	0	0
E_{1h}	0	$\cos\alpha_3$	$\cos\alpha_4$	$\cos\alpha_5$	0	0	W

Taking:

$$\alpha_3 = 60^\circ, \quad \alpha_4 = 45^\circ, \quad \alpha_5 = 30^\circ, \quad A = \text{const}, \quad l_n = l_{12}$$

$$S_{12} = 1, \quad S_{13} = 1/\sin\alpha_3 = 1{\cdot}1547, \quad S_{14} = 1/\sin\alpha_4 = 1{\cdot}4144,$$

$$S_{15} = 1/\sin\alpha_5 = 2{\cdot}0$$

the above matrix becomes:

	12	13	14	15	$r.\Delta 1v$	$r.\Delta 1h$	
C_{12}	+1	0	0	0	+1	0	0
C_{13}	0	+1·1547	0	0	0·866	0·5	0
C_{14}	0	0	+1·4144	0	0·707	0·707	0
C_{15}	0	0	0	+2·0	0·5	0·866	0
E_{1v}	+1	−0·866	−0·707	−0·5	0	0	0
E_{1h}	0	0·5	0·707	0·866	0	0	W

The resulting values are shown in Fig. 11.6.

11.3 Statically Determinate Trusses

In statically determinate trusses, such as shown in Fig. 11.7, internal forces can be determined by using the conditions of equilibrium of forces. The compatibility conditions (based on eqs. (11.1–11.3) are required only for the evaluation of deflections of joints.

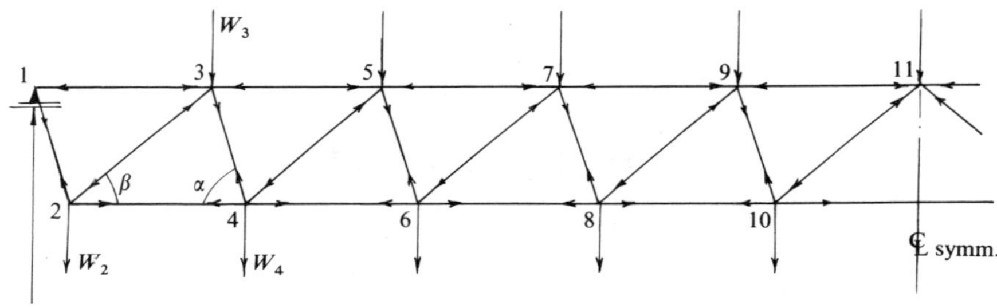

Fig. 11.7

Referring to Fig. 11.7, the equilibrium conditions at any junction, such as number 3, can be written as:

$$E_{3v}, \quad N_{23}.\sin\beta - N_{34}.\sin\alpha = W_3$$

$$E_{3h}, \quad N_{13} + N_{23}.\cos\beta + N_{34}.\cos\alpha - N_{35} = 0$$

These two equations form an element matrix, and Table 11.1 shows the general matrix for a ten-bay symmetrical truss for the evaluation of forces only.

When the deformations as well as forces are required, then the compatibility conditions must be included based on the relationships (11.1) to (11.3). For the joint 3 in Fig. 11.7 these are:

$$C_{34}, \quad r(\Delta 3h - \Delta 4h)\cos\alpha + r(\Delta 3v - \Delta 4v)\sin\alpha + N_{34}.S_{34} = 0$$

$$C_{35}, \quad r(\Delta 3h - \Delta 5h) + N_{35}.S_{35} = 0$$

The equilibrium conditions at joint 3 are as shown before.

The general matrix can thus be set out for a statically determinate truss as shown

Table 11.1 General matrix for evaluation of forces in statically determinate truss in Fig. 11.7

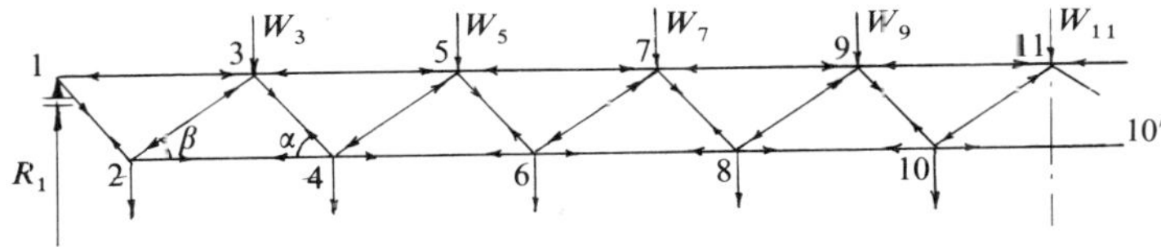

	1		2		3		4		5		6		7		8		9		10		
	1 2	1 3	2 3	2 4	3 4	3 5	4 5	4 6	5 6	5 7	6 7	6 8	7 8	7 9	8 9	8 10	9 10	9 11	10 11	10 10′	
E_{1v}	$\sin\alpha$	0																			R_1
E_{1h}	$\cos\alpha$	-1																			0
E_{2v}	$\sin\alpha$	0	$-\sin\beta$	0																	W_2
E_{2h}	$\cos\alpha$	0	$\cos\beta$	-1																	0
E_{3v}		0	$\sin\beta$	0	$-\sin\alpha$	0															W_3
E_{3h}		1	$\cos\beta$	0	$\cos\alpha$	-1															0
E_{4v}						F^e															W_4
E_{4h}																					0
E_{5v}								F^e													W_5
E_{5h}																					0
E_{6v}										F^e											W_6
E_{6h}																					0
E_{7v}												F^e									W_7
E_{7h}																					0
E_{8v}														F^e							W_8
E_{8h}																					0
E_{9v}																F^e					W_9
E_{9h}																					0
E_{10v}																		F^e			W_{10}
E_{10h}																					0

Note: The resulting forces are in terms of W_n.

in Fig. 11.7 for evaluation of forces and displacements. This is shown in Table 11.2. It should be noted that this matrix is set out in terms of the non-dimensional coefficients so that the resulting values must be multiplied by W_n for forces and by $(W_n.l_n)/EA_n$ for the displacements.

Table 11.2 General matrix for evaluation of forces and displacements in statically determinate truss shown in Fig. 11.7

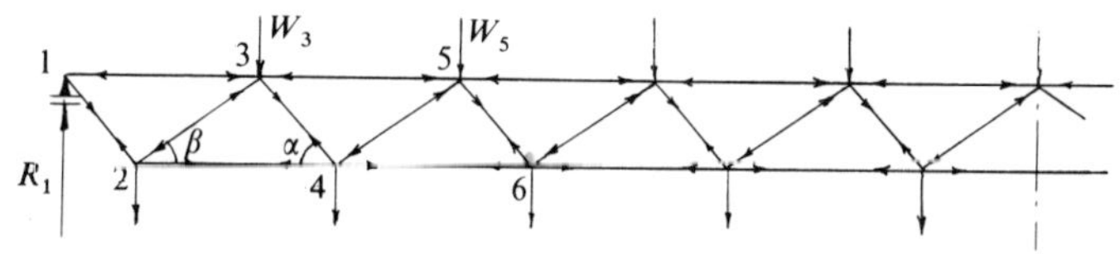

	1			2				3				4				5				6						
	1 2	1 3	$r.\Delta h$	2 3	2 4	$r.\Delta v$	$r.\Delta h$	3 4	3 5	$r.\Delta v$	$r.\Delta h$	4 5	4 6	$r.\Delta v$	$r.\Delta h$	5 6	5 7	$r.\Delta v$	$r.\Delta h$	6 7	6 8	$r.\Delta v$	$r.\Delta h$			
C_{12}	S_{12}		$\cos\alpha$			$-\sin\alpha$	$-\cos\alpha$																			0
C_{13}		S_{13}	$+1$								-1															0
E_{1v}	$\sin\alpha$																									R_1
E_{1h}	$\cos\alpha$	-1																								0
C_{23}				S_{23}		$\sin\beta$	$\cos\beta$			$-\sin\beta$	$-\cos\beta$															0
C_{24}					S_{24}		$+1$								-1											0
E_{2v}	$\sin\alpha$			$-\sin\beta$																						W_2
E_{2h}	$\sin\alpha$			$\cos\beta$	-1																					0
C_{34}								S_{34}		$\sin\alpha$	$\cos\alpha$			$-\sin\alpha$	$-\cos\alpha$											0
C_{35}									S_{35}		$+1$								-1							0
E_{3v}				$\sin\beta$				$-\sin\alpha$																		W_3
E_{3h}		$+1$		$\cos\beta$				$\cos\alpha$	-1																	0
C_{45}																										0
C_{46}														F^e												0
E_{4v}																										W_4
E_{4h}																										0
C_{56}																										0
C_{57}																		F^e								0
E_{5v}																										W_5
E_{5h}																										0
C_{67}																										0
C_{68}																					F^e					0
E_{6v}																										W_6
E_{6h}																										0

Note: Forces are in terms of $W_n.l_n$ and displacements in terms of $(W_n.l_n)/EA_n$.

Example 11.3

For the truss shown in Fig. 11.8, it is assumed that $\alpha=\beta=60^\circ$ and that all values of s are constant. The matrix for the evaluation of forces only based on Table 11.1 gives values as shown in Fig. 11.8(a).

To analyse forces as well as displacements in the above truss, the matrix shown in Table 11.2 is used. The resulting forces are identical to those shown in Fig. 11.8(a) and the displacements are shown in Fig. 11.8(b).

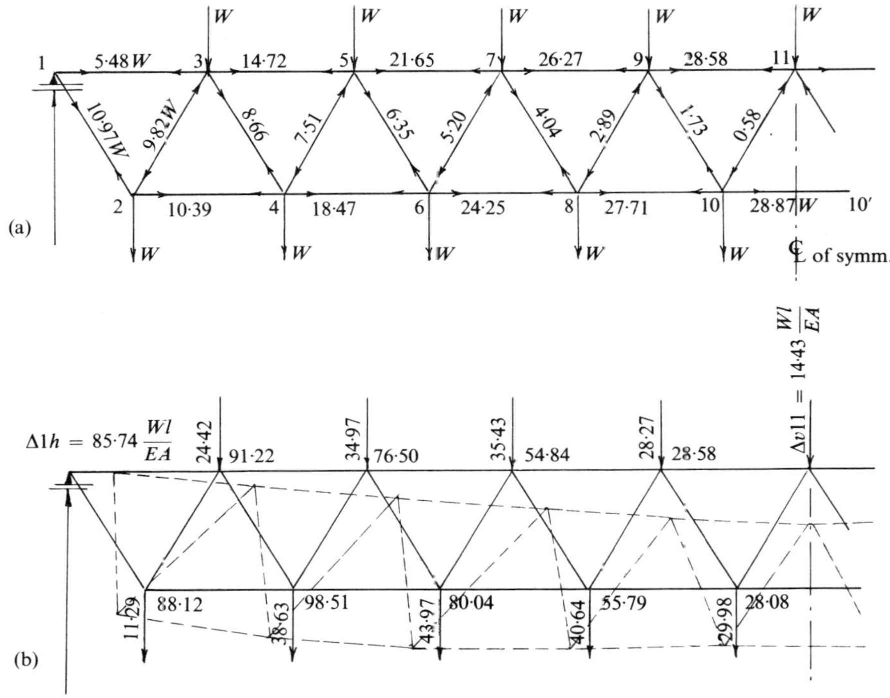

Fig. 11.8

It is interesting to note that trusses deflect slightly differently to beams in flexure, e.g., while the deflections of the upper chord in the above example increase towards the centre at an expected rate the deflections of the lower chord do not follow this pattern and are quite large at the joints near the supports. The maximum deflection may not always be at the centre of the lower chord.

Forcing members to meet at a joint during the fabrication of trusses, because of poor initial fit, is a well-known source of secondary stresses in trusses. If the magni-

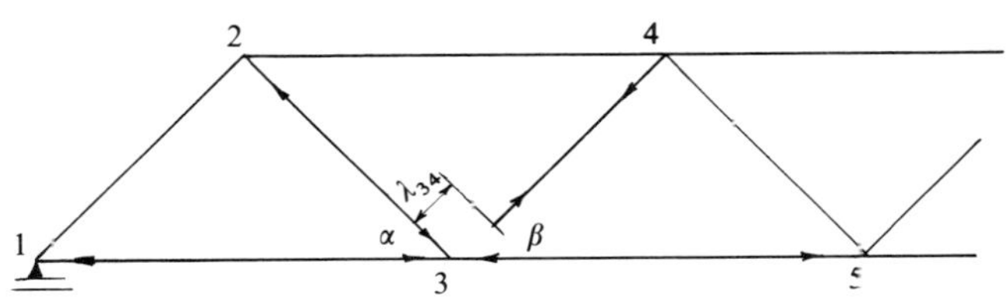

Fig. 11.9

tude of the lack of fit of a member is known, then the forces set in the truss and the deformations could be analysed based on the conditions of compatibility of deformations and the equilibrium of forces at each joint; e.g., from Fig. 11.9 we can write at joint 3:

$$C_{34}, \quad (\Delta 3h - \Delta 4h)\cos\beta + r(\Delta 3v - \Delta 4v)\sin\beta + \lambda_{34} = 0$$

while: $C_{35}, \quad r(\Delta 3h - \Delta 5h) + N_{35}.S_{35} = 0$

or $\quad C_{23}, \quad r(\Delta 2h - \Delta 3h)\cos\alpha + r(\Delta 3v - \Delta 5v)\sin\alpha + N_{35}.S_{35} = 0$

Similar equations of compatibility can be written for other members.

The conditions of the equilibrium of forces are written in the usual way and at joint 3 are:

$$E_{3v}, \quad N_{23}.\sin\beta - N_{34}.\sin\alpha = 0$$

$$E_{3h}, \quad N_{13} + N_{23}.\cos\beta + N_{34}.\cos\alpha - N_{35} = 0$$

The matrix can thus be set out for the forces and displacements as a function of given values of the lack of fit at the joints. However, statically determinate trusses are free from internal forces while indeterminate trusses develop forces in members and displacements of joints.

11.4 Internally Statically Indeterminate Trusses

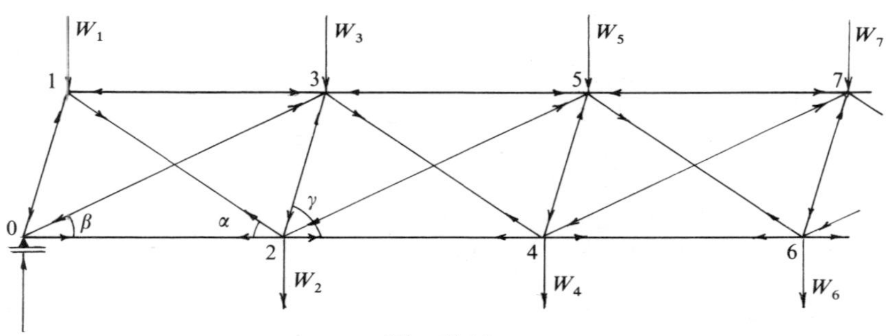

Fig. 11.10

The analysis of internally statically indeterminate trusses does not differ in principle from the procedure previously shown and eqs. (11.1) to (11.3) must be used in setting out the compatibility conditions. For the general case of an inclined truss as shown in Fig. 11.10, the element matrix at the joints 2 and 3 can be written as follows:

$$C_{23}, \quad r(\Delta 2h - \Delta 3h).\cos\gamma + r(\Delta 2v - \Delta 3v)\sin\gamma + N_{23}.S_{23} = 0$$

$$C_{24}, \quad r(\Delta 2h - \Delta 4h) + N_{24}.S_{24} = 0$$

$$C_{25}, \quad (\Delta 2h - \Delta 5h)\cos\beta + r(\Delta 2v - \Delta 5v)\sin\beta + N_{25}.S_{25} = 0$$

$$E_{2v}, \quad N_{12}.\sin\alpha - N_{23}.\sin\gamma - N_{25}.\sin\beta = W_2$$

$$E_{2h}, \quad N_{02} + N_{12}.\cos\alpha + N_{23}.\cos\gamma + N_{25}.\cos\beta - N_{24} = 0$$

$$C_{34},\quad r(\Delta 3h - \Delta 4h)\cos\alpha + r(\Delta 3v - \Delta 4v).\cos\alpha + N_{34}.S_{34} = 0$$

$$C_{35},\quad r(\Delta 3h - \Delta 5h) + N_{35}.S_{35} = 0$$

$$E_{3v},\quad N_{03}.\sin\beta + N_{23}.\sin\gamma - N_{34}.\sin\alpha = W_3$$

$$E_{3h},\quad N_{03}.\cos\beta + N_{13} + N_{34}.\cos\alpha - N_{35} = 0$$

In this way an overall matrix can be set out. Table 11.3 shows the general matrix for the evaluation of forces and displacements in the internally statically indeterminate truss shown in Fig. 11.10. For multi-bay trusses it is more convenient to use flow charts, similar to those used in the analysis of frames (see pages 16 and 17).

Example 4

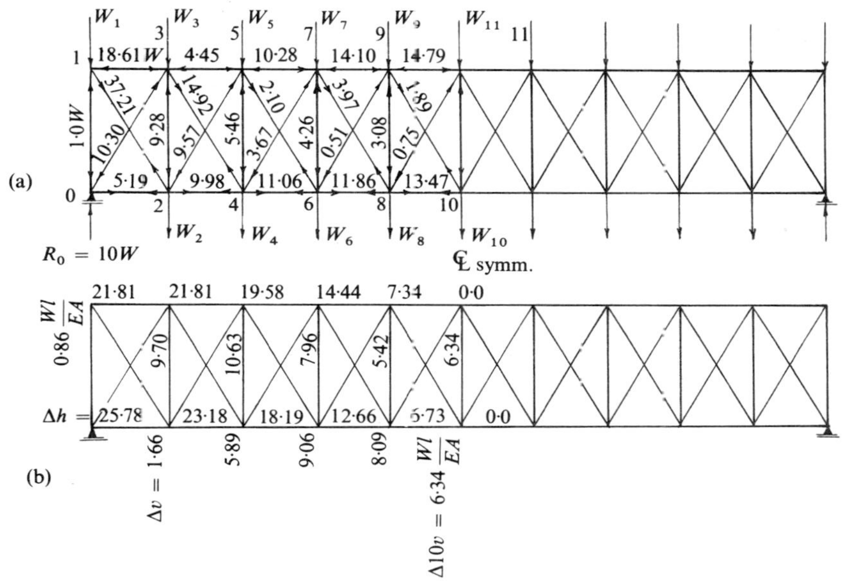

Fig. 11.11

For the truss shown in Fig. 11.11, the following data are assumed:

$$\alpha = \beta = 60^\circ, \qquad \gamma = 90^\circ$$

$$A_n = \text{constant throughout}$$

$$l_n = l_{03}$$

$$s_{01} = \frac{l_{01}}{A_{01}}\cdot\frac{A_n}{l_n} = 0{\cdot}866$$

$$s_{02} = 0{\cdot}5 \quad \text{and} \quad s_{03} = 1{\cdot}0$$

Introducing numerical values into the matrix shown in Table 11.3, the distribution of forces and displacements is as shown in Fig. 11.11(a) and (b).

Table 11.3 General matrix for the analysis of statically indeterminate truss shown in Fig. 11.10

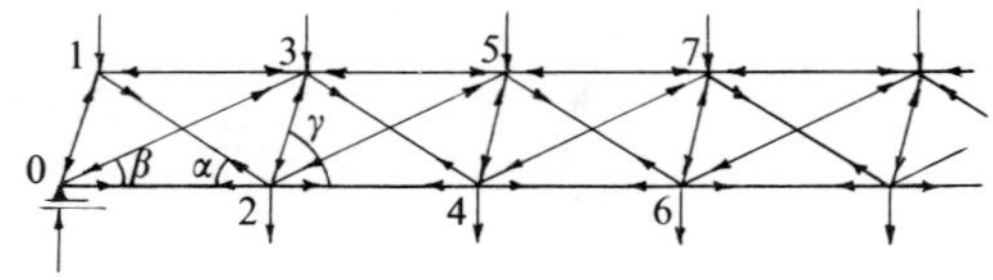

	0				1				2					
	0 1	0 2	0 3	$r.\Delta h$	1 2	1 3	$r.\Delta v$	$r.\Delta h$	2 3	2 4	2 5	$r.\Delta v$	$r.\Delta h$	3
C_{01}	S_{01}			$\cos\gamma$			$-\sin\gamma$	$-\cos\gamma$						
C_{02}		S_{02}		$+1$									-1	
C_{03}			S_{03}	$\cos\beta$										
E_{0v}	$\sin\gamma$		$\sin\beta$											
E_{0h}	$\cos\gamma$	-1	$\cos\beta$											
C_{12}					S_{12}		$\sin\alpha$	$\cos\alpha$				$-\sin\alpha$	$-\cos\alpha$	
C_{13}						S_{13}		$+1$						
E_{1v}	$\sin\gamma$				$-\sin\alpha$									
E_{1h}	$\cos\gamma$				$\cos\alpha$	-1								
C_{23}									S_{23}			$\sin\gamma$	$\cos\gamma$	
C_{24}										S_{24}			$+1$	
C_{25}											S_{25}	$\sin\beta$	$\cos\beta$	
E_{2v}					$\sin\alpha$				$-\sin\gamma$		$-\sin\beta$			
E_{2h}		$+1$			$\cos\alpha$				$\cos\gamma$	-1	$\cos\beta$			
C_{34}														S
C_{35}														
E_{3v}			$\sin\beta$						$\sin\gamma$					$-$si
E_{3h}			$\cos\beta$			$+1$								co
C_{45}														
C_{46}														
C_{47}														
E_{4v}														
E_{4h}														
C_{56}														
C_{57}														
E_{5v}														
E_{5h}														

3		4					5				6					
$r.\Delta v$	$r.\Delta h$	4 5	4 6	4 7	$r.\Delta v$	$r.\Delta h$	5 6	5 7	$r.\Delta v$	$r.\Delta h$	6 7	6 8	6 9	$r.\Delta v$	$r.\Delta h$	
																0
																0
$-\sin\beta$	$-\cos\beta$															0
																R_0
																0
																0
	-1															0
																W_1
																0
$-\sin\gamma$	$-\cos\gamma$															0
						-1										0
									$-\sin\beta$	$-\cos\beta$						0
																W_2
																0
$\sin\alpha$	$\cos\alpha$				$-\sin\alpha$	$-\cos\alpha$										0
	$+1$									-1						0
																W_3
																0
																0
																0
																0
				F^e												W_4
																0
																0
																0
																W_5
																0
									F^e							

11.5 Externally Indeterminate Trusses

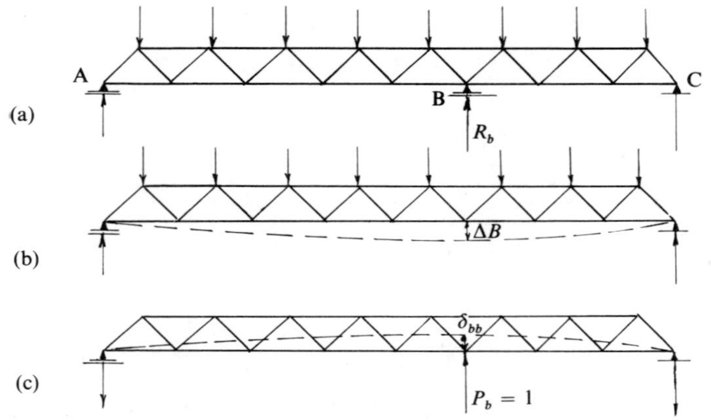

Fig. 11.12

The analysis of externally indeterminate trusses as shown in Fig. 11.12 can be made by using the general procedure of the flexibility method. At first, the redundant supports such as support B in Fig. 11.12, are removed and the structure is analysed for forces and displacements. At B the latter is denoted by ΔB. Now, the unit load $P=1$ is applied separately at each removed support and the corresponding deflections are calculated. At support B in Fig. 11.12 this is denoted by δb. The conditions of compatibility are now set out requiring that the displacements at the redundant supports be equal to zero. From these conditions the redundant reactions are calculated, e.g., in Fig. 11.12 this condition is written as follows:

$$R_b . \delta_{bb} = \Delta B$$

Once we have obtained the reactions, the rest of the procedure involves only the summing up of the relevant forces calculated at each step.

11.6 Bending Stresses in Trusses

Bending stresses in trusses are usually of secondary order and can occur from various sources, such as:

(a) loads placed between the joints (Fig. 11.13(a));
(b) joints being rigidly connected instead of being ideally hinged (Fig. 11.13(b));
(c) eccentric connections of members (Fig. 11.13(c)) or 'knee' bracing of frames (Fig. 11.13(d));
(d) fully continuous chords (Fig. 11.13(e) and (f));
(e) missing of some diagonal members (Fig. 11.13(g));
(f) temperature changes, etc.

These bending stresses are seldom evaluated, mainly because of the relatively large amount of work involved. However, some simple cases are considered here.

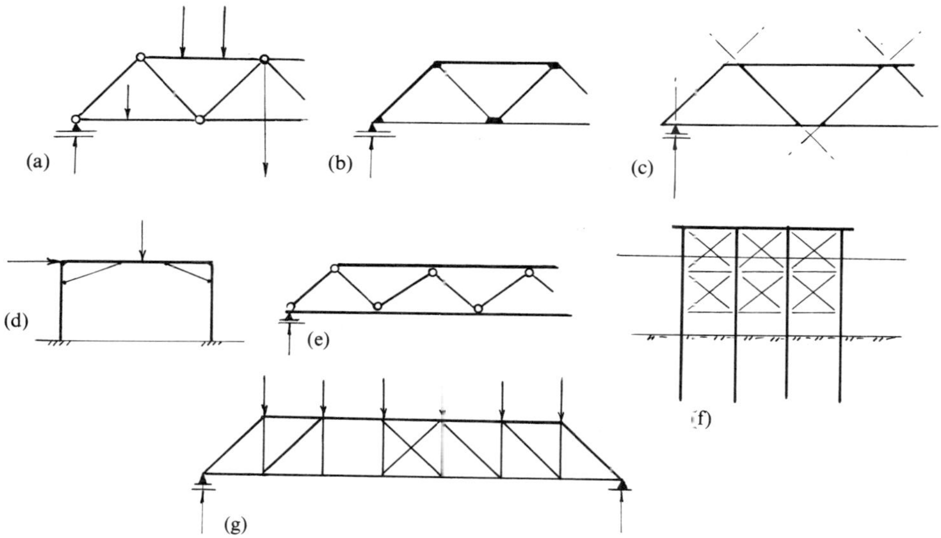

Fig. 11.13

Example 11.4

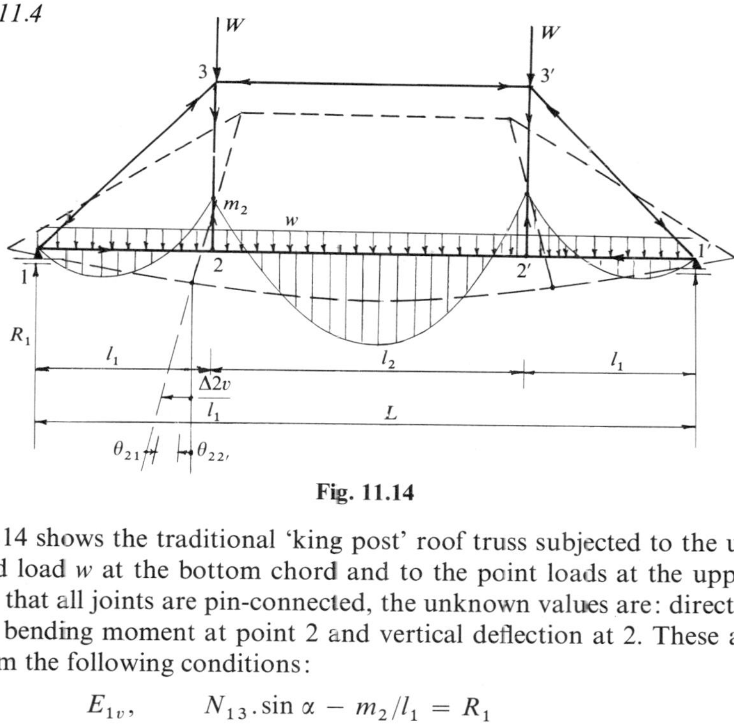

Fig. 11.14

Figure 11.14 shows the traditional 'king post' roof truss subjected to the uniformly distributed load w at the bottom chord and to the point loads at the upper joints. Assuming that all joints are pin-connected, the unknown values are: direct forces in members, bending moment at point 2 and vertical deflection at 2. These are determined from the following conditions:

$$E_{1v}, \qquad N_{13}.\sin\alpha - m_2/l_1 = R_1$$

$$E_{1h}, \qquad N_{13}.\cos\alpha - N_{12} = 0$$

$$Fl.\ 122', \quad m_2\left(\frac{l_1}{3EI} + \frac{l_2}{2EI}\right) = \frac{\Delta 2v}{l_1} - \theta_{21} - \theta_{22'}$$

$$E_{3v}, \qquad N_{13}.\sin\alpha - m_2/l_1 = v_3{}^0 + W$$

in which the symbol *Fl.* 122′ refers to the flexural compatibility condition at the joint 2.

Re-arranging the terms we obtain the following matrix:

	12	13	m_2/L	$r.\Delta 2v$	
E_{1v}	0	$\sin\alpha$	$-L/l_1$	0	$(n+\frac{1}{2})wL$
E_{1h}	-1	$\cos\alpha$	0	0	0
Fl. 122′	0	0	$3-4l_1/L$	$\dfrac{6I}{Ll_1A}$	$-\dfrac{wL}{4}\left[\left(\dfrac{l_1}{L}\right)^3+\left(\dfrac{l_2}{L}\right)^3\right]$
E_{3v}	0	$\sin\alpha$	L/l_1	0	$wL[n+\frac{1}{2}(1-l_1/L)]$

in which $n = W/wL$ is the ratio of the point load W to the total distributed load wL.

Taking:

$$n = W/wL = 2; \qquad A = \text{const}, \qquad \alpha = 45^\circ$$

$$bd = 100 \times 200, \qquad \frac{6I}{Ll_1A} = 0{\cdot}000\dot{5}$$

we obtain the following matrix:

	12	13	m_2/L	$r.\Delta 2v$	
E_{1v}	0	0·707	4·0	0	$2{\cdot}5wL$
E_{1h}	-1	0·707	0	0	0
Fl. 122′	0	0	2·0	$0{\cdot}000\dot{5}$	$-0{\cdot}0351wL$
E_{3v}	0	0·707	$-4{\cdot}0$	0	$2{\cdot}375wL$

The resulting values from the above matrix are shown in Fig. 11.15.

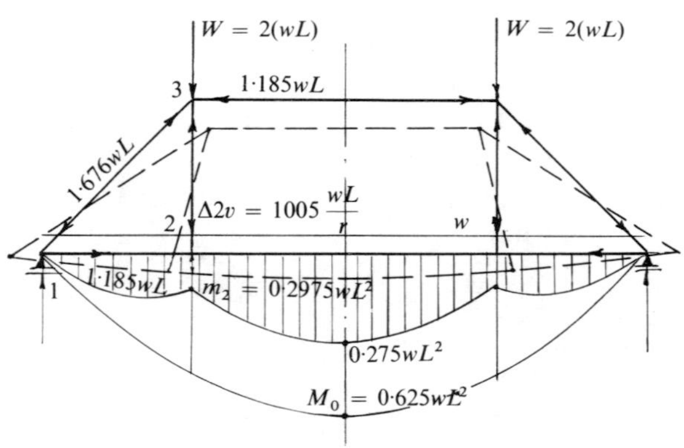

Fig. 11.15

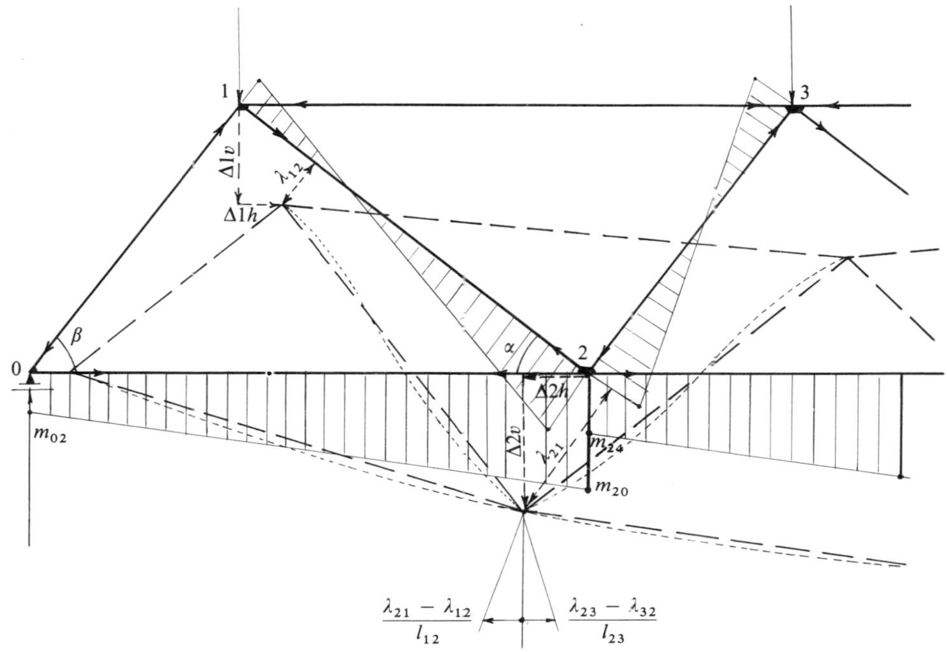

Fig. 11.16

Most trusses have rigid joints and these are the source of secondary bending stresses in members. These stresses are seldom evaluated because they are considered to be small and because of the large amount of work involved. However, if these are required the following procedure could be adopted based on the force-displacement method.

Referring to Fig. 11.16 the transverse displacements λ of the ends of members are first expressed in terms of the displacements Δv and Δh, e.g., for member 1–2 these are:

$$\lambda_{12} = \Delta 1v.\cos\alpha - \Delta 1h.\sin\alpha$$

$$\lambda_{21} = \Delta 2v.\cos\alpha + \Delta 2h.\sin\alpha$$

The relative transverse displacement of the ends of member 1–2 is therefore:

$$\lambda_{21} - \lambda_{12} = (\Delta 2v - \Delta 1v)\cos\alpha + (\Delta 2h - \Delta 1h)\sin\beta$$

Similarly, for the member 2–3:

$$\lambda_{23} - \lambda_{32} = (\Delta 2v - \Delta 3v)\cos\alpha + (\Delta 2h - \Delta 3h)\sin\beta$$

At release of joint 2 (see Fig. 11.16) there will be an angular discontinuity between the members 1–2 and 2–3 of the total value of:

$$\frac{\lambda_{21} - \lambda_{12}}{l_{12}/\sin\alpha} + \frac{\lambda_{23} - \lambda_{32}}{l_{23}/\sin\beta}$$

This discontinuity must be closed (simultaneously with all other discontinuities) by re-applying flexural bending moments in members. We can write that:

$$Fl.\ 123,\quad -m_{12}\frac{l_{12}}{6EI_{12}}+m_{21}\frac{l_{12}}{3EI_{12}}+m_{23}\frac{l_{23}}{3EI_{23}}-m_{32}\frac{l_{23}}{6EI_{23}}=$$

$$=\frac{\lambda_{21}-\lambda_{12}}{l_{12}}\sin\alpha+\frac{\lambda_{23}-\lambda_{32}}{l_{23}}\cdot\sin\beta$$

At joint 2 in Fig. 11.16 there are in all three independent flexural compatibility conditions denoted by 021, 123, and 324, one of which is as shown above and the remaining two are as follows:

$$Fl.\ 021,\quad m_{02}\frac{l_{02}}{6EI_{02}}+m_{20}\frac{l_{02}}{3EI_{02}}+m_{24}\frac{l_{24}}{3EI_{24}}+m_{42}\frac{l_{24}}{6EI_{24}}=$$

$$=\frac{\Delta 2v}{l_{02}}+\frac{\lambda_{21}-\lambda_{12}}{l_{12}}\sin\alpha$$

$$Fl.\ 324,\quad -m_{32}\frac{l_{32}}{6EI_{32}}+m_{23}\frac{l_{32}}{3EI_{32}}+m_{24}\frac{l_{24}}{3EI_{24}}+m_{42}\frac{l_{24}}{6EI_{24}}=$$

$$=\frac{\lambda_{23}-\lambda_{32}}{l_{23}}\sin\beta+\frac{\Delta 4v-\Delta 2v}{l_{24}}$$

The linear compatibility conditions at joint 2, i.e., C_{23} and C_{24}, are not affected by the rigidity of the joints and are written in the usual way based on the relationships (11.1) to (11.3).

However, the equilibrium conditions are affected by the rigidity of the joints, and the resulting bending moments and corresponding shear forces in members. At joint 2 these are as follows:

$$E_{2v},\quad N_{12}.\sin\alpha-N_{23}.\sin\beta+\left[\frac{m_{20}-m_{02}}{l_{02}}+\frac{m_{21}+m_{12}}{l_{12}}\cdot\cos\alpha+\right.$$

$$\left.+\frac{m_{23}+m_{32}}{l_{23}}\cos\beta-\frac{m_{42}-m_{24}}{l_{24}}\right]=W_2$$

in which, for equilibrium of bending moments at the joint: $m_{24}=m_{20}+m_{21}-m_{23}$.

Similarly:

$$E_{2h},\quad N_{02}+N_{12}.\cos\alpha+N_{23}\cos\beta-N_{24}+\left[-\frac{m_{12}+m_{21}}{l_{12}}\sin\alpha+\right.$$

$$\left.+\frac{m_{23}+m_{32}}{l_{23}}\cdot\sin\beta\right]=0$$

Summing up, the typical element force-displacement matrix at a rigid junction such as junction 2 in Fig. 11.16, consists of two linear compatibility conditions, three flexural conditions of compatibility of angular deformations and two conditions of the equilibrium of forces. These are given the following designation at joint 2: C_{23}, C_{24}, *Fl.* 021, *Fl.* 123, *Fl.* 324, E_{2v}, and E_{2h}. Such element is repetitive throughout the main matrix and this reduces the amount of work in the analysis.

11.8 Trusses with Continuous Chords

In most trusses the chords are continuous, while the diagonal members are usually separate, and are either hinged or rigidly connected to the chords. Because of the continuity of the chords, even if the diagonal members are ideally hinged, there will be some bending moments in the chords. These are usually omitted from the analysis and are considered to be small and of secondary effect.

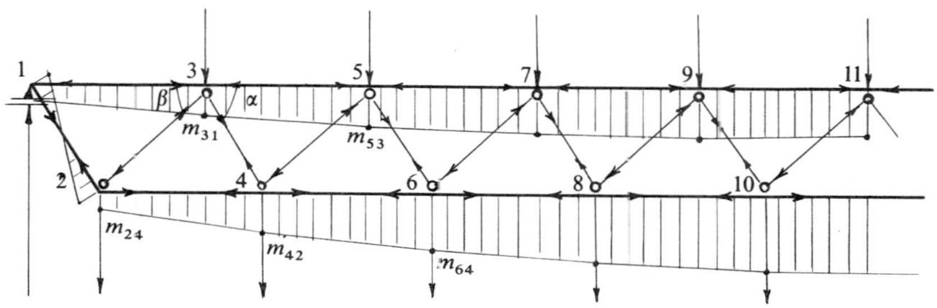

Fig. 11.17

However, if required, these bending moments can be taken into account with very little additional work in setting out of the matrices. Referring to Fig. 11.17 we can write that at joint 1:

$$C_{12}, \quad r(\Delta 1h - \Delta 2h)\cos\alpha + r(\Delta 1v - \Delta 2v)\sin\alpha + N_{12}.S_{12} = 0$$

$$C_{13}, \quad r(\Delta 1h - \Delta 3h) + N_{13}.S_{13} = 0$$

$$Fl.\ 213, \quad -m_2\frac{l_{12}}{6EI_{12}} + m_1\left(\frac{l_{12}}{3EI_{12}} + \frac{l_{13}}{3EI_{13}}\right) + m_3\frac{l_{13}}{6EI_{13}} = \frac{\Delta 2v}{l_{12}.\cos\alpha} - \frac{\Delta 3v}{l_{13}}$$

$$E_{1v}, \quad N_{12}.\sin\alpha + \frac{m_1 + m_2}{l_{12}}\cos\alpha + \frac{m_3 - m_1}{l_{13}} = R_0$$

$$E_{1h}, \quad N_{12}.\cos\alpha - N_{13} - \frac{m_1 + m_2}{l_{12}}.\sin\alpha = 0$$

Similarly at joint 2:

$$C_{23}, \quad r(\Delta 2h - \Delta 3h)\cos\beta + r(\Delta 2v - \Delta 3v)\sin\beta + N_{23}.S_{23} = 0$$

$$C_{24}, \quad r(\Delta 2h - \Delta 4h) + N_{24}.S_{24} = 0$$

$$Fl.\ 124, \quad -m_1\frac{l_{12}}{6EI_{12}} + m_2\left(\frac{l_{12}}{3EI_{12}} + \frac{l_{24}}{3EI_{24}}\right) + m_4\frac{l_{24}}{6EI_{24}} = \frac{\Delta 2v}{l_{12}.\cos\beta} - \frac{\Delta 4v - \Delta 2v}{l_{24}}$$

$$E_{2v}, \quad N_{12}.\sin\alpha - N_{23}.\sin\beta + \frac{m_1 + m_2}{l_{12}}\cos\alpha - \frac{m_4 - m_2}{l_{24}} = W_2$$

$$E_{2h}, \quad N_{12}.\cos\alpha + N_{23}.\cos\beta - N_{24} - \frac{m_1 + m_2}{l_{12}}.\sin\alpha = 0$$

Based on these equations, the overall force-displacement matrix can be set out.

Concluding Remarks

The design of modern tall buildings requires an increasingly higher degree of accuracy of analysis, because of the continuing tendency towards the design and construction of tall buildings of great slenderness and reduced mass, and the dependence on a high degree of continuity at joints.

While the aim of the analysis is to predict the response of a structure to loads and other environmental conditions, it is now known that this prediction, based on primary elastic analyses, can be insufficient. Both laboratory tests and observations of structures indicate that the overall performance of structures can differ considerably from that predicted by elastic analyses. Deflections and cracks are visible signs of this discrepancy, and it is expected that stresses also differ from those predicted by primary analyses. This is presumably also true in relation to the traditional forms, but these were less vulnerable because of their larger mass, shorter spans, smaller slenderness, and, above all, because traditional buildings did not possess or rely for their stability on a high degree of continuity. In modern tall buildings, a high degree of continuity of joints is essential for their satisfactory performance and stability. In addition, the standard of requirements has increased and, while in the past a crack, excessive deflection or deterioration were often accepted as a natural course of events, now more rigorous requirements are set with regard to serviceability of a structure within its envisaged life.

At present, there is no way of predicting with absolute certainty the response of a structure in all environmental conditions during its lifetime, although the risk involved can now be reduced to an acceptable level, and any structure can be designed to perform satisfactorily for all practical purposes.

In this book, efforts have been made to analyse the effects of various factors on the response of structures generally, with particular emphasis on tall buildings. Numerical examples have been deliberately restricted to a few types of frames to facilitate the comparison of various effects. No claim is made that all analyses are exact or rigorous, but it is hoped that the suggested methods may contribute to the understanding of the complex nature of various effects to which tall buildings can be subjected.

Most of this work is based on the concept of the force–displacement method, because the stiffness method, the only practical alternative, is well presented in technical literature and does not require elaboration. The force–displacement method on the other hand is less known and offers the advantage of having a clear physical meaning at all steps necessary in setting out of a matrix, is straightforward in application, and can provide a solution to a wide range of practical problems.

However, more work is needed and above all the verification of the assumptions and correlation with the actual response of structures. Although approximate mathematical formulation of various problems can now be made and problems solved with the aid of computers, closer solutions can only be made by further understanding of the properties of the materials, particularly of concrete, steel, and soil in actual environmental conditions.

Index

Axial deformations of columns, 204–210
Axial forces:
  analysis of effect, 201
  arising from temperature changes, 171
  arising from warping of tubes, 122
  effect on flexibility coefficients, 201

Balcony slab, stepped, 142–144
Beams:
  castellated, *see* Vierendeel girders
  continuous: on elastic foundation, 128
    foundation, 40
    temperature stresses in, 166
    with variable *I*, 28
  creep deformation of, 249–252
  inclined, effect of, 36
  parapet, *see* Parapet beams
  shrinkage stresses in, 262–267
  thermal stresses in, 150
  with variable *I*, 201
Bending:
  compatibility of deformations in, 6
  of cores, 117–122
  force-displacement matrix for, 7–23
  in frames, 9
  in frame-tubes, 124
  in trusses, 289–290
Brick, *see* Frames, infilled
Buckling, 226

Cantilever stairs, 139
Carry-over factor, 202, 203, 204
Castellated beam, *see* Vierendeel girders
Columns:
  axial deformations of, 204–210
  bending moments in, effect on foundations, 43
  eccentrically loaded, 227
  effect of creep on, 252–256
  elastic instability of, 227–229
  inclined, effect of, 35
Compatibility of deformations, 4
Concrete:
  creep effects, 247–252
  heat transfer in, 147–149
  modulus of elasticity, 241, 242
  Poisson's ratio for, 242
  properties of, 241
  shrinkage of, 242
    effect on structures, 259–267
  strains in, 241
  stress–strain diagram for, 241
  thermal expansion coefficient, 146, 242
  volume changes in, 240
    analysis of, 247–273
Contact pressure, 98
Continuum analysis, 49–53
Cores:
  analysis of, 111–126
  bending in, 117–122
  perforations in, 117
  shear lag in, 123–124
  torsion in, 112–117
  warping in, 122–123
Creep:
  in columns, 252–256
  of concrete, 244, 247–252
  deformations due to, 249
  effects on frames, 256–259
  effects on shear walls, 256–259
  specific, 245

Deformations:
  compatibility in bending, 6
  compatibility in direct forces, 276
  creep, 249
  effect on flexible joints, 204
  equilibrium of forces for large, 230
  of infilled frames, 82
  shearing, 210
  shrinkage, 268
  temperature, 156
  of trusses, 275
Diagrid floors, analysis of, 134
Distortion:
  maximum permissible, 44
  thermal, 156

Elastic instability, 225
  coefficients of, 202
  of columns, 227–229
  of frames, 229–238
Energy dissipation, *see* Vibrations, damping of

Flexible joints:
  coefficients of, 214
  effect of axial deformation, 204
  effect of axial forces, 201
  in frames, 214
Flexibility coefficients:
  effect of axial forces on, 201–202
  and stability functions, 202
Floor parameters, 77, 78
Floor systems:
  analysis of indeterminate, 127–144
  centre of rotation, 104–106
Floors:
  Diagrid, 127, 134–135
  interconnected beam, 128
  'waffle', 135–136
Force-displacement method, 2, 5
  analysis of trusses by, 274–295
  sign convention, 5
Foundations:
  continuous, 40
  effect of column bending moments on, 43
  settlement, 25–28
Frames:
  analysis of, 1–102
  analysis of gravity loads, 24
  analysis for lateral forces, 4–9
  bending in, 9
  discontinuities in, 33–38
  displacements in, 9
  effect of boundary conditions on analysis, 38
  effect of creep on, 256–259
  effect of horizontal forces, 4, 67
  effect of shear on tall, 210

Frames—*contd.*
elastic instability of, 225–239
equilibrium of forces in, 7
flexible joints in, 214
flow-charts for, 16, 17, 25
on four columns, 20
hinged, 39, 40
infilled, 79–89
analysis, 79
deformations of, 82
effect of contact pressure on analysis, 98–99
effect of incomplete filling, 80, 82, 84, 85
perforations in, 89–92
spaces in, 82
with inclined beams, 36
with inclined columns, 35
with local openings, approximate analysis of, 62
multi-bay, 18
multi-storey, 9, 14
out-of-plumb construction, 230
with parapet infills, 92–98
shearing deformations in, 210
shrinkage effects in, 270
stability of, 225
stress concentrations in, 99
space, *see* Space frames
temperature, 172
temperature effects in, 169
on three columns, 17, 62
three-dimensional, analysis of, 103–110
on two columns, 9, 14, 58
Frame-tube construction, 124
Frame-tubes:
analysis, 124
bending in, 124

Girders, *see* Trusses
Gravity loads on frames, 24
Grids, *see* Floors, interconnected beam

Heat transfer, effect on concrete structures, 147–149
Hysteresis, 199

Instability:
analysis, 229
causes, 225
defined, 225
elastic, *see* Elastic instability
Instability coefficients, 202
Instability functions, 203

Joints:
degrees of freedom, 3
importance of continuity, 1
three-dimensional, 104
Joint modulus, 215
Joint rotation factors, 101

Lateral forces:
analysis, 4–102
distribution, 67
Lift shafts, *see* Cores
Linear elastic response assumption, 3

Non-prismatic members, 28

Out-of-plumb construction, effect of, 230

Parapets:
analysis of, 92
temperature effect on, 165
Parapet beams, effects of shrinkage, 267–268
Parapet infill, effect of, 92
Poisson's ratio for concrete, 242

Rayleigh method, 187
Ritz method, 189

Second moment of area, variable, 28
'Secondary' effects, 1, 200–224
Secondary moments:
due to P/Δ effect, 229
effect of, 200
in trusses, 289
Settlement, 26, 28
Shape factor, 211
Shear, effect on tall frames, 210
Shear lag:
effect of, 123
in tubes, 124
Shear walls, 55
analysis, 55–102
analysis for temperature, 172
approximate analysis, 73
defined, 55
distribution of lateral forces, 67–73
effect of boundary condition on analysis, 39
effect of creep on, 256–259
effect of shearing forces on analysis, 210–214
local openings in, 62, 65–67
preliminary analysis without computers, 73–79
stress concentrations in, 100–102
temperature effect on, 161–172
with two rows of openings, 58
types of, 56
Shearing forces in analysis of shear walls, 210–214
Shrinkage:
analysis of stresses, 267
of concrete, 242
deformations due to, 268
effects in rectangular members, 267
general effects of, 259
in parapet beams, 267
stresses in beams, 263–265
Soil-structure interaction, 43–49
Space frames, 103
Stability functions, 202–204
Stairs:
cantilever, analysis of, 136
free-standing, 139
Statical indeterminacy, 2
expressing, 3
Statically determinate structures, 4
Steel:
modulus of elasticity, 242
strains in, 242
stress–strain diagram, 242
thermal expansion coefficient, 146, 242
Stiffness, including instability functions, 203
Stress concentrations:
in frames: 99
in shear walls, 100–102
Structural analysis:
assumptions in, 2
methods of, 2
requirements of, 2
Subsoil reaction modulus, 41, 44
Superposition principle, 3

Tall building defined, 1
Temperature, 145, *see also* Thermal stresses
analysis of frames for effects, 172, 176–180
analysis of shear walls for effects, 172–175
deformations, 156–161
effects on statically indeterminate structures, 166–171
unusual effects of, 181
Temperature distribution:
curvilinear, 159
linear, 160
in beams, 167–168
in cantilever shear wall, 161–164
in columns, 169
in frames, 169
parabolic, in cantilever shear wall, 164
thermal stresses due to non-symmetrical, 163–164
thermal stresses due to symmetrical, 151, 152
Thermal expansion coefficient, 146, 242
Thermal stresses, *see also* Temperature:
analysis of, 145–183
assumptions for, 146
in beams, 150

Thermal stresses—*contd.*
due to non-symmetrical temperature distribution, 163–164
due to symmetrical temperature distribution, 152
due to temperature change, 155
in parapet beams, 165
in plates, 149
in roof beams, 181
in statically determinate members, 149–156
Three-dimensional analysis, 103
Torsion:
constants, 106
in thin tubes, 112–117
Trusses:
bending in, 288, 289
compatibility relationship in, 274
deformation of, 275
displacements in, 275, 282
equilibrium of forces in, 280
externally statically indeterminate, 288
internally statically indeterminate, 284–287
lack of fit in, 283
rigid joints in, 291
secondary moments in, 289
statically determinate, 280–284
with continuous chords, 293
with rigid joints, 291
Tubes:
axial forces arising from warping, 122
torsion in thin, 112–117
warping in closed, 122–123

Unit Load Method, 28

Variable sections, 28
Vibrations, 184–199
analysis of single-degree-of-freedom systems, 185–186
damping of, 194
free, 185
frequency, 186
fundamental frequency, analysis, 187, 188, *see also* Rayleigh's method and Ritz method
period, 186, 187
Vierendeel girders, 128
analysis of, 131–134
Virtual Work principle, 5, 210

'Waffle' floors, 135
Walls, *see* Shear walls
Wind forces:
analysis of, 2–102
distribution, 67

MADE AND PRINTED IN GREAT BRITAIN BY
WILLIAM CLOWES & SONS LIMITED, LONDON, BECCLES AND COLCHESTER